Solutions Manual for
Quantitative Chemical Analysis

Fourth Edition

Daniel C. Harris

W. H. Freeman and Company
New York

Contents

Curve Fitting

The computer program on the following pages allows you to fit functions of your choice to experimental data pairs (x,y). The algorithm minimizes the sum of squares of deviations of y values from the curve, as in Figure 4-9. To alter the program for different functions, there are only two lines that you need to change: `nterms` gives the number of unknown coefficients to be fit and `value = ...` gives the new equation to be fit.

For example, in fitting data to the van Deemter equation, $y = A + B/x + Cx$ (Figure 22-13), there are three parameters to find (A, B and C). Therefore, we write

```
#define  nterms  3
```

and

```
value  =  c[2]*x[i]  +  c[1]/x[i]  +  c[0]
```

When you run the program, the computer will prompt you for data and initial estimates of the values of the three constants. If you have no idea of the value of the constants, try entering 1 for each of them. The program will probably home in on a solution from any reasonable starting point. Try different starting points to verify that the computer reaches the same solution. Some starting values for some problems can cause the program to "blow up" and other choices can lead the computer to a local minimum for the sum of squares of deviations that is not the true global minimum. Each problem is a matter of trial-and-error, but the procedure is quick.

For the data in Table 1, the best fit to the van Deemter equation is

$$A = 1.568\ 07 \pm 0.157\ 21$$
$$B = 26.727\ 86 \pm 2.179\ 23$$
$$C = 0.024\ 358\ 6 \pm 0.001\ 471\ 1$$

If you made replicate measurements of y for each value of x, you could use the standard deviation of y as an estimate of the uncertainty in y. The program allows you to enter the uncertainties and it then puts less weight on the less certain values, which will produce somewhat different results for A, B and C. If you do not know the uncertainties, the program weights all input equally.

Different algorithms for estimating uncertainty in the parameters to be fit give somewhat different results. The program described here does not give exactly the same error estimates as the equations in Chapter 4 for the slope and intercept of a straight line. We will always use the least squares equations in Chapter 4 for fitting straight lines.

Table 1. Data for van Deemter graph (y = A + B/x + Cx) in Figure 22-13[*]

Flow rate (mL/min) x	Plate height (mm) y
3.4	9.59
7.1	5.29
16.1	3.63
20.0	3.42
23.1	3.46
34.4	3.06
40.0	3.25
44.7	3.31
65.9	3.50
78.9	3.86
96.8	4.24
115.4	4.62
120.0	4.67

[*]From H. W. Moody, *J.Chem. Ed.* **1982**, *59*, 290.

The computer program below is based on the Marquardt procedure, with error estimation by the jackknife method. An article by M. S. Cacecci (*Anal. Chem.* **1989**, *61*, 2324) provides test data for debugging the error estimation routine. A least squares procedure that handles an arbitrary number of variables [such as w = f(x,y,z) with uncertainty in each variable] and carries out a more general analysis of uncertainty is given by W. E. Wentworth, *J. Chem. Ed.* **1965**, *42*, 96, 162.

/*Nonlinear Least Squares Curve Fitting Program*/

/*Marquardt algorithm from P.R. Bevington,"Data Reduction and Error
Analysis for the Physical Sciences," McGraw-Hill, 1969; Adapted by
Wayne Weimer & David Harris. Jackknife error algorithm of M.S. Caceci,
Anal. Chem. **1989**, *61*, 2327. Translated to ANSI C by Douglas Harris &
Tim Seufert 7-94 */

```
#include <stdio.h>
#include <stdlib.h>
#include <math.h>
#define maxnpts 50          /* Maximum data pairs - increase if desired */

/*Change nterms to the number of parameters to be fit in your equation*/
/*****************************************************************/
#define nterms 3            /* Number of parameters to be fit */
/*****************************************************************/

int param, iteration, nloops, n, cycle, nfree;
int npts;                                    /* Number of data pairs   */
long double x[maxnpts],y[maxnpts],sigmay[maxnpts]; /*x,y,y uncertainty*/
long double weight[maxnpts];                       /* Weighting factor */
long double yfit[maxnpts];                   /* Calculated values of y */
long double a[nterms], sigmaa[nterms], b[nterms]; /* a[i]=c[i] params */
long double beta[nterms], c[nterms];         /* To be fit by program */
long double finala[nterms], lastsigmaa[nterms];
long double alpha[nterms][nterms],  arry[nterms][nterms];
long double aug[nterms][nterms*2];           /* For matrix inversion */
long double deriv[maxnpts][nterms];          /* Derivatives          */
long double flambda;  /*Proportion of gradient search(=0.001 at start)*/
long double chisq;                 /* Variance of residuals in curve fit */
long double chisq1, fchisq, sy;
char errorchoice;
char filename[20], answer[100];
FILE *fp;

void readdata(void),                 unweightedinput(void);
void weightedinput(void),            chisquare(void);
void calcderivative(void),           matrixinvert(void);
void curvefit(int npoints),          display(void);
void uncertainties(void),            jackknifedata(char *filename, int k);
void print_data(void);

int main(void)              /* The main routine */
{
  int i;
  printf("Least Squares Curve Fitting. You must modify the constant\n");
  printf("'nterms' and the fuction 'Func' for new problems.\n");
  readdata();
  printf("\nEnter initial guesses for parameters:\n");
  printf("\t(Note: Parameters cannot be exactly zero.)\n");
  for (i=0; i<nterms; i++) {
    do {
      printf("Parameter #%d = ", i+1);   gets(answer);
    } while ( (a[i] = atof(answer)) == 0.0 );
  }
  flambda = 0.001;    iteration = 0;  cycle = 0;
  do {
    curvefit(npts);
    iteration++;
    display();
```

```
        printf("\n\tAnother iteration (Y/N)? ");    gets(answer);
    } while (answer[0] != 'N' && answer[0] != 'n');
    printf("\nDo you want to calculate uncertainty in parameters (Y/N)?");
    gets(answer);
    if (answer[0] == 'Y' || answer[0] == 'y') uncertainties();
    return 0;
}
                                        /*******************************/
long double func(int i)        /* The function you are fitting*/
{                                       /*******************************/
    int loop;       long double value;
    if (param==1) {
        for (loop=0; loop<nterms; loop++)      c[loop] = b[loop];
    } else {
        for (loop=0; loop<nterms; loop++)      c[loop] = a[loop];
    }

/***************************************************/
/*Enter the function to be fit:*/
/***************************************************/
    value = c[2]*x[i] + c[1]/x[i] + c[0];    /*van Deemter Equation*/
    /* x[i] is the independent variable */
    /* Values of c[n], c[n-1],  c[0] are determined by least squares */
    /* nterms must be set to equal n+1 */
 /* Example of another function: value = c[2]*x[i]*x[i]+c[1]*x[i]+c[0]*/
    return (value);
}

void  readdata(void)
{
    int n = 0;
    do {
        printf("\nDo you want to enter x,y values or read them from a
file?\n");
        printf("\tType E for enter and F for File: ");    gets(answer);
        answer[0] = toupper(answer[0]);
    } while (answer[0] != 'E' && answer[0] != 'F');
    if (answer[0] == 'F') {
        do {
            printf("\nPlease enter the name of the data file: ");
            gets(filename);
            printf("\n");
            fp = fopen(filename, "rb");
            if (fp == NULL) {
                printf("Fatal error: could not open file %s\n", filename);
                exit(1);
            }
            for(n=0; !feof(fp); n++) {
                fread(&x[n], sizeof(long double), 1, fp);
                fread(&y[n], sizeof(long double), 1, fp);
                fread(&sigmay[n], sizeof(long double), 1, fp);
                if (errorchoice == '1')   sigmay[n] = 1.0;
            }
            fclose(fp);
            npts = n - 1;
            print_data();
            printf("\nIs this data correct (Y/N)? ");   gets(answer);
        } while (answer[0] != 'Y' && answer[0] != 'y');
    } else {
        do {
            printf("\nChoices for error analysis:\n");
```

```
            printf("\t1. Let the program weight all points equally\n");
            printf("\t2. Enter estimated uncertainty for each point\n\n");
            printf("Choose option 1 or 2 now: ");   gets(answer);
        } while (answer[0] != '1' && answer[0] != '2');
        errorchoice = answer[0];
        do {
            if (errorchoice == '1')   unweightedinput();
            if (errorchoice == '2')   weightedinput();
            print_data();
            printf("Is this data correct (Y/N)? ");
            gets(answer);
        } while (answer[0] != 'y' && answer[0] != 'Y');
        printf("Enter name of file to save the data in: ");
        gets(filename);
        fp=fopen(filename, "wb");
        if (fp == NULL) {
            printf("Fatal error: could not open file %s\n", filename);
            exit(1);
        }
        for(n=0; n<npts; n++) {
            fwrite(&x[n], sizeof(long double), 1, fp);
            fwrite(&y[n], sizeof(long double), 1, fp);
            fwrite(&sigmay[n], sizeof(long double), 1, fp);
        }
        fclose(fp);
        printf("Data saved in file %s\n", filename);
    }
}


void   print_data(void)            /* Displays the data entered */
{
    int i;
    for (i=0; i<npts; i++) {
        printf("%d\tx = %- #12.8Lf\ty = %- #12.8Lf\t", i+1, x[i], y[i]);
        printf("Sigmay = %- #12.8Lf\n", sigmay[i]);
        weight[i] = 1 / ( sigmay[i] * sigmay[i] );
    }
}


void   unweightedinput(void)       /* Enter equal weighted data */
{
    int i, n;
    printf("List the data in the order: x y, with one set on each\n");
    printf("line and a space (not a comma) between the numbers.\n");
    printf("Type END to end input\n");
    for (n=0; ;n++) {
        gets(answer);
        if (answer[0]=='E' || answer[0]=='e') break;
        x[n] = atof(answer);   i = 0;
        while (answer[i] != ' ' && answer[i] != '\0') i++;
        y[n] = atof(answer+i);   sigmay[n]=1;
    }
    npts = n;
}

void   weightedinput(void)         /* Enter unequal weighted data */
{
    int i, n;
    printf("List the data in the order: x y sigmay, with one set on\n");
    printf("each line and a space (not a comma) between the numbers.\n");
```

```
    printf("Type END to end input\n");
    for (n=0; ;n++) {
      gets(answer);
      if (answer[0]=='E'  || answer[0]=='e') break;
      x[n] = atof(answer);   i = 0;
      while (answer[i] != ' ' && answer[i] != '\0') i++;
      y[n] = atof(answer+i);   i++;
      while (answer[i] != ' ' && answer[i] != '\0') i++;
      sigmay[n] = atof(answer+i);
    }
    npts = n;
}


void chisquare(void)    /* Sum of squares of differences between */
{                           /*    measured and calculated y values   */
    int i;
    fchisq = 0;
    for (i=0; i<npts; i++) {
      fchisq += weight[i] * ( y[i] - yfit[i] ) * ( y[i] - yfit[i] );
    }
    fchisq /= nfree;
}


void calcderivative(void)        /* Numerical derivative */
{
    int i, m;   long double atemp, delta;
    for (m=0; m<nterms; m++) {
      atemp = a[m];  delta = fabs( a[m] / 100000 );  a[m] = atemp + delta;
      for (i=0; i<npts; i++)  deriv[i][m] = ( func(i) - yfit[i] ) / delta;
      a[m] = atemp;
    }
}


void  matrixinvert(void)    /* Inverts the matrix arry[][] */
{                        /*A method called pivoting reduces rounding error*/
    int i, j, k, ik[nterms], jk[nterms];   long double rsave, amax;
    for (k=0; k<nterms; k++) {
      amax =0.0;
      for (i=k; i < nterms; i++) {
        for (j=k; j < nterms; j++) {
          if (abs(amax) <= abs(arry[i][j])) {
            amax = arry[i][j];
            ik[k] = i;   jk[k] = j;
          }
        }
      }
      i = ik[k];
      if (i>k) {
        for (j=0; j < nterms; j++) {
          rsave = arry[k][j];
          arry[k][j] = arry[i][j];   arry[i][j] = -1 * rsave;
        }
      }
      j = jk[k];
      if (j>k) {
        for (i=0; i < nterms; i++) {
          rsave = arry[i][k];
          arry[i][k] = arry[i][j];   arry[i][j] = -1 * rsave;
        }
      }
      for (i=0; i<nterms; i++) {
```

```
      if (i!=k) {
        arry[i][k] = -1 * arry[i][k] / amax;
      }
    }
    for (i=0; i<nterms; i++) {
      for (j=0; j<nterms; j++) {
        if (j!=k && i!=k) {
          arry[i][j] = arry[i][j] + arry[i][k] * arry[k][j];
        }
      }
    }
    for (j=0; j<nterms; j++) {
      if (j!=k) {
        arry[k][j] = arry[k][j]/amax;
      }
    }
    arry[k][k] = 1/amax;
  }
  for (k=nterms-1; k>-1; k--) {
    j = ik[k];
    if (j>k) {
      for (i=0; i<nterms; i++) {
        rsave = arry[i][k];
        arry[i][k] = -1 * arry[i][j];   arry[i][j] = rsave;
      }
    }
    i = jk[k];
    if (i>k) {
      for (j=0; j<nterms; j++) {
        rsave = arry[k][j];
        arry[k][j] = -1 * arry[i][j];   arry[i][j] = rsave;
      }
    }
  }
}

void curvefit(int npoints)                    /* Curve fitting algorthim */
{
  int i, j, k;
  nfree = npoints - nterms;
  for (j=0; j<nterms; j++) {
    b[j] = beta[j] = 0;
    for (k=0; k<=j; k++)   alpha[j][k] = 0;
  }
  param=0;
  for (i=0; i<npoints; i++)   yfit[i] = func(i);
  chisquare();
  chisq1 = fchisq;
  calcderivative();
  for (i=0; i<npoints; i++) {
    for (j=0; j<nterms; j++) {
      beta[j] += weight[i] * ( y[i] - yfit[i] ) * deriv[i][j];
      for (k=0; k<=j; k++)
        alpha[j][k] += (weight[i] * deriv[i][j] * deriv[i][k]);
    }
  }
  for (j=0; j<nterms; j++) {
    for (k=0; k<=j; k++)   alpha[k][j] = alpha[j][k];
  }
  nloops = 0;
  do {
```

```
      param = 1;
      for (j=0; j<nterms; j++) {
        for (k=0; k<nterms; k++)
          arry[j][k] = alpha[j][k] / sqrt( alpha[j][j] * alpha[k][k] );
        arry[j][j] = flambda + 1;
      }
      matrixinvert();
      for (j=0; j<nterms; j++) {
        b[j] = a[j];
        for (k=0; k<nterms; k++)
          b[j] += beta[k] * arry[j][k] / sqrt( alpha[j][j] * alpha[k][k]);
      }
      for (i=0; i<npoints; i++)   yfit[i] = func(i);
      chisquare();
      if ( (chisq1 - fchisq) < 0 )    flambda *= 10;
      nloops++;
    } while ( fchisq > chisq1 );
    for (j=0; j<nterms; j++) {
      a[j] = b[j];
      sigmaa[j] = sqrt( arry[j][j] / alpha[j][j] );
    }
    flambda /= 10;
}


void  display(void)              /* Prints result of curvefit */
{
  int i;
  printf("\nIteration #%d\n", iteration);
  for (i=0; i<nterms; i++) {
    printf("A[%3d] = %- #12.8Lf\n", i, a[i]);
    finala[i] = a[i];
  }
  printf("Sum of squares of residuals = %- #12.8Lf", fchisq * nfree);
  sy = sqrt(fchisq);
}


void uncertainties(void)/*Calculates uncertainties by removing one*/
                         /*data point and recalculating parameters*/
{
  int i,k;    long double ajack[nterms][maxnpts], avajack[nterms];

  do {
    cycle++;
    printf("Calculating uncertainties.  Let me think!\n");
    for (i=0; i<npts; i++) {
      jackknifedata(filename, i);
      for (k=0; k<=iteration; k++)   curvefit(npts-1);
      printf("Now playing with data point %d\n", i+1);
      for (k=0; k<nterms; k++)   ajack[k][i] = a[k];
    }
    printf("\n\n");
    for (k=0; k<nterms; k++)   avajack[k] = 0;
    for (k=0; k<nterms; k++) {
      for (i=0; i<npts; i++)   avajack[k] += ajack[k][i];
      avajack[k] = avajack[k]/npts;
    }
    for (k=0; k<nterms; k++)   sigmaa[k] = 0;
    for (k=0; k<nterms; k++) {
      for (i=0; i<npts; i++)
        sigmaa[k] += (ajack[k][i]-avajack[k])*(ajack[k][i]-avajack[k]);
      sigmaa[k] = sqrt( (npts-1) * sigmaa[k]/npts );
```

```
            printf("Parameter[%d] = %- 12.8Lf +/-  %- 12.8Lf\n",k, finala[k],
                    sigmaa[k]);
            if (cycle>1)
              printf("\t(Previous uncertainty = %- #12.8Lf)\n\n",
                      lastsigmaa[k]);
            lastsigmaa[k] = sigmaa[k];
        }
        printf("Standard deviation of y = %- #12.8Lf\n", sy);
        printf("Above result is based %d iterations\n", iteration);
        iteration += 5;
        printf("Iterations will now be increased to %d"
                " to see if the estimates of \n", iteration);
        printf("uncertainty change.  When consecutive cycles give\n");
        printf("similar results, it is time to stop.\n");
        printf("\tDo you want to try another cycle now (Y/N)? ");
        gets(answer);
    } while ( answer[0] == 'y' || answer[0] == 'Y' );
}


void jackknifedata(char *filename,  int k)
{                                          /* Removes one data point */

  int n = 0;
  fp = fopen(filename, "rb");
  while (!feof(fp)) {
    fread(&x[n], sizeof(long double), 1, fp);
    fread(&y[n], sizeof(long double), 1, fp);
    fread(&sigmay[n], sizeof(long double), 1, fp);
    if (errorchoice == '1')   sigmay[n] = 1.0;
    weight[n] = 1/(sigmay[n]*sigmay[n]);
    n++;
  }
  npts = n-1;
  fclose(fp);
  for (n=0; n<(npts-1); n++) {
    if (n>=k) {
      x[n] = x[n+1];    y[n] = y[n+1];   weight[n] = weight[n+1];
    }
  }
}
}
```

Spreadsheet Statistics with Pennies

Pennies minted in the United States after 1982 are *composites* with a core of zinc protected by a copper overlayer. (A composite does not have the same composition everywhere.) Prior to 1982, pennies had a uniform brass composition of 95% Cu and 5% Zn. During 1982, both the heavier brass coins and the lighter composite coins were made. In this experiment[†] we will weigh a large number of coins to gain concrete experience with statistics and least squares calculations and to answer the following questions:

1. Do the masses follow a Gaussian distribution? (This involves the χ^2 test.)
2. Do pennies from different years have the same mass?
3. Do pennies from different mints have the same mass?

Gathering Data

Sanity is best preserved by dividing this chore among the entire class. Each student should collect and weigh enough pennies to the nearest milligram so that the total set contains 300 to 500 values for the brass type and a similar number for the composite variety. Compile all of the class data in a spreadsheet. Each column should list the masses of pennies from one calendar year. There will be two columns for 1982, in which both brass and composite coins were made. Select a year other than 1982 for which you have many coins and divide them into those made in Denver (with a "D" beneath the year) and those minted in Philadelphia (with no mark beneath the year). Examine the number of coins made at each mint in 1982. Which mint made more composites in 1982?

Discrepant Data

For all calculations in this experiment, retain at least one extra digit beyond the milligram place to avoid round-off errors in your calculations. At the bottom of each column list the average and standard deviation for that year.

A damaged or corroded coin may have lower mass than the general population. Use the spreadsheet "sort" function to sort each column by increasing mass. This makes it easy to identify an outlier in any list. Discard grossly discrepant masses lying ≥ 4 standard deviations from the mean in any one year. (For example, if one column has an average mass of 3.000 g and a standard deviation of 0.030 g, the 4 standard deviation limit is 3.000 $\pm$ (4 $\times$ 0.030) = 3.000 $\pm$ 0.120 g. A mass that is ≤ 2.880 or ≥ 3.120 g may be discarded.) This is a conservative criterion in which only the most discrepant data will be rejected.

[†] T. H. Richardson, *J. Chem. Ed.* **1991**, *68*, 310.

After rejecting discrepant data, recompute the average and standard deviation for each column. Do not apply the same test again to the same data with the new standard deviation, because you could continue indefinitely until you have discarded much of your valid data.

Confidence Intervals and t Test

Select the two years in which the composite coins (≥1982) have the highest and lowest average mass. Compute the 95% and 99% confidence intervals for the masses. Use the t test to compare the two mean values at the 95 and 99% confidence levels. What can you conclude? Try the same for two years with the older brass coins (≤1982). Try the same for the one year in which you segregated coins made in Philadelphia from those in Denver. Do the two mints produce coins with the same mass?

Do the Masses Follow a Gaussian Distribution (χ^2 Test)

The χ^2 test (pronounced "ki squared") is used to compare an observed distribution to a theoretical distribution. We will examine the composite pennies minted since 1982 to see if their masses follow the Gaussian distribution.

Construct a bar graph of your data as in Figure 1. Group the data for all of the composite pennies into a single column sorted from lowest mass to highest mass. There should be at least 300 entries in this table. Divide the data into 0.01 g intervals and plot each category as a bar on the graph. For example, the bar at 2.485 g in Figure 1 shows that 69 coins had a mass between 2.480 and 2.489 g. Calculate the average, median and standard deviation for the entire set of coins in the graph. Indicate which bars (if any) lie beyond 3 standard deviations from the mean. In Figure 1 the two bars at the right are beyond 3 standard deviations. None of the bars at the left lie beyond 3 standard deviations.

Next, construct the smooth Gaussian curve that has the same mean, standard deviation and area as the observed data set. The equation for this curve is

$$y = \frac{\text{number of coins}/100}{s\sqrt{2\pi}}\, e^{-(x-\bar{x})^2/2s^2} \qquad (1)$$

where $\bar{x}$ is the average value and s is the standard deviation for the whole set of pennies in the bar chart. The factor (number of coins/100) in the numerator is required so that the area under the Gaussian curve is equal to the number of coins in the total set. The factor of 100 arises because we divided the bar chart into 0.01 g intervals. A values of y is calculated for the value of x for each bar in Figure 1.

Now we carry out a χ^2 test to see if the observed distribution (the bar chart in Figure 1) agrees with the Gaussian curve. The statistic χ^2 is given by

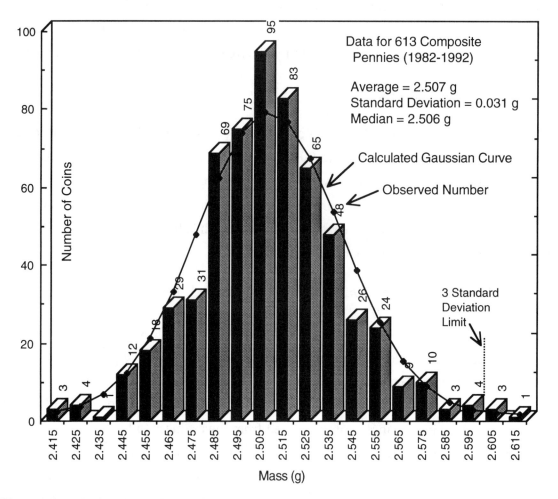

Figure 1. Distribution of penny masses measured in Dan's house by Jimmy Kusznir and Doug Harris in December 1992.

$$\chi^2 = \sum \frac{(y_{obs} - y_{calc})^2}{y_{calc}} \qquad (2)$$

where y_{obs} is the height of a bar on the chart, y_{calc} is the ordinate of the Gaussian curve (Equation 1), and the sum extends over all bars in the graph. The calculations for the data in Figure 1 are shown in Table 2.

At the bottom of Table 2 we see that χ^2 for all 21 bars is 43.231. In Table 3 we find a critical value of 31.4 for 20 degrees of freedom (degrees of freedom = one less than number of categories). Since χ^2 from Equation 2 exceeds the critical value, we conclude that *the distribution is not Gaussian.*

Table 2. Calculation of χ^2 for Figure 1

Mass (g) (x)	Observed number of coins (y_{obs})	Calculated ordinate of Gaussian curve (y_{calc})	$y_{obs} - y_{calc}$	$\dfrac{(y_{obs} - y_{calc})^2}{y_{calc}}$
2.415	3	1.060	1.940	3.550
2.425	4	2.566	1.434	0.801
2.435	1	5.611	-4.611	3.789
2.445	12	11.083	0.917	0.076
2.455	18	19.776	-1.776	0.159
2.465	29	31.875	-2.875	0.259
2.475	31	46.409	-15.409	5.116
2.485	69	61.039	7.961	1.038
2.495	75	72.519	2.481	0.085
2.505	95	77.829	17.171	3.788
2.515	83	75.453	7.547	0.755
2.525	65	66.077	-1.077	0.176
2.535	48	52.272	-4.272	0.349
2.545	26	37.354	-11.354	3.451
2.555	24	24.112	-0.112	0.001
2.565	9	14.060	-5.060	1.821
2.575	10	7.406	2.594	0.909
2.585	3	3.524	-0.524	0.078
2.595	4	1.515	2.485	4.076
2.605	3	0.588	2.412	9.894
2.615	1	0.206	0.794	3.060

$$\chi^2 \text{ (all 21 points)} = \mathbf{43.231}$$

$$\chi^2 \text{ (19 points - omitting bottom two points)} = \mathbf{30.277}$$

It would be reasonable to omit the smallest bars at the edge of the graph from the calculation of χ^2 because these bars contain the fewest observations but make large contributions to χ^2. Suppose that we reject bars lying >3 standard deviations from the mean. This removes the two bars at the right side of Figure 1 which give the last two entries in Table 2. Omitting these two points gives $\chi^2 = 30.277$, which is still greater than the critical value of 28.9 for 18 degrees of freedom in Table 3. Our conclusion is that at the 95% confidence level the observed distribution in Figure 1 is not quite Gaussian. It is possible that exceptionally light coins are nicked and exceptionally heavy coins are dirty or corroded. You would need to inspect these coins to very this hypothesis.

Table 3. Critical values of χ^2 that will be exceeded in 5% of experiments[*]

Degrees of freedom	Critical value	Degrees of freedom	Critical value	Degrees of freedom	Critical value
1	3.84	11	19.7	21	32.7
2	5.99	12	21.0	22	33.9
3	7.81	13	22.4	23	35.2
4	9.49	14	23.7	24	36.4
5	11.1	15	25.0	25	37.7
6	12.6	16	26.3	26	38.9
7	14.1	17	27.6	27	40.1
8	15.5	18	28.9	28	41.3
9	16.9	19	30.1	29	42.6
10	18.3	20	31.4	30	43.8

[*]*Example:* The value of χ^2 from 15 observations is 17.2. This value is less than the value of 23.7 listed for 14 (= 15-1) degrees of freedom. Since χ^2 does not exceed the critical value, the observed distribution is consistent with the theoretical distribution.

Is There a Systematic Change of Mass Over a Period of Years?

Make a graph in which the ordinate (y-axis) is the average mass of pennies minted each year since 1982 and the abscissa (x-axis) is the year. For simplicity, let 1982 be year 0, 1983 be year 1, and so on. Since the data have random variation, the slope of the curve will not be zero. The question we ask is "Does the slope have a value that is significantly different from zero?"

Suppose that you have data for 12 years. Enter all of the data into two columns of a linear least squares spreadsheet (as in Table 4-7). The first column (x_i) is the year (0 to 11) and the second column (y_i) is the mass of each penny. Your table will have several hundred entries. Calculate the slope and intercept of the best straight line through all points and find the uncertainties in slope and intercept. Find the 95% and 99% confidence interval for the slope. Do these confidence intervals include the value slope = 0? If they do not include 0, then the data tell us that the masses of pennies change as a function of time.

Analysis of Variance

When we analyze a *heterogeneous* material — one whose composition varies from place to place — the overall *variance* (s_o^2) is the sum of the variance of the sampling operation (s_s^2), and the variance of the analytical procedure (s_a^2):

Additivity of variance: $$s_o^2 = s_s^2 + s_a^2 \qquad (3)$$

We wish to dissect the observed variance s_o^2 into its component parts, $s_s^2 + s_a^2$. *Analysis of variance* is the statistical tool used to accomplish this task.

Variance Between Lab Samples *vs* Variance Between Aliquots from One Lab Sample

Consider the 100 m $\times$ 100 m square field in Figure 2 populated by sonflowers that are rich in copper. We want to know the average Cu content of the flowers. To do this, we divide the field into 10 000 ($= 100 \times 100$) one-square-meter parcels. We then select ten of the parcels at random (using a list of random numbers generated by a computer) and harvest the flowers in each parcel. These are chopped up into tiny pieces, mixed together and dried. The finely chopped dry material is our *laboratory sample*. From the lab sample, we take five 25-gram *aliquots* (portions) for analysis. The results are expressed as parts per thousand (ppt), which means grams of Cu per kg of dry mass of flower. Five aliquots gave values of 1.23, 1.27, 1.22, 1.26 and 1.28 ppt, which are listed for lab sample 1 in Table 4. We then repeat the whole procedure three more times to generate three different lab samples (each from 10 parcels) designated 2, 3 and 4 in Table 4.

The mean value for lab sample 1 is 1.25_2 ppt, with a standard deviation of 0.02_{588} ppt. This standard deviation arises from random errors in the analytical procedure and from residual heterogeneity in the finely chopped, well mixed lab sample. The mean values for lab samples 2, 3 and 4 in Data Set 1 are 1.21_2, 1.27_0 and 1.17_6, with standard deviations of 0.02_{280}, 0.02_{739} and 0.02_{680} ppt. The average standard deviation *within* lab samples is

Average standard deviation *within* lab samples

$$= \tfrac{1}{4}(0.02_{588} + 0.02_{280} + 0.02_{739} + 0.02_{660}) = 0.02_{572} \text{ ppt} \qquad (4)$$

$\uparrow$

16 degrees
of freedom

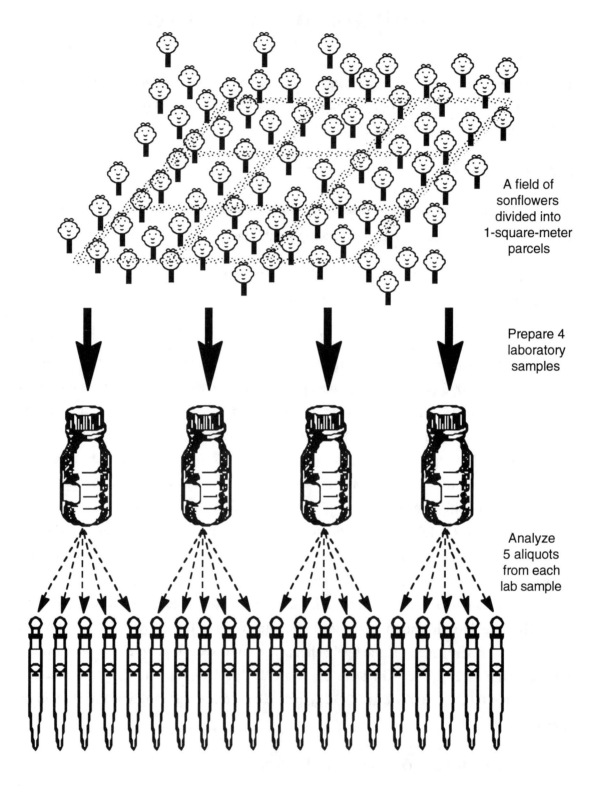

Figure 2. Experimental design for analysis of variance. A field of sonflowers is divided into 1-m² parcels and bushes from ten parcels chosen at random are chopped up, dried and mixed well to obtain Laboratory Sample 1. Other parcels are used to generate Laboratory Samples 2, 3 and 4. From each laboratory sample, five identical aliquots are analyzed.

Table 4.   Copper analysis of sonflowers

Lab sample	Cu content (ppt)	Mean $\bar{x}$	Standard deviation, s	Degrees of freedom
Data Set 1				
1	1.23, 1.27, 1.22, 1.26, 1.28	1.25_2	0.02_{588}	4
2	1.19, 1.19, 1.23, 1.24, 1.21	1.21_2	0.02_{280}	4
3	1.27, 1.26, 1.23, 1.29, 1.30	1.27_0	0.02_{739}	4
4	1.18, 1.16, 1.22, 1.16, 1.16	1.17_6	0.02_{680}	4
Average:			0.02_{572} ← 16	
Average:		1.22_{75} ← 3		
Standard deviation:		0.04_{203} ← 3		
Data Set 2				
5	1.23, 1.27, 1.22, 1.26, 1.28	1.25_2	0.02_{588}	4
6	1.16, 1.16, 1.20, 1.21, 1.18	1.18_2	0.02_{280}	4
7	1.27, 1.26, 1.23, 1.29, 1.30	1.27_0	0.02_{739}	4
8	1.15, 1.13, 1.19, 1.13, 1.13	1.14_6	0.02_{680}	4
Average:			0.02_{572} ← 16	
Average:		1.21_{25} ← 3		
Standard deviation:		0.05_{836} ← 3		

For five aliquots in one lab sample, there are 5-1 = 4 degrees of freedom associated with the mean and standard deviation. The average standard deviation in Equation 4 has 16 degrees of freedom, because it is composed of four measurements, each with 4 degrees of freedom.

The average copper content for Data Set 1 is 1.22_{75} ppt. The standard deviation from the mean of the four lab samples is

Standard deviation *between* lab samples

$$= \sqrt{\frac{(1.25_2-1.22_{75})^2 + (1.21_2-1.22_{75})^2 + (1.27_0-1.22_{75})^2 + (1.17_6-1.22_{75})^2}{4-1}}$$

$$= 0.04_{203} \text{ ppt} \qquad\qquad\qquad\qquad (5)$$

↑
3 degrees
of freedom

There are 3 degrees of freedom associated with Equation 5, because we are finding the standard deviation of four numbers from their own mean.

In analysis of variance, we seek to answer these questions:

1. *Can the variation among samples 1-4 be ascribed to random error in the analytical procedure (as opposed to real differences in the compositions of samples 1-4)?*

2. *If there are differences in sample composition, what portion of the variance in the analytical results can be ascribed to heterogeneity among the lab samples and what portion of the variance can be ascribed to irreproducibility in the analysis?*

The F Test

The first question is addressed by the **F test**. For two variances, s_1^2 and s_2^2 (with s_1 chosen to be the larger of the two), the quantity F is defined as

$$F = \frac{s_1^2}{s_2^2} \tag{6}$$

To decide whether s_1 is significantly greater than s_2, we compare F to the values in Table 5. *If the calculated value of F is greater than the value in Table 5, the difference is significant.*

When Variance Between Lab Samples is Not Significantly Different from Variance Between Aliquots

We want to know if the variance between lab samples in Equation 5 is significantly greater than the variance within replicate analyses of a single lab sample in Equation 4.

3 degrees of freedom
$$F = \frac{s_1^2}{s_2^2} = \frac{0.04_{203}^2}{0.02_{572}^2} = 2.70 < 3.239 \text{ in Table 5}$$
16 degrees of freedom

Since the calculated value of F (2.70) is less than the tabulated value of F (3.239 for 3 degrees of freedom in s_1 and 16 degrees of freedom in s_2), *the difference is not significant.* The variance among results for lab samples 1-4 in Table 4 should not be attributed to anything more than random variation in the analytical procedure.

When Variance Between Lab Samples is Significantly Greater than Variance Between Aliquots

Now consider Data Set 2 in Table 4. The standard deviations *within* each lab sample are identical in Data Sets 1 and 2. However, the differences between the mean values in

Table 5. Critical values of $F = \dfrac{s_1^2}{s_2^2}$ (at 95% confidence level)

Degrees of freedom for s_2	Degrees of freedom for s_1													
	2	3	4	5	6	7	8	9	10	12	15	20	30	∞
2	19.0	19.2	19.2	19.3	19.3	19.4	19.4	19.4	19.4	19.4	19.4	19.4	19.5	19.5
3	9.55	9.28	9.12	9.01	8.94	8.89	8.84	8.81	8.79	8.74	8.70	8.66	8.62	8.53
4	6.94	6.59	6.39	6.26	6.16	6.09	6.04	6.00	5.96	5.91	5.86	5.80	5.75	5.63
5	5.79	5.41	5.19	5.05	4.95	4.88	4.82	4.77	4.74	4.68	4.62	4.56	4.50	4.36
6	5.14	4.76	4.53	4.39	4.28	4.21	4.15	4.10	4.06	4.00	3.94	3.87	3.81	3.67
7	4.74	4.35	4.12	3.97	3.87	3.79	3.73	3.68	3.64	3.58	3.51	3.44	3.38	3.23
8	4.46	4.07	3.84	3.69	3.58	3.50	3.44	3.39	3.35	3.28	3.22	3.15	3.08	2.93
9	4.26	3.86	3.63	3.48	3.37	3.29	3.23	3.18	3.14	3.07	3.01	2.94	2.86	2.71
10	4.10	3.71	3.48	3.33	3.22	3.14	3.07	3.02	2.98	2.91	2.84	2.77	2.70	2.54
11	3.98	3.59	3.36	3.20	3.10	3.01	2.95	2.90	2.85	2.79	2.72	2.65	2.57	2.40
12	3.88	3.49	3.26	3.11	3.00	2.91	2.85	2.80	2.75	2.69	2.62	2.54	2.47	2.30
13	3.81	3.41	3.18	3.02	2.92	2.83	2.77	2.71	2.67	2.60	2.53	2.46	2.38	2.21
14	3.74	3.34	3.11	2.96	2.85	2.76	2.70	2.65	2.60	2.53	2.46	2.39	231	2.13
15	3.68	3.29	3.06	2.90	2.79	2.71	2.64	2.59	2.54	2.48	2.40	2.33	2.25	2.07
16	3.63	3.24	3.01	2.85	2.74	2.66	2.59	2.54	2.49	2.42	2.35	2.28	2.19	2.01
17	3.59	3.20	2.96	2.81	2.70	2.61	3.55	2.49	2.45	2.38	2.31	2.23	2.15	1.96
18	3.56	3.16	2.93	2.77	2.66	2.58	2.51	2.46	2.41	2.34	2.27	2.19	2.11	1.92
19	3.52	3.13	2.90	2.74	2.63	2.54	2.48	2.42	2.38	2.31	2.23	2.16	2.07	1.88
20	3.49	2.10	2.87	2.71	2.60	2.51	2.45	2.39	2.35	2.28	2.20	2.12	2.04	1.84
30	3.32	2.92	2.69	2.53	2.42	2.33	2.27	2.21	2.16	2.09	2.01	1.93	1.84	1.62
∞	3.00	2.60	2.37	2.21	2.10	2.01	1.94	1.88	1.83	1.75	1.67	1.57	1.46	1.00

Data Set 2 are greater than those in Data Set 1. The standard deviation *between* mean values in Data Set 2 is

Standard deviation *between* lab samples in Data Set 2

$$= \sqrt{\frac{(1.25_2-1.21_{25})^2 + (1.18_2-1.21_{25})^2 + (1.27_0-1.21_{25})^2 + (1.14_6-1.21_{25})^2}{4-1}}$$

$$= 0.05_{836} \text{ ppt} \tag{7}$$

↑

3 degrees
of freedom

To see if the variance between samples in Data Set 2 is significantly greater than the variance deviation within lab samples, we calculate F with Equation 6:

3 degrees of freedom

$$F = \frac{s_1^2}{s_2^2} = \frac{0.05_{836}^2}{0.02_{572}^2} = 5.15 > 3.239 \text{ in Table 5}$$

16 degrees of freedom

Since $F_{calculated} > F_{table}$, we conclude that there are real differences between samples, beyond what is expected from variations in the experimental procedure.

Two Contributions to Variance

Since there are real differences between the samples, what portion of the variance (s_s^2) can be ascribed to sample heterogeneity and what portion (s_a^2) can be ascribed to irreproducibility in the replicate analyses? We sort out the two sources of variation with the formula

$$\text{Variance between lab samples} = s_a^2 + n\, s_s^2 \tag{8}$$

where n is the number of replicate analyses (5) of each lab sample The variance between lab samples is the square of the standard deviation (0.05_{836}^2) calculated in Equation 7. The variance due to experimental error is the square of the average standard deviation (0.02_{572}^2) within lab samples in Data Set 2 in Table 4.

Substituting these variances into Equation 8 gives

$$0.05_{836}^2 = 0.02_{572}^2 + 5\, s_s^2$$

which we solve for the sampling variance attributable to heterogeneity among lab samples:

$$s_s^2 = 5.489 \times 10^{-4} \Rightarrow s_s^2 = 0.02_{343} \text{ ppt}$$

Differences among the 25 results in Data Set 2 in Table 4 can be broken down as follows:

$$\text{standard deviation due to sample heterogeneity} = s_s = 0.023 \text{ ppt}$$
$$\text{standard deviation due to analytical procedure} = s_a = 0.026 \text{ ppt}$$

Exercises

A. Is the difference between standard deviations for bromothymol blue and methyl red in Problem 17 of Chapter 4 significant?

B. In Table 4-3, is the standard deviation for the mass of nitrogen from air significantly different from the standard deviation for the mass of nitrogen from chemical sources?

C. A finely ground mixture of potassium nitrate and sodium chloride was shipped in a train car. To determine what fraction was potassium nitrate, samples were withdrawn at random from five locations, and each sample was subjected to four replicate analyses, with the results shown below:

Sample	Percent potassium	Mean	Standard Deviation
A	12.42, 12.28, 12.33, 12.36	12.34_{75}	0.05_{85}
B	12.27, 12.24, 12.19, 12.19	12.22_{25}	0.03_{95}
C	12.41, 12.48, 12.51, 12.39	12.44_{75}	0.05_{68}
D	12.42, 12.43, 12.47, 12.40	12.43_{00}	0.02_{94}
E	12.19, 12.28, 12.20, 12.32	12.24_{75}	0.06_{29}

(a) Show that the standard deviation among the five mean values is significantly greater than the average standard deviation within each sample.

(b) What standard deviation can be ascribed to variation in the sample composition (s_s) and what standard deviation can be ascribed to irreproducibility (s_a) in the replicate analyses of each sample?

D. Red blood cells from 6 mice were analyzed for iron, with the following results:

Mouse	Fe content (mM)	Mean	Standard Deviation
1	17.3, 16.8, 18.1, 18.4, 17.2	17.5_6	0.6_{66}
2	22.4, 20.4, 22.1, 21.3, 21.9	21.6_2	0.7_{92}
3	23.2, 22.0, 21.8, 22.5, 23.0	22.5_0	0.6_{08}
4	20.6, 19.1, 21.1, 20.4, 19.9	20.2_2	0.7_{60}
5	18.1, 18.4, 16.6, 17.3, 17.9	17.6_6	0.7_{16}
6	18.2, 20.4, 19.5, 18.7, 19.1	19.1_8	0.8_{35}
	average	19.7_9	0.7_{30}
	standard deviation	2.0_4	

(a) Show that the standard deviation between different mice is significantly greater than the average standard deviation for replicate measurements from each mouse.

(b) What standard deviation can be ascribed to variation between mice (s_s) and what standard deviation can be ascribed to irreproducibility (s_a) in replicate analyses from each mouse?

Solutions to Exercises in Analysis of Variance

A. $F = \dfrac{0.002\ 25^2 \ \leftarrow 27 \text{ degrees of freedom}}{0.000\ 98^2 \ \leftarrow 17 \text{ degrees of freedom}} = 5.27 > F_{table} = 2.2.$

The difference is significant.

B. $F = \dfrac{0.001\ 38^2 \ \leftarrow 6 \text{ degrees of freedom}}{0.000\ 143^2 \ \leftarrow 7 \text{ degrees of freedom}} = 93.1 > F_{table} = 3.87.$

The difference is significant.

C. (a)

Sample	Percent potassium	Mean	Standard Deviation
A	12.42, 12.28, 12.33, 12.36	12.34_{75}	0.05_{85}
B	12.27, 12.24, 12.19, 12.19	12.22_{25}	0.03_{95}
C	12.41, 12.48, 12.51, 12.39	12.44_{75}	0.05_{68}
D	12.42, 12.43, 12.47, 12.40	12.43_{00}	0.02_{94}
E	12.19, 12.28, 12.20, 12.32	12.24_{75}	0.06_{29}
		-------	-------
	average	12.33_{90}	0.04_{94}
	standard deviation	0.10_{26}	

0.10_{26} ↑ 4 degrees of freedom

0.04_{94} ↑ 15 degrees of freedom

$$F = \frac{0.10_{26}^2 \ \leftarrow 4 \text{ degrees of freedom}}{0.04_{94}^2 \ \leftarrow 15 \text{ degrees of freedom}} = 4.31 > F_{table} \approx 3.06$$

(b) Variance between samples $= s_a^2 + n s_s^2$

$$0.10_{26}^2 = 0.04_{94}^2 + 4 s_s^2$$

$$s_s = 0.04_{50}$$

$$s_a = 0.04_{94}$$

D. (a) $F = \dfrac{2.04^2 \ \leftarrow 5 \text{ degrees of freedom}}{0.730^2 \ \leftarrow 24 \text{ degrees of freedom}} = 7.81 > F_{table} \approx 2.6$

(b) Variance between mice $= s_a^2 + n s_s^2$

$$2.04^2 = 0.730^2 + 5 s_s^2$$

$$s_s = 0.8_5 \text{ mM}$$

$$s_a = 0.7_3 \text{ mM}$$

MEASUREMENTS

> *A note from Dan*: Don't worry if your numerical answers are slightly different from those in the *Solutions Manual*. You or I may have rounded intermediate results. In general, retain many extra digits for intermediate answers and save your roundoff until the end. We'll study this process in Chapter 3.

1. 1. Obtain a representative bulk sample. 2. Extract from it a smaller, homogeneous laboratory sample. 3. Transform the laboratory sample into a form suitable for analysis. 4. Remove or mask interfering species. 5. Measure the concentration of analyte in the sample. 6. Interpret the results and draw conclusions.

2. (a) A random heterogeneous material varies randomly from place to place. A segregated heterogeneous material has regions (such as layers or pockets) of distinct composition.

   (b) The random sample is divided into segments which are sampled at random using a list of random numbers to select segments.

   (c) A representative sample is constructed from the composite material. For example, if there are two regions of relative volume 2:1 we would take twice as many random samples from the larger region as from the smaller region.

3. (a) meter (m), kilogram (kg), second (s), ampere (A), kelvin (K), mole (mol)

   (b) hertz (Hz), newton (N), pascal (Pa), joule (J), watt (W)

4. See Table 1-3

5.

(a)	mW	=	milliwatt	=	10^{-3} watt
(b)	pm	=	picometer	=	10^{-12} meter
(c)	kΩ	=	kiloohm	=	10^3 ohm
(d)	μF	=	microfarad	=	10^{-6} farad
(e)	TJ	=	terajoule	=	10^{12} joule
(f)	ns	=	nanosecond	=	10^{-9} second
(g)	fg	=	femtogram	=	10^{-15} gram
(h)	dPa	=	decipascal	=	10^{-1} pascal

6. (a) 100 fJ or 0.1 pJ   (d) 0.1 nm or 100 pm

   (b) 43.172 8 nF   (e) 21 TW

   (c) 299.79 THz   (f) 0.483 amol or 483 zmol

7. (a) $(150 \times 10^{-15} \frac{\text{mol}}{\text{cell}}) / (2.5 \times 10^4 \frac{\text{vesicles}}{\text{cell}}) = 6.0$ amol/vesicle

(b) $(6.0 \times 10^{-18} \text{ mol}) (6.022 \times 10^{23} \frac{\text{molecules}}{\text{mol}}) = 3.6 \times 10^6$ molecules

8.   Table 1-4 tells us that 1 horsepower = 745.700 W = 745.700 J/s.

Therefore, 100 horsepower $= 7.457 \times 10^4$ J/s.

$$\frac{7.457 \times 10^4 \frac{\text{J}}{\text{s}}}{4.184 \frac{\text{J}}{\text{cal}}} \times 3\,600 \frac{\text{s}}{\text{h}} = 6.416 \times 10^7 \frac{\text{cal}}{\text{h}} .$$

9.   (a) $\dfrac{(2.2 \times 10^6 \frac{\text{cal}}{\text{day}})(4.184 \frac{\text{J}}{\text{cal}})(\frac{1 \text{ day}}{24 \text{ h}})(\frac{1 \text{ h}}{3\,600 \text{ s}})}{(120 \text{ pound}) (0.453\,6 \frac{\text{kg}}{\text{pound}})} = 2.0$ J/(s·kg)

$= 2.0$ W/kg

Similarly, $3.4 \times 10^3 \frac{\text{kcal}}{\text{day}} \Rightarrow 3.0$ J/(s·kg) $= 3.0$ W/kg.

(b)   The office worker's power output is

$(2.2 \times 10^6 \frac{\text{cal}}{\text{day}}) (4.184 \frac{\text{J}}{\text{cal}}) (\frac{1 \text{ day}}{24 \text{ h}}) (\frac{1 \text{ h}}{3\,600 \text{ s}}) = 107 \frac{\text{J}}{\text{s}} = 107$ W.

The person consumes more energy than the 100 W light bulb.

10.   (a) $\dfrac{0.025\,4 \text{ m}}{1 \text{ inch}} = \dfrac{1 \text{ m}}{x \text{ inch}} \Rightarrow x = 39.37$ inches

(b) $(345 \frac{\text{m}}{\text{s}}) (\frac{1 \text{ inch}}{0.025\,4 \text{ m}}) (\frac{1 \text{ foot}}{12 \text{ inch}}) (\frac{1 \text{ mile}}{5\,280 \text{ foot}}) = 0.214 \frac{\text{mile}}{\text{s}}$

$(0.214 \frac{\text{mile}}{\text{s}}) (3\,600 \frac{\text{s}}{\text{h}}) = 770$ mile/h

(c) $(3.00 \text{ s})(345 \frac{\text{m}}{\text{s}}) = 1.04 \times 10^3$ m $= 1.04$ km

$(1.04 \times 10^3 \text{ m}) (\frac{1 \text{ inch}}{0.025\,4 \text{ m}}) (\frac{1 \text{ foot}}{12 \text{ inch}}) (\frac{1 \text{ mile}}{5\,280 \text{ foot}}) = 0.643$ mile

11. $(5.00 \times 10^3 \frac{\text{Btu}}{\text{h}}) (1\,055.06 \frac{\text{J}}{\text{Btu}}) (\frac{1 \text{ h}}{3\,600 \text{ s}}) = 1.47 \times 10^3 \frac{\text{J}}{\text{s}} = 1.47 \times 10^3$ W

12.   Newton $=$ force $=$ mass $\times$ acceleration $= $ kg $\cdot (\frac{\text{m}}{\text{s}^2})$

Joule $=$ energy $=$ force $\times$ distance $=$ kg $(\frac{\text{m}}{\text{s}^2}) \cdot$ m $=$ kg $(\frac{\text{m}^2}{\text{s}^2})$

Pascal $=$ pressure $=$ force $/$ area $=$ kg $(\frac{\text{m}}{\text{s}^2}) / \text{m}^2 = \dfrac{\text{kg}}{\text{m} \cdot \text{s}^2}$

13.   molality $= \dfrac{\text{mol KI}}{\text{kg solvent}}$ 20.0 wt% KI $= \dfrac{200 \text{ g KI}}{1\,000 \text{ g solution}} = \dfrac{200 \text{ g KI}}{800 \text{ g } H_2O}$

To find the grams of KI in 1 kg of H_2O, we set up a proportion:

$$\frac{200 \text{ g KI}}{800 \text{ g } H_2O} = \frac{x \text{ g KI}}{1\,000 \text{ g } H_2O} \Rightarrow x = 250 \text{ g KI}$$

But 250 g KI = 1.51 mol KI, so the molality is 1.51 m.

14. (a) molarity = moles of solute / liter of solution

(b) molality = moles of solute / kilogram of solvent

(c) density = grams of substance / milliliter of substance

(d) weight percent = $100 \times$ (mass of substance / mass of solution or mixture)

(e) volume percent = $100 \times$ (volume of substance / volume of solution or mixture)

(f) parts per million = $10^6 \times$ (grams of substance / grams of sample)

(g) parts per billion = $10^9 \times$ (grams of substance / grams of sample)

(h) formal concentration = moles of formula / liter of solution

15. Acetic acid (CH_3CO_2H) is a weak electrolyte that is partially dissociated. When we dissolve 0.01 mol in a liter, there is less than 0.01 mol of CH_3CO_2H plus some amount of $CH_3CO_2^-$.

16. 32.0 g / [(22.989 768 + 35.452 7) g/mol] = 0.547 5 mol NaCl

0.547 5 mol / 0.500 L = 1.10 M

17. $(1.71 \frac{\text{mol } CH_3OH}{\text{L solution}})$ (0.100 L solution) = 0.171 mol CH_3OH

$(0.171 \text{ mol } CH_3OH)$ $(\frac{32.042 \text{ g}}{\text{mol } CH_3OH})$ = 5.48 g

18. $1 \text{ ppm} = \frac{1 \text{ g solute}}{10^6 \text{ g solution}}$. Since 1 L of dilute solution $\approx 10^3$ g, 1 ppm = 10^{-3} g solute/L (= 10^{-3} g solute / 10^3 g solution). Since 10^{-3} g = 10^3 μg, 1 ppm = 10^3 μg/L or 1 μg/mL. Since 10^{-3} g = 1 mg, 1 ppm = 1 mg/L.

19. 0.2 ppb means 0.2×10^{-9} g of $C_{20}H_{42}$ per g of rainwater

$= 0.2 \times 10^{-6} \frac{\text{g } C_{20}H_{42}}{1\ 000 \text{ g rainwater}} \approx \frac{0.2 \times 10^{-6} \text{ g } C_{20}H_{42}}{\text{L rainwater}}$.

MW of $C_{20}H_{42}$ = $20 \times 12.011 + 42 \times 1.007\ 94$ = 282.55 g/mol

$\frac{0.2 \times 10^{-6} \text{ g/L}}{282.55 \text{ g/mol}}$ = 7×10^{-10} M

20. $(0.705 \frac{\text{g } HClO_4}{\text{g solution}})$ (37.6 g solution) = 26.5 g $HClO_4$

37.6 g solution - 26.5 g $HClO_4$ = 11.1 g H_2O

21. (a) $(1.67 \frac{\text{g solution}}{\text{mL}})$ (1 000 $\frac{\text{mL}}{\text{L}}$) = 1 670 g solution

(b) $(0.705 \frac{\text{g HClO}_4}{\text{g solution}})(1\,670 \text{ g solution}) = 1.18 \times 10^3 \text{ g HClO}_4$

(c) $(1.18 \times 10^3 \text{ g}) / (100.458 \text{ g/mol}) = 11.7 \text{ mol}$

22. (a) Volume $= \frac{4}{3} \pi (200 \times 10^{-9} \text{ m})^3 = 3.35 \times 10^{-20} \text{ m}^3$;

$$\frac{3.35 \times 10^{-20} \text{ m}^3}{10^{-3} \text{ m}^3 / \text{ L}} = 3.35 \times 10^{-17} \text{ L}$$

(b) $\dfrac{10 \times 10^{-18} \text{ mol}}{3.35 \times 10^{-17} \text{ L}} = 0.30 \text{ M}$

23. $\dfrac{80 \times 10^{-3} \text{ g}}{180.2 \text{ g/mol}} = 4.4 \times 10^{-4} \text{ mol}$; $\dfrac{4.4 \times 10^{-4} \text{ mol}}{100 \times 10^{-3} \text{ L}} = 4.4 \times 10^{-3} \text{ M}$;

Similarly, $120 \text{ mg/L} = 6.7 \times 10^{-3} \text{ M}$

24. (a) Mass of 1 L $= 1.046 \frac{\text{g}}{\text{mL}} \times 1\,000 \frac{\text{mL}}{\text{L}} = 1\,046 \text{ g}$

Grams of $C_2H_6O_2$ per liter $= 6.067 \frac{\text{mol}}{\text{L}} \times 62.07 \frac{\text{g}}{\text{mol}} = 377 \frac{\text{g}}{\text{L}}$

(b) 1 L contains 377 g of $C_2H_6O_2$ and $1\,046 - 377 = 669$ g of $H_2O = 0.669$ kg

Molality $= \dfrac{6.067 \text{ mol } C_2H_6O_2}{0.669 \text{ kg } H_2O} = 9.07 \dfrac{\text{mol } C_2H_6O_2}{\text{kg } H_2O} = 9.07 \, m$

25. Shredded wheat: 1 g contains 0.099 g protein + 0.799 g carbohydrate

$0.099 \text{ g} \times 4.0 \frac{\text{Cal}}{\text{g}} + 0.799 \text{ g} \times 4.0 \frac{\text{Cal}}{\text{g}} = 3.6 \text{ Cal}$

Doughnut: 1 g contains 0.046 g protein + 0.514 g carbohydrate + 0.186 g fat

$0.046 \text{ g} \times 4.0 \frac{\text{Cal}}{\text{g}} + 0.514 \text{ g} \times 4.0 \frac{\text{Cal}}{\text{g}} + 0.186 \text{ g} \times 9.0 \frac{\text{Cal}}{\text{g}} = 3.9 \text{ Cal}$

In a similar manner, we find $2.8 \frac{\text{Cal}}{\text{g}}$ for hamburger and $0.48 \frac{\text{Cal}}{\text{g}}$ for apple.

Table 1-4 says that 16 ounces $= 453.592\,37$ g $\Rightarrow 28.35 \frac{\text{g}}{\text{ounce}}$.

To convert Cal/g to Cal/ounce, multiply by 28.35:

	Shredded Wheat	Doughnut	Hamburger	Apple
Cal/g	3.6	3.9	2.8	0.48
Cal/ounce	102	111	79	14

26. Mass of water $= \pi (225 \text{ m})^2 (10 \text{ m}) (\frac{1\,000 \text{ kg}}{\text{m}^3}) = 1.59 \times 10^9 \text{ kg}$

$1.6 \text{ ppm} = \dfrac{1.6 \times 10^{-3} \text{ g } F^-}{\text{kg } H_2O}$,   Mass of F^- required =

$(1.6 \times 10^{-3} \frac{\text{g } F^-}{\text{kg } H_2O})(1.59 \times 10^9 \text{ kg } H_2O) = 2.54 \times 10^6 \text{ g } F^-$.

$$\frac{\text{atomic weight F}}{\text{FW NaF}} = \frac{18.998\ 4}{41.988} = \frac{2.54 \times 10^6 \text{ g F}}{x \text{ g NaF}} \Rightarrow x = 5.6 \times 10^6 \text{ g NaF}$$

27. (a) $PV = nRT$

$(1 \text{ atm})(5.24 \times 10^{-6} \text{ L}) = n\ (0.082\ 06\ \frac{\text{L·atm}}{\text{mol·K}})(298.16 \text{ K})$

$\Rightarrow n = 2.14 \times 10^{-7} \text{ mol} \Rightarrow 2.14 \times 10^{-7} \text{ M}$

(b) Ar: 0.934% means 0.009 34 L of Ar per L of air

$(1 \text{ atm})(0.009\ 34 \text{ L}) = n\ (0.082\ 06\ \frac{\text{L·atm}}{\text{mol·K}})(298.16 \text{ K})$

$\Rightarrow n = 3.82 \times 10^{-4} \text{ mol} \Rightarrow 3.82 \times 10^{-4} \text{ M}$

Kr: 1.14 ppm $\Rightarrow$ 1.14 µL Kr per L of air $\Rightarrow 4.66 \times 10^{-8}$ M

Xe: 87 ppb $\Rightarrow$ 87 nL Xe per L of air $\Rightarrow 3.6 \times 10^{-9}$ M

28. $2.00 \text{ L} \times 0.050\ 0\ \frac{\text{mol}}{\text{L}} \times 61.83\ \frac{\text{g}}{\text{mol}} = 6.18 \text{ g in a 2 L volumetric flask}$

29. Weigh out $2 \times 0.050\ 0 \text{ mol} = 0.100 \text{ mol} = 6.18 \text{ g B(OH)}_3$ and dissolve in 2.00 kg H_2O.

30. $M_{con} \cdot V_{con} = M_{dil} \cdot V_{dil}$

$(0.80\ \frac{\text{mol}}{\text{L}})\ (1.00 \text{ L}) = (0.25\ \frac{\text{mol}}{\text{L}})\ V_{dil} \Rightarrow V_{dil} = 3.2 \text{ L}$

31. We need $1.00 \text{ L} \times 0.10\ \frac{\text{mol}}{\text{L}} = 0.10 \text{ mol NaOH} = 4.0 \text{ g NaOH}$

$\frac{4.0 \text{ g NaOH}}{0.50\ \frac{\text{g NaOH}}{\text{g solution}}} = 8.0 \text{ g solution}$

32. (a) $V_{con} = V_{dil}\frac{M_{dil}}{M_{con}} = 1\ 000 \text{ mL} (\frac{1.00 \text{ M}}{18.0 \text{ M}}) = 55.6 \text{ mL}$

(b) One liter of 98.0% H_2SO_4 contains $(18.0 \text{ mol})(98.073 \text{ g/mol}) = 1\ 765 \text{ g of}$ H_2SO_4. Since the solution contains 98.0 wt% H_2SO_4, and the mass of H_2SO_4 per mL is 1.765 g, the mass of solution per liter is

$\frac{1.765 \text{ g H}_2\text{SO}_4\text{/mL}}{0.980 \text{ g H}_2\text{SO}_4\text{/g solution}} = 1.80 \text{ g solution/mL}$

33. 2.00 L of 0.169 M NaOH = 0.338 mol NaOH = 13.5 g NaOH

$\text{density} = \frac{\text{g solution}}{\text{mL solution}}$

$= \frac{13.5 \text{ g NaOH}}{(16.7 \text{ mL solution})(0.534\ \frac{\text{g NaOH}}{\text{g solution}})} = 1.51\ \frac{\text{g}}{\text{mL}}$

1. The primary rule is not to do something you consider to be dangerous. This implies that you are responsible for evaluating the safety of what you are doing before you do it.

3. $PbSiO_3$ is insoluble and will not leach into ground water.

4. The lab notebook must: (1) state what was done; (2) state what was observed; and (3) be understandable to a stranger.

5. See Section 2.3.

6. The buoyancy correction is zero when the substance being weighed has the same density as the weight used to calibrate the balance.

7. $$m = \frac{(14.82 \text{ g})\left(1 - \dfrac{0.001\ 2 \text{ g/mL}}{8.0 \text{ g/mL}}\right)}{\left(1 - \dfrac{0.001\ 2 \text{ g/mL}}{0.626 \text{ g/mL}}\right)} = 14.85 \text{ g}$$

8. Smallest: lead dioxide; Largest: lithium

9. $\Delta F = (2.3 \times 10^{-10} \dfrac{m^2}{g \cdot Hz})(8.1 \times 10^6 \text{ Hz})^2 (200 \times 10^{-9} \text{ g}) / (16 \times 10^{-6} \text{ m}^2) = 190 \text{ Hz}$

10. (a) One mol of He (= 4.003 g) occupies a volume of

$$V = \frac{nRT}{P} = \frac{(1 \text{ mol})\left(0.082\ 06 \dfrac{L \cdot atm}{mol \cdot K}\right)(293.15 \text{ K})}{1 \text{ atm}} = 24.06 \text{ L}$$

Density = 4.003 g / 24.06 L = 0.166 g/L = 0.000 166 g/mL

(b) $$m = \frac{(0.823 \text{ g})\left(1 - \dfrac{0.000\ 166 \text{ g/mL}}{8.0 \text{ g/mL}}\right)}{\left(1 - \dfrac{0.000\ 166 \text{g/mL}}{0.97 \text{ g/mL}}\right)} = 0.823 \text{ g}$$

11. (a) (0.42) (2 330 Pa) = 979 Pa

(b) Using the note in the margin near Equation 2-1, we find

$$\text{density} = \frac{(0.003\ 485)(94\ 000) - (0.001\ 318)(979)}{293.15} = 1.11 \text{ g/L} = 0.001\ 1 \text{ g/mL}$$

(c) $$\text{mass} = 1.000\ 0 \text{ g}\left(\frac{1 - \dfrac{0.001\ 1 \text{ g/mL}}{8.0 \text{ g/mL}}}{1 - \dfrac{0.001\ 1 \text{ g/mL}}{1.00 \text{ g/mL}}}\right) = 1.001\ 0 \text{ g}$$

12. TD means "to deliver" and TC means "to contain."

13. Dissolve $(0.250\ 0\ L)(0.150\ 0\ mol/L) = 0.037\ 50$ mol of K_2SO_4 in less than 250 mL of water in a 250-mL volumetric flask. Add more water and mix. Dilute to the 250.0 mL mark and invert the flask 20 times for complete mixing.

14. The plastic flask is needed for trace analysis on analytes at ppb levels that might be lost by adsorption on the glass surface.

15. See Section 2.6.

16. Transfer pipet.

17. The last liquid should be blown out of the serological pipet, because it is calibrated down to the tip. With the measuring pipet, simply drain liquid from the 0 mL mark to the 1 mL mark.

18. The trap prevents tap water from backing up into the suction flask. The watchglass keeps dust out of the sample.

19. Phosphorus pentoxide

20. $20.214\ 4\ g - 10.263\ 4\ g = 9.951\ 0\ g$. Using column 3 of Table 2-6 tells us that the true volume is $(9.951\ 0\ g)(1.002\ 9\ mL/g) = 9.979\ 9$ mL.

21. $\text{Expansion} = \dfrac{0.999\ 102\ 6}{0.997\ 047\ 9} = 1.002\ 060\ 8 \approx 0.2\%$.

Densities were taken from Table 2-6. The 0.500 0 M solution at 25° would be $(0.500\ 0\ M)\ /\ (1.002) = 0.499\ 0$ M.

22. Using column 2 of Table 2-6,

mass in vacuum $= (50.037\ mL)(0.998\ 207\ 1\ g/mL) = 49.947$ g.

Using column 3, mass in air $= \dfrac{50.037\ mL}{1.002\ 9\ mL/g} = 49.892$ g.

23. The concentration needed at 24° will be $(1.000\ M)\left(\dfrac{0.997\ 299\ 5\ g/mL}{0.998\ 207\ 1\ g/mL}\right)$

$= 0.999\ 1$ M (using the quotient of densities from Table 2-6). The true mass of KNO_3 needed is $(0.500\ 0\ L)(0.999\ 1\ mol/L)(101.103\ g/mol) = 50.506$ g.

$$m' = \frac{(50.506\ g)\left(1 - \dfrac{0.001\ 2\ g/mL}{2.109\ g/mL}\right)}{\left(1 - \dfrac{0.001\ 2\ g/mL}{8.0\ g/mL}\right)} = 50.484\ g$$

CHAPTER 3
EXPERIMENTAL ERROR

1.   (a) 5   (b) 4   (c) 3

2.   (a) 1.237   (b) 1.238   (c) 0.135   (d) 2.1   (e) 2.00

3.   (a) 0.217   (b) 0.216   (c) 0.217

4.   (b) 1.18 (3 significant figures)   (c) 0.71 (2 significant figures)

5.   (a) 3.71   (b) 10.7   (c) 4.0×10^1 (d) 2.85×10^{-6}
       (e) 12.625 1   (f) 6.0×10^{-4} (g) 242

6.   (a) $BaCl_2 = 208.232$ (b) $C_{31}H_{32}O_8N_2 = 560.604$

7.   (a) 12.3   (b) 75.5   (c) 5.520×10^3 (d) 3.04
       (e) 3.04×10^{-10} (f) 11.9   (g) 4.600   (h) 4.9×10^{-7}

8.   Since all measurements have some uncertainty, there is no way to know the true value.

9.   See Section 3-4.

10.   After all these years, we still don't know.

11.   (a) Carmen   (b) Cynthia   (c) Chastity   (d) Cheryl

12.   3.124 (± 0.005), 3.124 ($\pm 0.2\%$). It would also be reasonable to keep an additional digit : 3.123_6 ($\pm 0.005_2$), 3.123_6 ($\pm 0.1_7\%$)

13.   (a)   6.2 (± 0.2)

       <u>- 4.1 (± 0.1)</u>

         2.1 $\pm e$ $e^2 = 0.2^2 + 0.1^2 \Rightarrow e = 0.2_2$

       Answer: 2.1 ± 0.2 (or $2.1 \pm 11\%$)

       (b)   9.43 (± 0.05)   9.43 ($\pm 0.53\%$)

         <u>$\times$ 0.016 (± 0.001)</u> $\Rightarrow$ <u>$\times$ 0.016 ($\pm 6.25\%$)</u> $\%e^2 = 0.53^2 + 6.25^2$

                          0.150 88 ($\pm \%e$) $\Rightarrow$ $\%e = 6.272$

       Relative uncertainty = 6.27%

       Absolute uncertainty = $0.150\ 88 \times 0.062\ 7 = 0.009\ 46$

       Answer: 0.151 ± 0.009 (or $0.151 \pm 6\%$)

       (c)   The first term in brackets is the same as part (a), so we can rewrite the problem as $\quad 2.1\ (\pm 0.2_{24}) \div 9.43\ (\pm 0.05) = 2.1\ (\pm 10.6\%) \div 9.43\ (\pm 0.53\%)$

$\%e = \sqrt{10.6^2 + 0.53^2} = 10.6\%$

Absolute uncertainty $= 0.106 \times 0.222 = 0.023\ 6$

Answer: $0.22_3 \pm 0.02_4\ (\pm 11\%)$

(d) The term in brackets is

$\qquad 6.2\ (\pm 0.2) \times 10^{-3} \qquad\qquad e = \sqrt{0.2^2 + 0.1^2} \Rightarrow e = 0.2_{24}$

$\underline{\quad + 4.1\ (\pm 0.1) \times 10^{-3}\quad}$

$\qquad 10.3\ (\pm\ 0.2_{24}) \times 10^{-3} = 10.3 \times 10^{-3}\ (\pm 2.2\%)$

$9.43\ (\pm 0.53\%) \times 0.010\ 3\ (\pm 2.2\%) = 0.097\ 13 \pm 2.23\% = 0.097\ 13 \pm 0.002\ 17$

Answer: $\quad 0.097_1 \pm 0.002_2\ (\pm 2._2\%)$

14. (a) $10.18\ (\pm 0.07)\ (\pm 0.7\%)$

(b) $174\ (\pm 3)\ (\pm 2\%)$

(c) $0.147\ (\pm 0.003)\ (\pm 2\%)$

(d) $2.016\ 4\ (\pm 0.000\ 8)$

$\quad 1.233\ (\pm 0.002)$

$\underline{+\ 4.61\quad (\pm 0.01)\qquad}$

$\quad 7.85_{94}\ \sqrt{(0.000\ 8)^2 + (0.002)^2 + (0.01)^2}\ = 0.01_{02}$

$\quad 7.86\ (\pm 0.01)(\pm 0.1\%)$

(e) $2\ 016.4 (\pm 0.8)$

$\quad +123.3\ (\pm 0.2)$

$\underline{+\ 46.1\ (\pm 0.1)\qquad}$

$\quad 2\ 185.8\ \sqrt{(0.8)^2 + (0.2)^2 + (0.1)^2}\ = 0.8$

$\quad 2\ 185.8\ (\pm 0.8)\ (\pm 0.04\%)$

(f) For $y = x^a$, $\%e_y = a\%e_x$

$x = 3.14 \pm 0.05 \Rightarrow \%e_x = (0.05\ /\ 3.14) \times 100 = 1.592\%$

$\%e_y = \frac{1}{3}(1.592\%) = 0.531\%$

Answer: $1.464_3 \pm 0.007_8\ (\pm 0.5_3\%)$

(g) For $y = \log x$, $e_y = 0.434\ 29\ \frac{e_x}{x}$

$x = 3.14 \pm 0.05 \Rightarrow e_y = 0.434\ 29\left(\frac{0.05}{3.14}\right) = 0.006\ 915$

Answer: $0.496_9 \pm 0.006_9\ (\pm 1.3_9\%)$

15. (a) $y = x^{1/2} \Rightarrow \%e_y = \frac{1}{2}\left(100 \times \frac{0.001\,1}{3.141\,5}\right) = 0.017\,5\%$

$(1.75 \times 10^{-4})\sqrt{3.141\,5} = 3.1 \times 10^{-4}$

Answer: $1.772\,4_3 \pm 0.000\,3_1$

(b) $y = \log x \Rightarrow e_y = 0.434\,29\left(\frac{0.001\,1}{3.141\,5}\right) = 1.52 \times 10^{-4}$

Answer: $0.497\,1_4 \pm 0.000\,1_5$

(c) $y = \text{antilog } x = 10^x \Rightarrow e_y = y \times 2.302\,6\,e_x$

$= (10^{3.141\,5})(2.302\,6)(0.001\,1) = 3.51$

Answer: $1\,385._2 \pm 3._5$

(d) $y = \ln x \Rightarrow e_y = \frac{0.001\,1}{3.141\,5} = 3.5 \times 10^{-4}$

Answer: $1.144\,7_0 \pm 0.000\,3_5$

(e) Numerator of log term: $y = x^{1/2} \Rightarrow e_y = \frac{1}{2}\left(\frac{0.006}{0.104} \times 100\right) = 2.88\%$

$\frac{0.322\,5 \pm 2.88\%}{0.051\,1 \pm 0.000\,9} = \frac{0.322\,5 \pm 2.88\%}{0.051\,1 \pm 1.76\%}$

$= 6.311 \pm 3.375\% = 6.311 \pm 0.213$

For $y = \log x$, $e_y = 0.434\,29\,\frac{e_x}{x} = 0.434\,29\left(\frac{0.213}{6.311}\right) = 0.015$

Answer: $0.80_0 \pm 0.01_5$

16. (a) $Na = 22.989\,768 \pm 0.000\,006$ g/mol

$Cl = \underline{35.452\,7 \quad \pm 0.000\,9 \quad}$ g/mol

$58.442\,468 \quad \sqrt{(6 \times 10^{-6})^2 + (9 \times 10^{-4})^2} = 9 \times 10^{-4}$

$58.442\,5 \pm 0.000\,9$ g/mol

(b) molarity $= \frac{\text{mol}}{L} = \frac{[2.634\,(\pm 0.002)\text{g}]\,/\,[58.442\,5\,(\pm 0.000\,9)\text{g/mol}]}{0.100\,00\,(\pm 0.000\,08)\,L}$

$= \frac{2.634\,(\pm 0.076\%)\,/\,[58.442\,5\,(\pm 0.001\,5\%)]}{0.100\,00\,(\pm 0.08\%)}$

relative error $= \sqrt{(0.076\%)^2 + (0.001\,5\%)^2 + (0.08\%)^2} = 0.11\%$

molarity $= 0.450\,7\,(\pm 0.000\,5)$ M

17. $1.054\,572\,67\,(\pm 0.000\,000\,64) \times 10^{-34}$ J·s

(Both h and the uncertainty were divided by 2π.)

18.

$$m = \frac{m'\left(1 - \dfrac{d_a}{d_w}\right)}{1 - \dfrac{d_a}{d}}$$

$$m = \frac{[1.034\,6\ (\pm0.000\,2)\ \text{g}]\left(1 - \dfrac{0.001\,2(\pm0.000\,1)\ \text{g/mL}}{8.0\ (\pm0.5)\ \text{g/mL}}\right)}{1 - \dfrac{0.001\,2(\pm0.000\,1)\ \text{g/mL}}{0.997\,299\,5\ \text{g/mL}}}$$

$$m = \frac{[1.034\,6\ (\pm0.019\,3\%)]\left(1 - \dfrac{0.001\,2\ (\pm8.33\%)}{8.0\ (\pm6.25\%)}\right)}{1 - \dfrac{0.001\,2\ (\pm8.33\%)}{0.997\,299\,5\ (\pm0\%)}}$$

$$m = \frac{[1.034\,6\ (\pm0.019\,3\%)][1 - 0.000\,150\ (\pm10.4\%)]}{[1 - 0.001\,203\ (\pm8.33\%)]}$$

$$m = \frac{[1.034\,6\ (\pm0.019\,3\%)]\ [1 - 0.000\,150\ (\pm0.000\,015\,6)]}{[1 - 0.001\,203\ (\pm0.000\,100)]}$$

$$m = \frac{[1.034\,6\ (\pm0.019\,3\%)]\ [0.999\,850\,0\ (\pm0.000\,015\,6)]}{[0.998\,797\ (\pm0.000\,100)]}$$

$$m = \frac{[1.034\,6\ (\pm0.019\,3\%)]\ [0.999\,850\,0\ (\pm0.001\,56\%)]}{[0.998\,797\ (\pm0.010\,0\%)]}$$

$$m = 1.035\,7\ (\pm0.021\,8\%) = 1.035\,7\ (\pm0.000\,2)\ \text{g}$$

CHAPTER 4
STATISTICS

1. The smaller the standard deviation, the greater the precision. There is no necessary relationship between standard deviation and accuracy. The statistics that we do in this chapter pertains to precision, not accuracy.

2. (a) $\mu \pm \sigma$ corresponds to $z = -1$ to $z = +1$. The area from $z = 0$ to $z = +1$ is 0.341 3. The area from $z = 0$ to $z = -1$ is also 0.341 3. Total area (= fraction of population) from $z = -1$ to $z = +1 = 0.682$ 6.

   (b) $z = -2$ to $z = +2 \Rightarrow$ area $= 2 \times 0.477\ 3 = 0.954\ 6$

   (c) $z = 0$ to $z = +1 \Rightarrow$ area $= 0.341\ 3$

   (d) $z = 0$ to $z = 0.5 \Rightarrow$ area $= 0.191\ 5$

   (e) Area from $z = -1$ to $z = 0$ is 0.341 3. Area from $z = -0.5$ to $z = 0$ is 0.191 5. Area from $z = -1$ to $z = -0.5$ is $0.341\ 3 - 0.191\ 5 = 0.149\ 8$.

3. (a) Mean $= \frac{1}{8}$ (1.526 60 + 1.529 74 + 1.525 92 + 1.527 31 + 1.528 94 + 1.528 04 + 1.526 85 + 1.527 93) = 1.527 67

   (b) Standard deviation =

   $$\sqrt{\frac{(1.526\ 60 - 1.527\ 67)^2 + \cdots + (1.527\ 93 - 1.527\ 67)^2}{8 - 1}} = 0.001\ 26$$

   (c) Variance $= (0.001\ 26)^2 = 1.59 \times 10^{-6}$

4. (a) 1000 hours corresponds to $z = (1000 - 845.2)/94.2 = 1.643$.
   Area from $\bar{x}$ to $z = 1.643$ is 0.449 6.
   Area beyond $z = 1.643$ is $0.500\ 0 - 0.449\ 6 = 0.050\ 4$

   (b) 800 to 845.2: $z = -0.479\ 8 \Rightarrow$ area $= 0.184\ 2$
   845.2 to 900: $z = 0.581\ 7 \Rightarrow$ area $= 0.219\ 5$
   Total area from 800 to 900 $= 0.403\ 7$

5. $y = \dfrac{4\ 768 \times 20}{94.2\sqrt{2\pi}}\, e^{-(x\ -\ 845.2)^2/2(94.2)^2} = 104.7$ when $x = 1000$.

6. The values 14.55 to 14.60 correspond to the range ($z = 0.504\ 7$) to ($z = 0.972\ 0$). Interpolating in Table 4-1, the area between the two is $0.334\ 2 - 0.193\ 1 = 0.141\ 1$.

7. Your spreadsheet should look like the one in Figure 4-11, with data covering 500 to 1 200 hours.

8.   Use the same spreadsheet as in the previous problem, but vary the standard
deviation in cell A4. Here are the results:

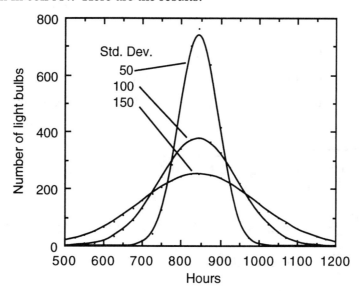

9.   A confidence interval is a region around the measured mean in which the true mean
is likely to lie.

10.   Since the bars are drawn at a 50% confidence level, 50% of them ought to include
the mean value if many experiments are performed.

11.   Case 1: Comparing a measured result to a "known" value. (Use Equation 4-6.)

Case 2: Comparing replicate measurements. (Use Equation 4-7 and 4-8.)

Case 3: Comparing individual differences. (Use Equations 4-9 and 4-10.)

12. $\bar{x} = 0.14_8, \quad s = 0.03_4$

90% confidence: $\mu = 0.14_8 \pm \dfrac{(2.015)(0.03_4)}{\sqrt{6}} = 0.14_8 \pm 0.02_8$

99% confidence: $\mu = 0.14_8 \pm \dfrac{(4.032)(0.03_4)}{\sqrt{6}} = 0.14_8 \pm 0.05_6$

13.   (a) 99% confidence interval: $\bar{x} \pm \dfrac{(3.707)(0.000\ 07)}{\sqrt{7}} = \bar{x} \pm 0.000\ 10$

(1.527 83 to 1.528 03).

(b) All of the samples lie outside the 99% confidence interval for sample 8.
There is a real variation among samples.

14.

Sample	Method 1	Method 2	d_i	$d_i - \bar{d}$	$(d_i - \bar{d})^2$
A	0.013 4	0.013 5	-0.000 1	+0.000 6	3.6×10^{-7}
B	0.014 4	0.015 6	-0.001 2	-0.000 5	2.5×10^{-7}
C	0.012 6	0.013 7	-0.001 1	-0.000 4	1.6×10^{-7}
D	0.012 5	0.013 7	-0.001 2	-0.000 5	2.5×10^{-7}
E	0.013 7	0.013 6	+0.000 1	+0.000 8	6.4×10^{-7}

$$\bar{d} = -0.000\ 70 \qquad \text{sum} = 16.6 \times 10^{-7}$$

$$s_d = \sqrt{\frac{\Sigma(d_i - \bar{d})^2}{n-1}} = \sqrt{\frac{16.6 \times 10^{-7}}{4}} = 6.4 \times 10^{-4}$$

$$t = \frac{0.000\ 70}{0.000\ 64}\sqrt{5} = 2.4_5 < 2.776 \text{ (Student's t for 4 degrees of freedom)}$$

The difference is <u>not</u> significant.

15. Sample 1: $\bar{x}_1 = 0.013\ 4_{00}$ $s_1 = 0.000\ 3_{937}$

Sample 2: $\bar{x}_2 = 0.013\ 9_{60}$ $s_2 = 0.000\ 3_{435}$

$$s_{pooled} = \sqrt{\frac{4s_1^2 + 4s_2^2}{5+5-2}} = 0.000\ 369\ 5$$

$$t = \frac{0.013\ 960 - 0.013\ 400}{0.000\ 369\ 5}\sqrt{\frac{5 \cdot 5}{5+5}} = 2.40 > 2.306 \text{ (Student's t for 8}$$

degrees of freedom)   The difference <u>is</u> significant.

16. $\mu = \bar{x} \pm \dfrac{(2.353)(1\%)}{\sqrt{4}} = \bar{x} \pm 1.1_8 \% < 1.2\%.$ The answer is yes.

17. For indicators 1 and 2: $s_{pooled} = \sqrt{\dfrac{0.002\ 25^2 \cdot 27 + 0.000\ 98^2 \cdot 17}{28 + 18 - 2}}$

$$= 0.001\ 864\ 8$$

$$t = \frac{0.095\ 65 - 0.086\ 86}{0.001\ 864\ 8}\sqrt{\frac{28 \cdot 18}{28+18}} = 15.60$$

This is much greater than t for 44 degrees of freedom, which is ~2.02.
The difference <u>is</u> significant.

For indicators 2 and 3: $s_{pooled} = 0.001\ 075\ 8$

$t = 1.39 < 2.02 \Rightarrow$ difference is not significant.

18. $t = \dfrac{255 - 238}{14}\sqrt{\dfrac{4 \cdot 5}{4+5}} = 1.81 < 2.365$ (Student's t for 7 degrees of freedom)

Difference is <u>not</u> significant.

19. Average blank value = 1.34 ppm

Corrected measurements are 4.7 - 1.34 = 3.36, 4.06, 4.86, 4.66, 3.26, 4.26,

3.86, 4.46

Measurements have $\bar{x} = 4.0_{98}$ and $s = 0.5_{80}$

For 7 degrees of freedom, t = 2.998 in Table 4-2 for 98% confidence

Detection limit = (2.998)(0.5_{80}) = 1.7_4 ppm.

20. A control chart tracks the performance of a process to see if it remains within expected bounds. Three indications that a process might be out of control are (1) a reading outside the action lines, (2) two consecutive readings outside the warning lines and (3) systematic drift away from the target value.

21. Q = (216 - 204) / (216 - 192) = 0.50 < 0.64. Retain 216.

22. Hopefully, the negative concentration lies within experimental error of the value zero. If so, there is no detectable analyte in the unknown. If the negative concentration is beyond the experimental error, there is something wrong with your analysis. The same is true for a value above 100% of the theoretical maximum concentration of an analyte.

23. Your spreadsheet should be similar to the one in Figure 4-12.

24.

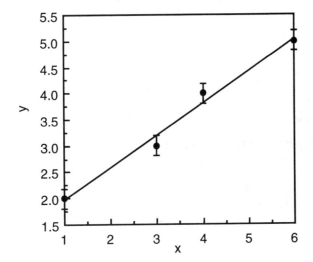

25. Slope $= -1.298\ 72 \times 10^4\ (\pm 0.001\ 319\ 0 \times 10^4) = -1.299\ (\pm 0.001) \times 10^4$

Intercept $= 256.695\ (\pm 323.57) = 3\ (\pm 3) \times 10^2$

26. 1σ: $x = \dfrac{y(\pm\sigma_y) - b(\pm\sigma_b)}{m(\pm\sigma_m)} = \dfrac{-40.00(\pm0.039) \times 10^4 - 0.03(\pm0.03) \times 10^4}{-1.298\,7(\pm0.001\,3) \times 10^4}$

$= \dfrac{-40.03(\pm0.049\,2)}{-1.298\,7(\pm0.001\,3)} = \dfrac{-40.03(\pm0.123\%)}{-1.298\,7(\pm0.10\%)} = 30.82 \pm 0.159\%$

$= 30.82 \pm 0.05$

For 95% confidence, we multiply the uncertainties by 3.182, which is student's t

for 3 degrees of freedom

$e_y = 3.182\,\sigma_y = 0.125 \times 10^4\,; \quad e_b = 3.182\,\sigma_b = 0.103 \times 10^4\,;$

$e_m = 3.182\,\sigma_m = 0.004\,1 \times 10^4$

95%: $x = \dfrac{-40.00(\pm0.125) \times 10^4 - 0.03(\pm0.103) \times 10^4}{-1.298\,7(\pm0.004\,1) \times 10^4} = 30.82 \pm 0.16$

27. (a) $x = \dfrac{y-b}{m} = \dfrac{2.58\,(\pm0.196) - 1.3_5\,(\pm0.2_1)}{0.61_5\,(\pm0.05_4)} = 2.0_0 \pm 0.5_0$

The uncertainty in y is $\sqrt{s_y^2}$.

(b) $\sigma_x^2 = \dfrac{\sigma_y^2}{m^2}\left(1 + \left(\dfrac{y-b}{m}\right)^2 \dfrac{n}{D} + \dfrac{\Sigma(x_i^2)}{D} - 2\left(\dfrac{y-b}{m}\right)\dfrac{\Sigma(x_i)}{D}\right)$

$= \dfrac{0.038\,462}{(0.615\,38)^2}\left(1 + (2.00)^2\,\dfrac{4}{52} + \dfrac{6\,2}{52} - 2\,(2.00)\,\dfrac{1\,4}{52}\right)$

$= 0.144\,5 \Rightarrow \sigma_x = 0.3_8$

(c) Replace the 1 in brackets by $\dfrac{1}{4} \Rightarrow \sigma_x = 0.2_6$

28. Slope $= -\dfrac{1}{K} \Rightarrow K = -\dfrac{1}{\text{slope}}$

Intercept $= \dfrac{nC_t}{K} \Rightarrow n = \dfrac{(K)(\text{intercept})}{C_t} = -\dfrac{(\text{intercept})}{(\text{slope})(C_t)}$

$n = \dfrac{-3.138\,6(\pm0.289\,9) \times 10^{-3}}{[-1.364\,5(\pm0.132\,8) \times 10^6][4.13(\pm0.09) \times 10^{-13}]}$

$= 5.6\,(\pm0.8) \times 10^3$

29. (a) $y(\pm 0.17) = -0.159\,75(\pm0.019\,15)x + 308._95(\pm38._00)$

(b) $y(\text{at } 2\,010) = (-0.159\,75)(2\,010) + 308.95 = -12.15$

temperature $= 10^y = 7 \times 10^{-13}$ K.

Extrapolations like this are nonsense.

There is no way to predict experimental progress.

1. Concentrations in an equilibrium constant are really dimensionless <u>ratios</u> of actual concentrations divided by standard state concentrations. Since standard states are 1 M for solutes, 1 atm for gases and pure substances for solids and liquids, these are the units we must use. A solvent is approximated as a pure liquid.

2. All concentrations in equilibrium constants are expressed as dimensionless ratios of actual concentrations divided by standard state concentrations.

3. Predictions based on free energy or Le Châtelier's principle tell us which way a reaction will go (thermodynamics), but not how long it will take (kinetics). A reaction could be over instantly or it could take forever.

4. (a) $K = 1/[Ag^+]^3 [PO_4^{3-}]$ (b) $K = P_{CO_2}^6 / P_{O_2}^{15/2}$

5. $K = \dfrac{P_E^3}{P_A^2 [B]} = \dfrac{[3.6 \times 10^4 \text{ torr}/(760 \text{ torr/atm})]^3}{\left(\dfrac{2.8 \times 10^3 \text{ Pa}}{1.013 \times 10^5 \text{ Pa/atm}}\right)^2 (1.2 \times 10^{-2} \text{ M})} = 1.2 \times 10^{10}$

6.
$$HOBr + OCl^- \rightleftarrows HOCl + OBr^- \qquad K_1 = 1/15$$
$$HOCl \rightleftarrows H^+ + OCl^- \qquad K_2 = 3.0 \times 10^{-8}$$
$$\overline{\qquad\qquad\qquad\qquad\qquad}$$
$$HOBr \rightleftarrows H^+ + OBr^- \qquad K = K_1 K_2 = 2 \times 10^{-9}$$

7. (a) Decrease (b) give off (c) negative

8. $K = e^{-(59.0 \times 10^3 \text{ J/mol})/(8.314\,510 \text{ J/(K·mol)})(298.15 \text{ K})} = 5 \times 10^{-11}$

9. (a) Right (b) right (c) neither (d) right (e) smaller

10. (a) $K = P_{H_2O} = e^{-(\Delta H° - T\Delta S°)/RT}$
$= e^{-\{[(63.11 \times 10^3 \text{ J/mol}) - (298.15K)(148 \text{ J/K/mol})]/(8.314\,510 \text{ J/K/mol})(298.15 \text{ K})\}}$
$= 4.7 \times 10^{-4} \text{ atm}$

(b) $P_{H_2O} = 1 = e^{-(\Delta H° - T\Delta S°)/RT} \Rightarrow \Delta H° - T\Delta S°$ must be zero.
$\Delta H° - T\Delta S° = 0 \Rightarrow T = \dfrac{\Delta H°}{\Delta S°} = 426 \text{ K} = 153°C$

11. (a) $K_1 = e^{-\Delta H°/RT_1} \cdot e^{\Delta S°/R}$
$K_2 = e^{-\Delta H°/RT_2} \cdot e^{\Delta S°/R}$
Dividing K_1 / K_2 gives $\dfrac{K_1}{K_2} = e^{-\Delta H°/R (1/T_1 - 1/T_2)}$

$$\Delta H° = \left(\frac{1}{T_2} - \frac{1}{T_1}\right)^{-1} R \ln \frac{K_1}{K_2}$$

Putting in $K_1 = 1.479 \times 10^{-5}$ at $T_1 = 278.15$ K and

$K_2 = 1.570 \times 10^{-5}$ at $T_2 = 283.15$ K gives $\Delta H° = +7.82$ kJ / mol.

(b) $K = e^{-\Delta H°/RT} \cdot e^{\Delta S°/R}$

$$\underset{y}{\ln K} = \underset{m}{-\frac{\Delta H°}{R}} \underset{x}{\left(\frac{1}{T}\right)} + \underset{b}{\frac{\Delta S°}{T}}$$

A graph of ln K vs 1/T will have a slope of $-\Delta H°/R$

12.   (a) $Q = \left(\dfrac{48.0 \text{ Pa}}{1.013 \times 10^5 \text{ Pa/atm}}\right)^2 \Big/ \left(\dfrac{1\,370}{1.013 \times 10^5}\right)\left(\dfrac{3\,310}{1.013 \times 10^5}\right)$

$= 5.08 \times 10^{-4} < K$ The reaction will go to the right. Note that it was not necessary to convert Pa to atm, since the units cancel.

(b) H_2 + Br_2 $\rightleftarrows$ $2HBr$

Initial pressure: 1 370   3 310   48.0

Final pressure: 1 370 - x   3 310 - x   48.0 + 2x

Note that 2x Pa of HBr are formed when x Pa of H_2 are consumed.

$$\frac{(48.0 + 2x)^2}{(1\,370-x)(3\,310-x)} = 7.2 \times 10^{-4} \Rightarrow x = 4.50 \text{ Pa}$$

$P_{H_2} = 1\,366$ Pa, $P_{Br_2} = 3\,306$ Pa, $P_{HBr} = 57.0$ Pa

(c) Neither, since Q is unchanged.

(d) HBr will be formed, since $\Delta H°$ is positive.

13.   (a) $[Cu^+][Br^-] = K_{sp}$

(x)   (x) $= 5 \times 10^{-9} \Rightarrow [Cu^+] = 7._1 \times 10^{-5}$ M

(b) $(143.45$ g/mol$)(7.1 \times 10^{-5}$ mol/L$)(0.100$ L/100 mL$) = 1._0 \times 10^{-3}$ g/100 mL

14.   (a) $[Ag^+]^4[Fe(CN)_6^{4-}] = K_{sp}$

$(4x)^4$ (x) $= 8.5 \times 10^{-45} \Rightarrow x = 5.0_6 \times 10^{-10}$ M

(b) $(643.43$ g/mol$)(5.0_6 \times 10^{-10}$ mol/L$)(0.100$ L$) = 3.3 \times 10^{-8}$ g/100 mL

(c) $[Ag^+] = 4x = 2.02 \times 10^{-9}$ M $= 2.18 \times 10^{-7}$ g/L $= 2.18 \times 10^{-7}$ mg/mL

$= 0.218$ ng/mL $= 0.22$ ppb

15. $[Ag^+] = \sqrt{K_{sp}}$ $= 1.34 \times 10^{-5}$ M for AgCl $= 1.4 \times 10^{-3}$ g/L $= 1\,400$ ppb

$= 7.07 \times 10^{-7}$ M for AgBr $= 7.6 \times 10^{-5}$ g/L $= \boxed{76 \text{ ppb}}$

$= 9.11 \times 10^{-9}$ M for AgI $= 9.8 \times 10^{-7}$ g/L $= 0.98$ ppb

16. $K = [Cu^{2+}] \, [OH^-]^{3/2} \, [SO_4^{2-}]^{1/4} = (x) \, (\frac{3}{2}x)^{3/2} \, (\frac{1}{4}x)^{1/4} = 6.9 \times 10^{-18}$

   $\Rightarrow x = [Cu^{2+}] = 5.2 \times 10^{-7} \, M.$

17. True. Two moles of F^- are lost for each mole of Ba^{2+}.

18. (a) $CaSO_4(s) \overset{K_{sp}}{\rightleftharpoons} Ca^{2+} + SO_4^{2-}$
   FW 136.14   x   x

   $x^2 = 2.4 \times 10^{-5} \Rightarrow x = 4.9_0 \times 10^{-3} \, M = 0.66_7 \, g/L$

   (b) $[Ca^{2+}] \, [SO_4^{2-}] = [0.50] \, [SO_4^{2-}] = 2.4 \times 10^{-5} \Rightarrow [SO_4^{2-}] = 4.8 \times 10^{-5} \, M$
      $= 6.5 \times 10^{-3} \, g \, CaSO_4/L$

19. $\dfrac{[Ag^+](\text{in } 0.010 \, M \, IO_3^-)}{[Ag^+](\text{in } H_2O)} = \dfrac{K_{sp}/(0.010)}{\sqrt{K_{sp}}} = 0.018$

20. (a) $[Zn^{2+}]^2[Fe(CN)_6^{4-}] = (2x)^2 \, (x) = 2.1 \times 10^{-16}$
      $\Rightarrow x = 3.74 \times 10^{-6} \, M = 1.3 \, mg/L$

   (b) $[Zn^{2+}]^2[Fe(CN)_6^{4-}] = (2x + 0.040)^2 \, (x) \approx (0.040)^2 \, (x) = 2.1 \times 10^{-16}$
      $\Rightarrow x = 1.3 \times 10^{-13} \, M.$

   (c) $[Zn^{2+}]^2[Fe(CN)_6^{4-}] = (5.0 \times 10^{-7})^2[Fe(CN)_6^{4-}] = 2.1 \times 10^{-16}$
      $\Rightarrow [Fe(CN)_6^{4-}] = 8.4 \times 10^{-4} \, M.$

21.

	A	B	C	D	E
1	y (positive)	y(3y+0.10)^3 - (1E-8)		y (negative)	y(3y+0.10)^3 - (1E-8)
2	0.1	2.979E-03		-0.002	-1.013E-08
3	0.01	6.300E-07		-0.007	-6.830E-10
4	0.001	-7.803E-09		-0.007 1	2.446E-10
5	0.002	-1.808E-09		-0.007 07	-4.053E-11
6	0.002 2	6.345E-11		-0.007 075	6.576E-12
7	0.002 19	-3.651E-11		-0.007 074 3	-2.872E-14
8	0.002 193	-6.592E-12			
9	0.002 193 6	-6.014E-13			
10	0.002 193 66	-2.243E-15			
11					
12	B2 = A2*(3*A2+0.01)^3 - 1E-8			E2 = D2*(3*D2+0.01)^3 - 1E-8	

22.

	A	B	C
1	4.236 067 98		A1 = Sqrt(4*A1+1)

23. BX_2 coprecipitates with AX_3. This means that some BX_2 is trapped in the AX_3 during precipitation of AX_3.

24. 99.9% precipitation means $[Zn^{2+}] = (0.001)(0.10) = 1.0 \times 10^{-4} \, M$

$$[Zn^{2+}][CO_3^{2-}] = K_{sp}$$
$$(1.0 \times 10^{-4})\,(x) = 1.0 \times 10^{-10} \Rightarrow x = 1.0 \times 10^{-6}\ M$$

25.

Salt	K_{sp}	$[Ag^+]$ (M, in equilibrium with 0.1 M anion)		
AgCl	1.8×10^{-10}	$K_{sp}/0.10$	=	1.8×10^{-9}
AgBr	5.0×10^{-13}	$K_{sp}/0.10$	=	5.0×10^{-12}
AgI	8.3×10^{-17}	$K_{sp}/0.10$	=	8.3×10^{-16}
Ag_2CrO_4	1.2×10^{-12}	$\sqrt{K_{sp}/0.10}$	=	3.5×10^{-6}

Order of precipitation : $I^- < Br^- < Cl^- < CrO_4^{2-}$

26. For $CaSO_4$, $K_{sp} = 2.4 \times 10^{-5}$. For Ag_2SO_4, $K_{sp} = 1.5 \times 10^{-5}$.

It appears that Ca^{2+} will precipitate first. Removing 99% of the Ca^{2+} reduces $[Ca^{2+}]$ to 0.000 500 M. The concentration of $[SO_4^{2-}]$ needed to accomplish this is $[SO_4^{2-}] = 2.4 \times 10^{-5}/0.000\,500 = 0.048\ M$.

This much $SO_4{}^{2-}$ <u>will</u> precipitate Ag_2SO_4, because $Q = [Ag^+]^2 [SO_4] = (0.030\,0)^2$ $(0.048) = 4.3 \times 10^{-5} > K_{sp}$. The separation is not feasible.

When Ag^+ first precipitates, $[SO_4^{2-}] = 1.5 \times 10^{-5}/(0.030\,0)^2 = 1.67 \times 10^{-2}\ M$.

$[Ca^{2+}] = 2.4 \times 10^{-5}/1.67 \times 10^{-2} = 0.001\,4\ M$. 97% of the Ca^{2+} has precipitated.

27. At low I^- concentration, $[Pb^{2+}]$ decreases with increasing $[I^-]$ because of the reaction $Pb^{2+} + 2I^{2-} \rightarrow PbI_2(s)$. Concentrations of other $Pb^{2+}-I^-$ species are negligible. At high I^- concentration, complex ions form by reactions such as $PbI_2(s) + I^- \rightarrow PbI_3^-$.

28. (a) BF_3 (b) AsF_5

29. $\dfrac{[SnCl_2(aq)]}{[Sn^{2+}][Cl^-]^2} = \beta_2 \Rightarrow [SnCl_2(aq)] = \beta_2[Sn^{2+}][Cl^-]^2$

$$= (12)(0.20)(0.20)^2 = 0.096\ M$$

30. $[Zn^{2+}] = K_{sp}/[OH^-]^2 = 2.9 \times 10^{-3}\ M$

$[ZnOH^+] = \beta_1[Zn^{2+}][OH^-] = \beta_1 K_{sp}/[OH^-] = 2.3 \times 10^{-5}\ M$

$[Zn(OH)_3^-] = \beta_3[Zn^{2+}][OH^-]^3 = \beta_3 K_{sp}[OH^-] = 6.9 \times 10^{-7}\ M$

$[Zn(OH)_4^{2-}] = \beta_4[Zn^{2+}][OH^-]^4 = \beta_4 K_{sp}[OH^-]^2 = 8.6 \times 10^{-14}\ M$

31.

	Na^+	$+$	OH^-	$\rightleftarrows$	$NaOH(aq)$
Initial concentration:	1		1		0
Final concentration:	1-x		1-x		x

$\dfrac{x}{(1-x)^2} = 0.2 \Rightarrow x = 0.15\ M$. 15% is in the form $NaOH(aq)$.

32.

	A	B	C	D	E	F	G	H
1	Ksp =	Log[I-]	[I-]	[Ag+]	[AgI(aq)]	[AgI2]	[AgI3]	[AgI4]
2	4.5E-17	-8	1.0E-08	4.5E-01	5.8E-01	4.0E-06	2.5E-11	1.1E-18
3	B1 =	-7	1.0E-07	4.5E-03	5.8E-02	4.0E-06	2.5E-10	1.1E-16
4	1.3E+08	-1	1.0E-01	4.5E-15	5.8E-08	4.0E-06	2.5E-04	1.1E-04
5	B2 =	0	1.0E+00	4.5E-17	5.8E-09	4.0E-06	2.5E-03	1.1E-02
6	9.0E+10							
7	B3 =		C2 = 10^B2				L2 = Log10(D2)	
8	5.6E+13		D2 = $A$2/C2^2				M2 = Log10(E2)	
9	B4 =		E2 = $A$4*D2*C2				N2 = Log10(F2)	
10	2.5E+14		F2 = $A$6*D2*C2^2				O2 = Log10(G2)	
11	K26 =		G2 = $A$8*D2*C2^3				P2 = Log10(H2)	
12	7.6E+29		H2 = $A$10*D2*C2^4				Q2 = Log10(I2)	
13	K38 =		I2 = $A$12*D2^2*C2^6				R2 = Log10(J2)	
14	2.3E+46		J2 = $A$14*D2^3*C2^8				S2 = Log10(K2)	
15			K2 = D2+E2+F2+G2+H2+2*I2+3*J2					

	I	J	K	L	M	N	O
1	[Ag2I6]	[Ag3I8]	Ag(total)	log[Ag+]	log[AgI]	log[AgI2]	log[AgI3]
2	1.5E-19	2.1E-19	1.04E+00	-3.5E-01	-2.3E-01	-5.4E+00	-1.1E+01
3	1.5E-17	2.1E-17	6.30E-02	-2.3E+00	-1.2E+00	-5.4E+00	-9.6E+00
4	1.5E-05	2.1E-05	4.62E-04	-1.4E+01	-7.2E+00	-5.4E+00	-3.6E+00
5	1.5E-03	2.1E-03	2.31E-02	-1.6E+01	-8.2E+00	-5.4E+00	-2.6E+00

	P	Q	R	S
1	log[AgI4]	log[Ag2I6]	log[Ag3I8]	log[Ag(Tot)]
2	-1.8E+01	-1.9E+01	-1.9E+01	0.015
3	-1.6E+01	-1.7E+01	-1.7E+01	-1.201
4	-3.9E+00	-4.8E+00	-4.7E+00	-3.335
5	-1.9E+00	-2.8E+00	-2.7E+00	-1.636

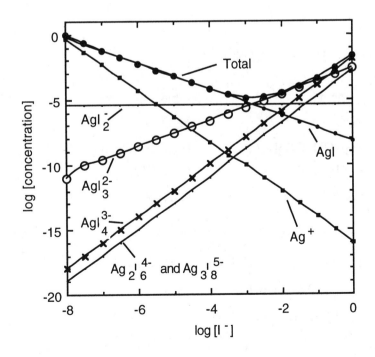

33. Lewis acids and bases are electron pair acceptors and donors, respectively:

$$F_3B \ + \ :\ddot{O}(CH_3)_2 \ \rightarrow \ F_3\overset{-}{B} - \overset{+}{O}(CH_3)_2$$

  Lewis   Lewis   Adduct
  acid   base

Brønsted acids and bases are proton donors and acceptors, respectively:

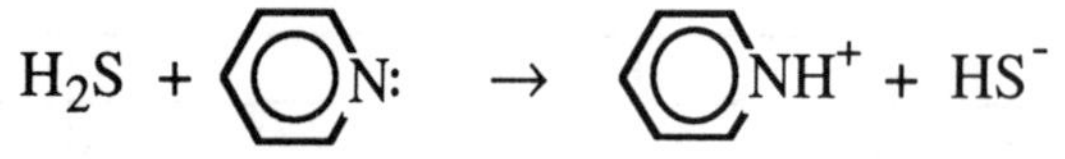

$$H_2S \ + \ \langle \rangle N: \ \rightarrow \ \langle \rangle NH^+ + HS^-$$

  Brønsted   Brønsted
  acid   base

34. (a) An adduct   (b) dative or coordinate covalent

     (c) conjugate   (d) $[H^+] > [OH^-]$; $[OH^-] > [H^+]$

35. Dissolved CO_2 from the atmosphere lowers the pH by making carbonic acid. Water can be distilled under an inert atmosphere to exclude CO_2, or most CO_2 can be removed by boiling the distilled water.

36. SO_2 in the atmosphere reacts with moisture to make H_2SO_3, which is a weak acid.

37.

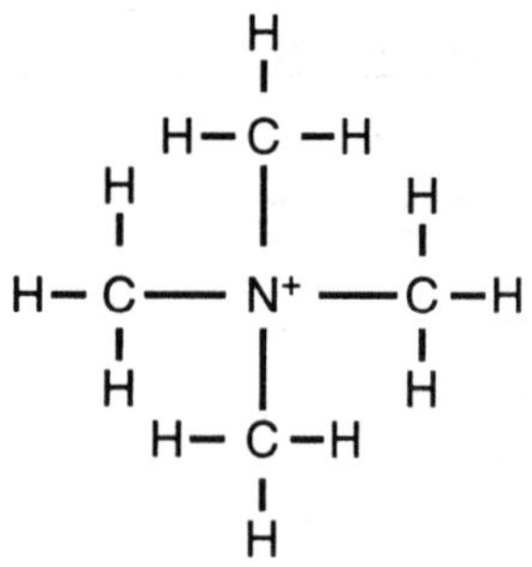

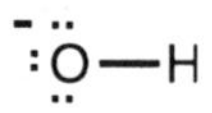

There is no place for OH^- to bond to $(CH_3)_4N^+$

38. (a) HI   (b) H_2O

39. $2H_2SO_4 \ \rightleftarrows \ HSO_4^- \ + \ H_3SO_4^+$

40.

	acid	base
(a)	H_3O^+	H_2O
(a)	$H_3\overset{+}{N}CH_2CH_2\overset{+}{N}H_3$	$H_3\overset{+}{N}CH_2CH_2NH_2$
(b)	$C_6H_5CO_2H$	$C_6H_5CO_2^-$
(b)	$C_5H_5NH^+$	C_5H_5N

41. (a) $[H^+] = 0.010 \ M \Rightarrow pH = -\log [H^+] = 2.00$

     (b) $[OH^-] = 0.035 \ M \Rightarrow [H^+] = K_w / [OH^-] = 2.86 \times 10^{-13} \ M \Rightarrow pH = 12.54$

     (c) $[H^+] = 0.030 \ M \Rightarrow pH = 1.52$

     (d) $[H^+] = 3.0 \ M \Rightarrow pH = -0.48$

     (e) $[OH^-] = 0.010 \ M \Rightarrow [H^+] = 1.0 \times 10^{-12} \ M \Rightarrow pH = 12.00$

42. $K_w = [H^+] [OH^-] = 1.14 \times 10^{-15}$ at $0°$

            x   x

$x^2 = 1.14 \times 10^{-15} \Rightarrow x = [H^+] = 3.38 \times 10^{-8}$ M $\Rightarrow$ pH = $-\log [H^+] = 7.472$

At $25°$, pH = 6.998 and at $100°$ C, pH = 6.132

43. Since $[H^+] [OH^-] = 1.0 \times 10^{-14}$, $K = [H^+]^4 [OH^-]^4 = 1.0 \times 10^{-56}$

44. $[La^{3+}] [OH^-] = K_{sp} = 2 \times 10^{-21}$

      $[OH^-]^3 = K_{sp} / (0.010) \Rightarrow [OH^-] = 5.8 \times 10^{-7}$ M $\Rightarrow$ pH = 7.8

45. (a) At $25°$ C, K_w increases as temperature increases $\Rightarrow$ endothermic

     (b) At $100°$ C, K_w increases as temperature increases $\Rightarrow$ endothermic

     (c) At $300°$ C, K_w decreases as temperature increases $\Rightarrow$ exothermic

46. See Table 5-2.

47. Weak acids: RCO_2H $R_3NH^+X^-$
                  Carboxylic Ammonium
                    acids   ions

    Weak bases: R_3N: $RCO_2^-M^+$
                  Amines   Carboxylate
                           ions

48. $Cl_3CCO_2H \rightleftarrows Cl_3CCO_2^- + H^+$ $C_6H_5\overset{+}{N}H_3 \rightleftarrows C_6H_5-NH_2 + H^+$

49. $C_5H_5N: + H_2O \rightleftarrows C_5H_5NH^+ + OH^-$

    $HOCH_2CH_2S^- + H_2O \rightleftarrows HOCH_2CH_2SH + OH^-$

50. K_a: $HCO_3^- \rightleftarrows H^+ + CO_3^{2-}$ K_b: $HCO_3^- + H_2O \rightleftarrows H_2CO_3 + OH^-$

51. (a) $H_3\overset{+}{N}CH_2CH_2\overset{+}{N}H_3 \overset{K_{a1}}{\rightleftarrows} H_2NCH_2CH_2\overset{+}{N}H_3 + H^+$

        $H_2NCH_2CH_2\overset{+}{N}H_3 \overset{K_{a2}}{\rightleftarrows} H_2NCH_2CH_2NH_2 + H^+$

    (b) $^-O_2CCH_2CO_2^- + H_2O \overset{K_{b1}}{\rightleftarrows} HO_2CCH_2CO_2^- + OH^-$

        $HO_2CCH_2CO_2^- + H_2O \overset{K_{b2}}{\rightleftarrows} HO_2CCH_2CO_2H + OH^-$

52. (a) , (c)

53. $CN^- + H_2O \rightleftarrows HCN + OH^-$ $K_b = \dfrac{K_w}{K_a} = 1.6 \times 10^{-5}$

54. $H_2PO_4^- \overset{K_{a2}}{\rightleftarrows} HPO_4^{2-} + H^+$ $HC_2O_4^- + H_2O \overset{K_{b2}}{\rightleftarrows} H_2C_2O_4 + OH^-$

55. $K_{a1} = \dfrac{K_w}{K_{b3}} = 7.04 \times 10^{-3}$ $K_{a2} = \dfrac{K_w}{K_{b2}} = 6.29 \times 10^{-8}$

$K_{a3} = \dfrac{K_w}{K_{b1}} = 7.1 \times 10^{-13}$

56. Add the two reactions and multiply their equilibrium constants to get $K = 3.0 \times 10^{-6}$.

57. (a) $Ca(OH)_2 (s) \rightleftarrows Ca^{2+} + 2\,OH^-$

　　　　　　　　　　　　　　　 x　　　 2x

$x(2x)^2 = K_{sp} = 10^{-5.19} \Rightarrow x = 1.2 \times 10^{-2}\,M$

(b) Since some Ca^{2+} reacts with OH^-, the K_{sp} reaction will be drawn to the right, and the solubility of $Ca(OH)_2$ will be greater than we expect just on the basis of K_{sp}.

58. Reversing the first reaction and then adding the four reactions gives

$Ca^{2+} + CO_2(g) \rightleftarrows CaCO_3(s) + 2H^+$

$K = (3.4 \times 10^{-2})(4.4 \times 10^{-7})(4.7 \times 10^{-11})/(6.0 \times 10^{-9})$

$\dfrac{[H^+]^2}{[Ca^{2+}]P_{CO_2}} = \dfrac{(1.8 \times 10^{-7})^2}{[Ca^{2+}][0.10]} = K = 1.17 \times 10^{-10}$

$[Ca^{2+}] = 2.76 \times 10^{-3}\,M = 0.22\,g/2.00\,L$

A FIRST LOOK AT SPECTROPHOTOMETRY

1. (a) Double   (b) halve   (c) double

2. (a) $E = h\nu = hc/\lambda = (6.626\ 2 \times 10^{-34}\ \text{J s})(2.997\ 9 \times 10^8\ \text{m s}^{-1})/(650 \times 10^{-9}\ \text{m})$

        $= 3.06 \times 10^{-19}\ \text{J/photon} = 184\ \text{kJ/mol}$

   (b) For $\lambda = 400$ nm, $E = 299$ kJ/mol.

3. $\nu = c/\lambda = 2.997\ 9 \times 10^8\ \text{m s}^{-1}/562 \times 10^{-9}\ \text{m} = 5.33 \times 10^{14}\ \text{Hz}$

   $\tilde{\nu} = 1/\lambda = 1.78 \times 10^6\ \text{m}^{-1} \times (1\ \text{m}/100\ \text{cm}) = 1.78 \times 10^4\ \text{cm}^{-1}$

   $E = h\nu = (6.626\ 2 \times 10^{-34}\ \text{J s})(5.33 \times 10^{14}\ \text{s}^{-1}) = 3.53 \times 10^{-19}\ \text{J/photon}$

        $= 213$ kJ/mol (after multiplication by Avogadro's number).

4. Microwave energies correspond to molecular rotation energies. Infrared corresponds to vibrations. Visible light can promote electrons to excited states (in colored compounds). Ultraviolet light also promotes electrons and can even break chemical bonds.

5. From the definition of index of refraction we can write

   $c_{\text{vacuum}} = n \cdot c_{\text{air}}$

   $\lambda_{\text{vacuum}} \cdot \nu = n \cdot \lambda_{\text{air}} \cdot \nu$

   $\lambda_{\text{air}} = \lambda_{\text{vacuum}}/n$

   $\nu = c/\lambda_{\text{vacuum}} = 5.088\ 491\ 0$ and $5.083\ 335\ 8 \times 10^{14}$ Hz

   $\lambda_{\text{air}} = \lambda_{\text{vacuum}}/n = 588.985\ 54$ and $589.582\ 86$ nm

   $\tilde{\nu}_{\text{air}} = 1/\lambda_{\text{air}} = 1.697\ 834\ 5$ and $1.696\ 114\ 4 \times 10^4\ \text{cm}^{-1}$

6. Transmittance (T) is the fraction of incident light that is transmitted by a substance: $T = P/P_0$, where P_0 is incident power and P is transmitted power. Absorbance is logarithmically related to transmittance: $A = -\log T$. When all light is transmitted, absorbance is zero. When no light is transmitted, absorbance is infinite. Absorbance is proportional to concentration. Molar absorptivity is the constant of proportionality between absorbance and the product cb, where c is concentration and b is pathlength.

7. An absorption spectrum is a graph of absorbance vs wavelength.

8. The color of transmitted light is the complement of the color that is absorbed. If blue green light is absorbed, red light is transmitted.

9. Luminescence is light given off after a molecule absorbs light. Chemiluminscence is light given off by a molecule created in an excited state in a chemical reaction.

10. A single-beam spectrophotometer has one beam of light, into which a sample or reference cell can be placed. In the double-beam instrument, a rapidly chopped beam is alternately passed through sample and reference. The double-beam spectrophotometer continuously compensates for drift in source intensity, since the sample and reference are compared many times each second.

11. If absorbance is too high, too little light reaches the detector for accurate measurement. If absorbance is too low, there is too little difference between sample and reference for accurate measurement.

12. $\varepsilon = A/bc = 0.822/[(1.00 \text{ cm})(2.31 \times 10^{-5} \text{ M})] = 3.56 \times 10^4 \text{ M}^{-1} \text{ cm}^{-1}$

13. Violet blue, according to Table 6-1.

14. [Fe] in reference cell $= \left(\dfrac{10.0}{50.0}\right)(6.80 \times 10^{-4}) = 1.36 \times 10^{-4}$ M. Setting the absorbances of sample and reference equal to each other gives $\varepsilon_s b_s c_s = \varepsilon_r b_r c_r$. But $\varepsilon_s = \varepsilon_r$, so $(2.48 \text{ cm})c_s = (1.00 \text{ cm})(1.36 \times 10^{-4} \text{ M}) \Rightarrow c_s = 5.48 \times 10^{-5}$ M. This is a 1/4 dilution of runoff, so [Fe] in runoff $= 2.19 \times 10^{-4}$ M.

15. (a) Measured from graph:

$\sigma \approx 1.3 \times 10^{-20} \text{ cm}^2$ at 325 nm $\qquad\qquad \sigma \approx 3.5 \times 10^{-19} \text{ cm}^2$ at 300 nm

at 325 nm: $T = e^{-(8 \times 10^{18} \text{ cm}^{-3})(1.3 \times 10^{-20} \text{ cm}^2)(1 \text{ cm})} = 0.90$

$A = -\log T = 0.045$

at 300 nm: $T = e^{-(8 \times 10^{18} \text{ cm}^{-3})(3.5 \times 10^{-19} \text{ cm}^2)(1 \text{ cm})} = 0.061$

$A = -\log T = 1.22$

(b) $T = e^{-n\sigma b} \qquad 0.14 = e^{-(8 \times 10^{18} \text{ cm}^{-3})\sigma (1 \text{ cm})} \Rightarrow \sigma = 2.4_{576} \times 10^{-19} \text{ cm}^2$

If n is decreased by 1%, $T = e^{-(7.92 \times 10^{18} \text{ cm}^{-3})(2.457 6 \times 10^{-19} \text{ cm}^2)(1 \text{ cm})}$

$= 0.142 8$

Increase in transmittance is $\dfrac{0.142 8 - 0.14}{0.14} = 2.0\%$.

Note that the fractional increase in transmittance is greater than the fractional decrease in ozone concentration.

(c) $T_{\text{winter}} = e^{-(290 \text{ D.U.})(2.69 \times 10^{16} \text{ molecules/cm}^3/\text{D.U.})(2.5 \times 10^{-19} \text{ cm}^2)(1 \text{ cm})}$

$= 0.142$

$T_{\text{summer}} = e^{-(350)(2.69 \times 10^{16})(2.5 \times 10^{-19})} = 0.095$

Fractional increase in transmittance is $(0.142 - 0.095) / (0.095) = 49\%$.

16. Neocuproine reacts with Cu(I) and prevents it from forming a complex with ferrozine that would give a false positive result in the analysis of iron.

17. (a) $c = A/\varepsilon b = 0.427/[(6\ 130\ M^{-1}\ cm^{-1})(1.000\ cm)] = 6.97 \times 10^{-5}\ M$

(b) The sample had been diluted $\times 10 \Rightarrow 6.97 \times 10^{-4}\ M$

(c) $\dfrac{x\ g}{(292.16\ g/mol)(5 \times 10^{-3}\ L)} = 6.97 \times 10^{-4}\ M \Rightarrow x = 1.02\ mg$

18. Yes.

19. (a) The absorbance due to the colored product in sample C is $0.967 - 0.622 = 0.345$. The concentration of colored product due to added nitrite in sample C is $\dfrac{(7.50 \times 10^{-3}\ M)(10.0 \times 10^{-6}\ L)}{0.054\ L} = 1.389 \times 10^{-6}\ M$

$\varepsilon = A/bc = 0.345/[(1.389 \times 10^{-6})(5.00)] = 4.97 \times 10^{4}\ M^{-1}\ cm^{-1}$

(b) 7.50×10^{-8} mol of nitrite (from 10.0 μL added to sample C) gives $A = 0.345$. In sample B, x mole of nitrite in food extract gives $A = 0.622 - 0.153 = 0.469$.

$\dfrac{x\ mol}{7.50 \times 10^{-8}\ mol} = \dfrac{0.469}{0.345} \Rightarrow x = 1.020 \times 10^{-7}\ mol\ NO_2^- = 4.69\ \mu g$

20.

Curve	Absorbance Measured on Graph	$[Fe^{3+}]$ (ng/mL)
a	0.084	0
b	0.278	0.40
c	0.472	0.80
d	0.666	1.20
e	0.860	1.60

(a) A graph of absorbance vs $[Fe^{3+}]$(ng/mL) has a slope (k) of 0.485 and an intercept (b) of 0.084.

(b) $[Fe^{3+}]$ in final solution $= \dfrac{1}{k}(A_t - b) = \left(\dfrac{1}{0.485}\right)(0.515 - 0.084)$

$= 0.889$ ng/mL. But the unknown was diluted from 5.00 to 26.00 mL.

Therefore $[Fe^{3+}]$ in unknown $= \left(\dfrac{26.00}{5.00}\right)(0.889) = 4.6_2$ ng/mL $= 82.7$ nM.

21. At some point in almost every analytical method, a primary standard whose composition is presumed to be known is used to standardize other reagents and, ultimately, is compared to the composition of the unknown. If the primary standard is not what we think it is, the analysis of the unknown will be incorrect.

22. Corrected absorbance = 0.264 - 0.095 = 0.169

$0.169 = 0.016\ 30\ x + 0.004\ 7 \Rightarrow x = 10.1\ \mu g$

23. (a) Absorbance = $0.266_0 \pm 0.006_1$;   blank = $0.097_0 \pm 0.004_3$

(b) corrected absorbance = $0.266_0\ (\pm 0.006_1) - 0.097_0\ (\pm 0.004_3)$

$= 0.169_0 \pm 0.007_5$

(c) $x = \dfrac{0.169_0\ (\pm 0.007_5) - 0.004_7\ (\pm 0.002_6)}{0.016\ 3_0\ (\pm 0.000\ 2_2)} = \dfrac{0.164_3\ (\pm 0.007_9)}{0.016\ 3_0\ (\pm 0.000\ 2_2)}$

$= 10.1 \pm 0.5\ \mu g$

(d) $x = \dfrac{0.169_0\ (\pm 0.005_9) - 0.004_7\ (\pm 0.002_6)}{0.016\ 3_0\ (\pm 0.000\ 2_2)} = 10.1 \pm 0.4\ \mu g$

24. (a) $m = 869 \pm 11,\ b = -22.1 \pm 8.9$

(b) unknown signal = 154.0 ± 1.4 mV, blank signal = 9.0 ± 1.3

(c) corrected signal = $154.0\ (\pm 1.4) - 9.0\ (\pm 1.3) = 145.0 \pm 1.9$

(d) $[CH_4] = \dfrac{145.0\ (\pm 1.9) + 22.1\ (\pm 8.9)}{869\ (\pm 11)} = \dfrac{167.1\ (\pm 9.1)}{869\ (\pm 11)} = 0.192\ (\pm 0.011)$

25. (a)

As(III) (μM)	Current (nA)	As(III) (μM)	Current (nA)
20	319	40	633
20	319	40	636
20	319	40	630
20	319	40	630
20	317	40	628
20	317	40	632
30	472	50	775
30	475	50	772
30	473	50	772
30	472	50	764
30	474	50	766
30	475	50	770

(b) $m = 15.12\ (\pm 0.10),\quad b = 18.9\ (\pm 3.8),\quad s_y = 5.7$

(c) $x = \dfrac{501\ (\pm 5.7) - 18.9\ (\pm 3.8)}{15.12\ (\pm 0.10)} = \dfrac{482.1\ (\pm 6.9)}{15.12\ (\pm 0.10)} = 31.9\ (\pm 0.5)\ \mu M$

26. $0.350 = -1.17 \times 10^{-4}\ x^2 + 0.018\ 58\ x - 0.000\ 7$

$1.17 \times 10^{-4}\ x^2 - 0.018\ 58\ x + 0.350\ 7 = 0$

$x = \dfrac{+0.018\ 58 \pm \sqrt{0.018\ 58^2 - 4\ (1.17 \times 10^{-4})\ (0.350\ 7)}}{2\ (1.17 \times 10^{-4})} = 137$ or $21.9\ \mu g$

Correct answer is 21.9 μg

27. (a) The graph is linear over the entire range.

(b) log (current, nA) = 0.969 2 log (concentration, μg/mL) + 1.339

28. In the simplest case, the colorimetric reagent in the glass particles reacts completely with analyte in the water. For a given volume of water, the quantity of analyte is proportional to the concentration of analyte. The amount of colorimetric reagent reacting with analyte is proportional to the length of column that has changed color. Even if only a fraction of colorimetric reagent reacts with analyte, the length of column that changes color is still directly proportional to the quantity of analyte in the water.

29. Standard addition is appropriate when the sample matrix is unknown or complex and hard to duplicate, and unknown matrix effects are anticipated. An internal standard can be added to an unknown at the start of a procedure in which uncontrolled losses of sample will occur. The relative amounts of unknown and standard remain constant.

30. A small volume of standard will not change the sample matrix very much, so matrix effects remain nearly constant. If large, variable volumes of standard are used, the matrix is different in every mixture and the matrix effects will be different in every sample.

31. (a) $[Cu^{2+}]_f = [Cu^{2+}]_i \dfrac{V_i}{V_f} = 0.950\,[Cu^{2+}]_i$

(b) $[S]_f = [S]_i \dfrac{V_i}{V_f} = (100.0\ \text{ppm})\left(\dfrac{1.00\ \text{mL}}{100.0\ \text{mL}}\right) = 1.00\ \text{ppm}$

(c) $\dfrac{[Cu^{2+}]_i}{1.00\ \text{ppm} + 0.950[Cu^{2+}]_i} = \dfrac{0.262}{0.500} \Rightarrow [Cu^{2+}]_i = 1.04\ \text{ppm}$

32. (a) $[S] = (8.47\ \text{mM})\left(\dfrac{1.00\ \text{mL}}{10.0\ \text{mL}}\right) = 0.847\ \text{mM}$

(b) $\dfrac{[X]/[S]\ \text{in unknown}}{[X]/[S]\ \text{in standard}} = \dfrac{\text{signal ratio in unknown}}{\text{signal ratio in standard mixture}}$

$\dfrac{[X]/(0.847)}{(3.47/1.72)} = \dfrac{5\ 428/4\ 431}{3\ 473/10\ 222} \Rightarrow [X] = 6.16\ \text{mM}$

(c) The original concentration of [X] was twice as great as the diluted concentration, so [X] = 12.3 mM.

33. We seek the value of x (= concentration) when y (= signal) is exactly zero:

$0 = 0.371\ 3\ (\pm 0.005\ 4)\ x + 4.22\ (\pm 0.13)$

$$x = \left| \frac{-4.22 \ (\pm 0.13)}{0.371 \ 3 \ (\pm 0.005 \ 4)} \right| = 11.3_7 \ (\pm 0.3_9) \ \text{ppm}$$

s_y is not used in the calculation, because y is exactly zero at the x intercept.

34.

Volume of added x	[X] (mM)	Counts/min
0 μL	0	1 084
100.0	1.060	1 844
200.0	2.116	2 473
300.0	3.167	3 266
400.0	4.214	4 010

(a)-(b) Plotting counts/min vs [X](mM) gives the straight line

counts/min (± 45) = 690 (± 13) [X] + 1 078 (± 35) whose x intercept is

obtained by setting counts/min to exactly zero:

$$[X] = \left| \frac{-1 \ 078 \ (\pm 35)}{690 \ (\pm 13)} \right| = 1.56_2 \ (\pm 0.05_9) \ \text{mM}$$

35.   In the standard mixture $\dfrac{[DDT]}{[CHCl_3]} = \dfrac{0.800 \ \text{mM}}{0.500 \ \text{mM}}$ and the signal ratio is $\dfrac{10.1 \ \mu A}{15.3 \ \mu A}$

Chloroform added to unknown = $(10.2 \times 10^{-6} \ \text{L}) (1 \ 484 \ \text{g/L}) = 0.015 \ 1_4 \ \text{g} =$
0.126$_8$ mmol in 0.100 L = 1.26$_8$ mM

$$\frac{[DDT]/1.268}{0.800/0.500} = \frac{8.7/29.4}{10.1/15.3} \Rightarrow [DDT] = 0.909 \ \text{mM}$$

[DDT] in unknown = $(0.909 \ \text{mM}) \left(\dfrac{100 \ \text{mL}}{10.0 \ \text{mL}} \right) = 9.09 \ \text{mM}$

VOLUMETRIC ANALYSIS

1. The concentrations of reagents used in an analysis are determined either by weighing out supposedly pure primary standards or by reaction with such standards. If the standards are not pure, none of the concentrations will be correct.

2. The equivalence point occurs when the exact stoichiometric quantities of reagents have been mixed. The end point, which comes near the equivalence point, is marked by a sudden change in a physical property brought about by the disappearance of a reactant or appearance of a product.

3. In a blank titration, the quantity of titrant required to reach the end point in the absence of analyte is measured. This is subtracted from the quantity of titrant needed in the presence of analyte.

4. In a direct titration, titrant reacts directly with analyte. In a back titration a known excess of reagent that reacts with analyte is used. The excess is then titrated.

5. 40.0 mL of 0.040 0 M $Hg_2(NO_3)_2$ = 1.60 mmol of Hg_2^{2+}, which will require 3.20 mmol of KI. This is contained in volume = $\dfrac{3.20 \text{ mmol}}{0.100 \text{ mmol/mL}}$ = 32.0 mL

6. 108.0 mL of 0.165 0 M oxalic acid = 17.82 mmol, which requires $\left(\dfrac{2}{5}\right)(17.82)$ = 7.128 mmol of MnO_4^-. 7.128 mmol / (0.165 0 mmol/mL) = 43.2 mL of $KMnO_4$. An easy way to see this is to note that the reagents are both 0.165 0 M. Volume of $MnO_4^- = \dfrac{2}{5}$ (volume of oxalic acid). For the second part of the question, volume of oxalic acid $= \dfrac{5}{2}$ (volume of MnO_4^-) = 270.0 mL.

7. 1.69 mg of NH_3 = 0.099 2 mmol of NH_3. This will react with $\dfrac{3}{2}(0.099 \ 2)$ = 0.149 mmol of OBr^-. The molarity of OBr^- is 0.149 mmol/1.00 mL = 0.149 M

8. 5.00 mL of 0.033 6 M HCl = 0.168 0 mmol. 6.34 mL of 0.010 0 M NaOH = 0.063 4 mmol. HCl consumed by NH_3 = 0.168 0 - 0.063 4 = 0.104 6 mmol = 1.465 mg of nitrogen. 256 μL of protein solution contains 9.702 mg protein. 1.465 mg of N/9.702 mg protein = 15.1 wt%.

9. (a) Theoretical molarity = 3.214/158.034 = 0.020 34 M.

    (b) 25.00 mL of 0.020 34 M $KMnO_4$ = 0.508 5 mmol. But 2 moles of MnO_4^- react with 5 moles of H_3AsO_3, which comes from $\dfrac{5}{2}$ moles of As_2O_3. The moles of As_2O_3 needed to react with 0.508 5 mmol of MnO_4^- =

$(^1/_2)(^5/_2)(0.508\ 5)$ = 0.635 6 mmol = 0.125 7 g of As_2O_3.

(c) $\dfrac{0.508\ 5\ \text{mmol KMnO}_4}{0.125\ 7\ \text{g As}_2\text{O}_3} = \dfrac{x\ \text{mmol KMnO}_4}{0.146\ 8\ \text{g As}_2\text{O}_3} \Rightarrow x = 0.593\ 9$ mmol

KMnO$_4$ in (29.98 - 0.03) = 29.95 mL $\Rightarrow$ [KMnO$_4$] = 0.019 83 M.

10. FW of NaCl = 58.443. FW of KBr = 119.002. 48.40 mL of 0.048 37 M Ag$^+$ = 2.341 1 mmol. This must equal the mmol of (Cl$^-$+ Br$^-$). Let x = mass of NaCl and y = mass of KBr.   x + y = 0.238 6 g.

$$\underbrace{\frac{x}{58.443}}_{\text{moles of Cl}^-} + \underbrace{\frac{y}{119.002}}_{\text{moles of Br}^-} = 2.341\ 1 \times 10^{-3}\ \text{mol}$$

Substituting x = 0.238 6 -y gives y = 0.200 0 g of KBr = 1.681 mmol of KBr = 1.681 mmol of Br = 0.134 3 g of Br = 56.28% of the sample.

11. Let x = mg of FeSO$_4 \cdot$(NH$_4$)$_2$SO$_4 \cdot$6H$_2$O and (54.85 -x) = mg of FeCl$_2 \cdot$6H$_2$O

mmol of Ce^{4+} = mmol FeSO$_4 \cdot$(NH$_4$)$_2$SO$_4 \cdot$6H$_2$O + mmol FeCl$_2 \cdot$6H$_2$O

$(13.39\ \text{mL})(0.012\ 34\ \text{M}) = \dfrac{x\ \text{mg}}{392.13\ \text{mg/mmol}} + \dfrac{(54.85 - x)}{234.84\ \text{mg/mmol}}$

$\Rightarrow$ x = 40.01 mg FeSO$_4 \cdot$(NH$_4$)$_2$SO$_4 \cdot$6H$_2$O

mass of FeCl$_2 \cdot$6H$_2$O = 14.84 mg = 0.063 19 mmol = 4.48 mg Cl

wt% Cl = $\dfrac{4.48\ \text{mg}}{54.85\ \text{mg}} \times 100$ = 8.17%

12. 30.10 mL of Ni^{2+} reacted with 39.35 mL of 0.013 07 M EDTA. Therefore, the Ni^{2+} molarity is

$$[\text{Ni}^{2+}] = \frac{(39.35\ \text{mL})(0.013\ 07\ \text{mol/L})}{30.10\ \text{mL}} = 0.017\ 09\ \text{M}.$$

25.00 mL of Ni^{2+} contains 0.427 2 mmol of Ni^{2+}. 10.15 mL of EDTA = 0.132 7 g of EDTA. The Ni^{2+} which must have reacted with CN$^-$ was 0.427 2 - 0.132 7 = 0.294 5 mmol. The cyanide which reacted with Ni^{2+} must have been (4)(0.294 5) = 1.178 mmol. [CN$^-$] = 1.178 mmol/12.73 mL = 0.092 54 M.

13. Prior to the equivalence point, all added Fe(III) binds to the protein to form a red colored complex whose absorbance is measured in the Figure. After the equivalence point, there are no more protein binding sites available. The slight increase in absorbance arises from the color of the iron reagent in the titrant.

14. A small quantity of precipitate scatters little light, so the signal seen in nephelometry is small. However, increasing the intensity of incident light increases the intensity of scattered light. Therefore, the nephelometric signal can be increased by using a stronger light source. Turbidimetry measures the fraction of incident light that is transmitted. This fraction is independent of incident light intensity.

15. (a) 163×10^{-6} L of 1.43×10^{-3} M Fe(III) $= 2.33 \times 10^{-7}$ mol Fe(III)

(b) 1.17×10^{-7} mol transferrin in 2.00×10^{-3} L $\Rightarrow 5.83 \times 10^{-5}$ M transferrin.

16. Theoretical equivalence point =

$$\frac{\left(2 \dfrac{\text{mol Ga}}{\text{mol transferrin}}\right)\left(\dfrac{0.003\ 57 \text{ g transferrin}}{81\ 000 \text{ g transferrin/mol transferrin}}\right)}{0.006\ 64 \dfrac{\text{mol Ga}}{\text{L}}} = 13.3 \text{ } \mu\text{L}$$

Observed end point ≈ intersection of lines taken from first 6 points and last 4 points in the graph below = 12.2 μL, corresponding to $\dfrac{12.2}{13.3} = 91.7\%$ of 2 Ga/transferrin = 1.83 Ga/transferrin. In the absence of oxalate, there is no evidence for specific binding of Ga to the protein, since the slope of the curve is small and does not change near 1 or 2 Ga/transferrin.

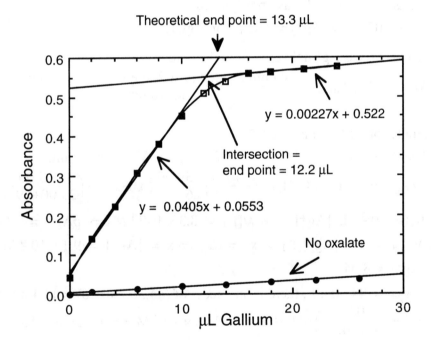

17. In nephelometry the scatter would increase until the equivalence point, when all sulfate has precipitated. In turbidimetry, the transmittance would decrease until the equivalence point, when all of the sulfate has been consumed. The end points would not be as sharp as shown in the schematic diagram below.

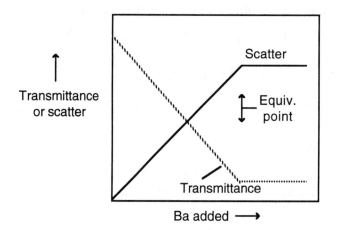

18. (i) I^-(excess) + Ag^+ → AgI (s) $[Ag^+]$ = K_{sp} (for AgI) / $[I^-]$

     (ii) A stoichiometric quantity of Ag^+ has been added that would be just equivalent
           to I^-, if no Cl^- were present. Instead, a tiny amount of $AgCl$ precipitates and a
           slight amount of I^- remains in solution.

     (iii) Cl^-(excess) + Ag^+ → $AgCl$ (s) $[Ag^+]$ = K_{sp} (for AgCl) / $[Cl^-]$

     (iv) Virtually all I^- and Cl^- have precipitated.
           $$[Ag^+] \approx [Cl^-] \Rightarrow [Ag^+] = \sqrt{K_{sp} \text{ (for AgCl)}}$$

     (v)   There is excess Ag^+ delivered from the buret.
           $$[Ag^+] = [Ag^+]_{titrant} \cdot \frac{\text{volume added past 2nd equivalence point}}{\text{total volume}}$$

19. At V_e, moles of Ag^+ = moles of I^-

     $(V_e$ mL)(0.051 1 M) = (25.0 mL)(0.082 3 M) $\Rightarrow V_e$ = 40.26 mL

     When V_{Ag^+} = 39.00 mL, $[I^-] = \dfrac{40.26 - 39.00}{40.26}(0.082\ 30)\left(\dfrac{25.00}{25.00+39.00}\right)$

     = 1.006×10^{-3} M. $[Ag^+]$ = $K_{sp}/[I^-]$ = 8.3×10^{-14} M $\Rightarrow$ pAg^+ = 13.08.

     When V_{Ag^+} = V_e, $[Ag^+][I^-]$ = x^2 = $K_{sp} \Rightarrow x$ = $[Ag^+]$ = 9.1×10^{-9} M
           $\Rightarrow$ pAg^+ = 8.04.

     When V_{Ag^+} = 44.30 mL, there is an excess of (44.30 - 40.26) = 4.04 mL of Ag^+.

     $[Ag^+] = \left(\dfrac{4.04}{25.00+44.30}\right)(0.051\ 10)$ = 2.98×10^{-3} M $\Rightarrow$ pAg^+ = 2.53.

20. (a) moles of La^{3+} = $\frac{2}{3}$ (moles of $C_2O_4^{2-}$)

          $(V_e)(0.025\ 7$ M) = $\frac{2}{3}$(25.00 mL) (0.031 1 M) $\Rightarrow$ V_e = 20.17 mL

     (b) The fraction of $C_2O_4^{2-}$ remaining when 10.00 mL of La^{3+} have been added
          is (20.17 - 10.00)/(20.17) = 0.504 2. The concentration of $C_2O_4^{2-}$ is

          $[C_2O_4^{2-}]$ = (0.504 2)(0.031 1 M) $\left(\dfrac{25.00}{35.00}\right)$ = 0.011 2 M

$$[La^{3+}]^2 = K_{sp}/[C_2O_4^{2-}]^3 \Rightarrow [La^{3+}] = 2.7 \times 10^{-10} \Rightarrow pLa^{3+} = 9.57$$

(c) $[La^{3+}]^2 [C_2O_4^{2-}]^3 = (x)^2 (\frac{3}{2}x)^3 = K_{sp} \Rightarrow x = 7.84 \times 10^{-6}$ M

$$pLa^{3+} = 5.11$$

(d) $[La^{3+}] = (0.025\ 7$ M$) \left(\dfrac{25.00 - 20.17}{50.00}\right) = 0.002\ 48$ M. $pLa^{3+} = 2.61$

21. (a) moles of $Th^{4+} = \frac{1}{4}$ (moles of F^-)

$(V_e) (0.010\ 0$ M$) = \frac{1}{4} (10.00$ mL$)(0.100$ M$) \Rightarrow V_e = 25.0$ mL

(b) $[Th^{4+}][F^-]^4 = K_{sp} \Rightarrow [Th^{4+}] = (5 \times 10^{-29})/(0.087\ 3)^4 =$
$8.6 \times 10^{-25} \Rightarrow pTh^{4+} = 24.07$

22. The reaction is $3Hg_2^{2+} + 2Co(CN)_6^{3-} \rightarrow (Hg_2)_3[Co(CN)_6]_2(s)$, with $V_e = 75.0$ mL.
Excess $Hg_2(NO_3)_2 = 15.0$ mL. The first 75.0 mL was used up precipitating product.

$$[Hg_2^{2+}] = \left(\frac{15.0}{140.0}\right)(0.010\ 0) = 1.071 \times 10^{-3}\ M$$

$$[Hg_2^{2+}]^3 [Co(CN)_6^{3-}]^2 = K_{sp} \Rightarrow [Co(CN)_6^{3-}] = \sqrt{\frac{K_{sp}}{[Hg_2^{2+}]^3}} = \sqrt{\frac{1.9 \times 10^{-37}}{(1.071 \times 10^{-3})^3}}$$

$$= 1.2 \times 10^{-14} \Rightarrow pCo(CN)_6^{3-} = 13.91$$

23. Titration of 40.00 mL of 0.050 2 M KI + 0.050 0 M KCl with 0.084 5 M $AgNO_3$

$I^- + Ag^+ \rightarrow AgI(s)$ $V_{e1} = (40.00$ mL$) \left(\dfrac{0.050\ 2\ M}{0.084\ 5\ M}\right) = 23.76$ mL

$Cl^- + Ag^+ \rightarrow AgCl(s)$ $V_{e2} = (40.00$ mL$) \left(\dfrac{0.050\ 2 + 0.050\ 0\ M}{0.084\ 5\ M}\right) = 47.43$ mL

The figure gives $V_{e2} = 47.41$ mL, which we will use as a more accurate value.

(a) 10.00 mL: A fraction of the I^- has reacted.

$$[I^-] = \left(\frac{23.76 - 10.00}{23.76}\right) (0.050\ 2\ M) \left(\frac{40.00}{50.00}\right) = 0.023\ 3\ M$$

$$\underbrace{\hspace{3cm}}_{\substack{\text{Fraction remaining}}} \quad \underbrace{\hspace{1.5cm}}_{\substack{\text{Initial} \\ \text{concentration}}} \quad \underbrace{\hspace{1.5cm}}_{\substack{\text{Dilution} \\ \text{factor}}}$$

$$[Ag^+] = \frac{K_{sp}(AgI)}{[I^-]} = \frac{8.3 \times 10^{-17}}{0.023\ 3} = 3.57 \times 10^{-15}\ M \Rightarrow$$

$pAg^+ = -\log [Ag^+] = 14.45$

(b) 20.00 mL: A fraction of the I^- has reacted.

$$[I^-] = \left(\frac{23.76 - 20.00}{23.76}\right) (0.050\ 2\ M) \left(\frac{40.00}{60.00}\right) = 0.005\ 30\ M$$

$$[Ag^+] = \frac{8.3 \times 10^{-17}}{0.005\ 30} = 1.57 \times 10^{-14}\ M \Rightarrow pAg^+ = 13.80$$

(c) 30.00 mL: I^- has been consumed and a fraction of Cl^- has reacted.

$$[Cl^-] = \underbrace{\left(\frac{47.41 - 30.00}{47.41 - 23.76}\right)}_{\text{Fraction remaining}} \underbrace{(0.050\ 0\ M)}_{\substack{\text{Initial} \\ \text{concentration}}} \underbrace{\left(\frac{40.00}{70.00}\right)}_{\substack{\text{Dilution} \\ \text{factor}}} = 0.021\ 0\ M$$

$$[Ag^+] = \frac{K_{sp}(AgCl)}{[Cl^-]} = \frac{1.8 \times 10^{-10}}{0.021\ 0} = 8.56 \times 10^{-9}\ M \Rightarrow pAg^+ = 8.07$$

(d) $[Ag^+][Cl^-] = x^2 = 1.8 \times 10^{-10} \Rightarrow [Ag^+] = 1.3 \times 10^{-5} \Rightarrow pAg^+ = 4.87$

(e) 50.00 mL: There is excess Ag^+.

$$[Ag^+] = (0.084\ 5\ M) \underbrace{\left(\frac{50.00 - 47.41}{90.00}\right)}_{} = 0.002\ 43\ M \Rightarrow pAg^+ = 2.61$$

$$\underbrace{}_{\substack{\text{Initial} \\ \text{concentration}}} \qquad \underbrace{\phantom{\left(\frac{50.00-47.41}{90.00}\right)}}_{\substack{\text{Dilution} \\ \text{factor}}}$$

24. $Hg_2^{2+} + 2CN^- \rightarrow Hg_2(CN)_2(s)$ $\qquad$ $K_{sp} = 5 \times 10^{-40}$

$\quad$ $Ag^+ + CN^- \rightarrow AgCN(s)$ $\qquad\qquad$ $K_{sp} = 2.2 \times 10^{-16}$

The Hg_2^{2+} will precipitate first and the equivalence point occurs at 20.00 mL. The second equivalence point is at 30.00 mL.

At 5.00, 10.00, 15.00 and 19.90 mL, there is excess, unreacted Hg_2^{2+}.

At 5.00 mL, $[Hg_2^{2+}] = \left(\frac{20.00-5.00}{20.00}\right)(0.100\ 0\ M)\left(\frac{10.00}{10.00+5.00}\right) = 0.050\ 00\ M$

$[CN^-] = \sqrt{K_{sp}\ (\text{for } Hg_2(CN)_2)/([Hg^{2+}])} = 1.0 \times 10^{-19} \Rightarrow pCN^- = 19.00$

By similar calculations we find $\qquad\qquad$ 10.00 mL: $\quad$ $pCN^- = 18.85$

$\qquad\qquad\qquad\qquad\qquad\qquad\qquad\qquad$ 15.00 mL: $\quad$ $pCN^- = 18.65$

$\qquad\qquad\qquad\qquad\qquad\qquad\qquad\qquad$ 19.90 mL: $\quad$ $pCN^- = 17.76$

At 20.10 mL, AgCN has begun to precipitate. The Ag^+ remaining is

$$[Ag^+] = \left(\frac{30.00-20.10}{10.00}\right)(0.100\ 0\ M)\left(\frac{10.00}{10.00+20.10}\right) = 0.032\ 9\ M$$

$$[CN^-] = K_{sp}\ (\text{for AgCN})/[Ag^+] = 6.7 \times 10^{-15}\ M \Rightarrow pCN^- = 14.17$$

By similar reasoning we find $pCN^- = 13.81$ at 25.00 mL.

At the second equivalence point (30.00 mL), $[Ag^+] = [CN^-] = x \Rightarrow x^2 =$

$\qquad$ K_{sp} (for AgCN) $\Rightarrow [CN^-] = 1.5 \times 10^{-8}\ M \Rightarrow pCN^- = 7.83$

At 35.00 mL, there are 5.00 mL of excess CN^-.

$$[CN^-] = \left(\frac{5.00}{10.00 + 35.00}\right)(0.100\ 0\ M) = 0.011\ 1\ M \Rightarrow pCN^- = 1.95$$

The last question is "Will Ag^+ precipitate when 19.90 mL of CN^- have been added?" We calculated above that $pCN^- = 17.76$ ($\Rightarrow [CN^-] = 1.7 \times 10^{-18}$ M) at 19.90 mL if no Ag^+ had precipitated. We can check to see whether the solubility product of AgCN is exceeded if $[CN^-] = 1.7 \times 10^{-18}$ M and $[Ag^+] = \left(\dfrac{10.00}{10.00+19.90}\right)$ $\times$ (0.100 0 M) = 0.033 4 M. $[Ag^+][CN^-] = 5.7 \times 10^{-20} < K_{sp}$ (for AgCN). The Ag^+ will not precipitate at 19.90 mL.

25. mmol of $BrCH_2CH_2CH_2CH_2Cl = \dfrac{82.67 \text{ mg}}{171.464 \text{ mg/mmol}} = 0.482\ 1$ mmol

There will be 0.482 1 mmol of Cl^- and 0.482 1 mmol of Br^- liberated by reaction with $CH_3O^-Na^+$.

$$Ag^+ \text{ required for } Br^- = \dfrac{0.482\ 1 \text{ mmol}}{0.025\ 70 \text{ mmol/mL}} = 18.76 \text{ mL}$$

The same amount of Ag^+ is required to react with Cl^-, so the second equivalence point is at $18.76 + 18.76 = 37.52$ mL.

26. Derivation of Equation 7-20:

Mass balance for M: $\quad C_M^o V_M = [M^{m+}](V_M + V_X^o) + x\{\text{mol } M_xX_m(s)\}$

Mass balance for X: $\quad C_X^o V_X^o = [X^{x-}](V_M + V_X^o) + m\{\text{mol } M_xX_m(s)\}$

Equating mol M_xX_m from the two equations gives

$$\tfrac{1}{x}\{C_M^o V_M - [M^{m+}](V_M + V_X^o)\} = \tfrac{1}{m}\{C_X^o V_X^o - [X^{x-}](V_M + V_X^o)\}$$

which can be rearranged to give Equation 7-20.

27. Derivation of Equation 7-24:

From Equation 7-22, mol MX (s) $= C_X^o V^o - [X^-](V_M + V^o)$

From Equation 7-23, mol MY (s) $= C_Y^o V^o - [Y^-](V_M + V^o)$

Substituting these into Equation 7-21 gives

$$C_M^o V_M = [M^+](V_M + V^o) + C_X^o V^o - [X^-](V_M + V^o) + C_Y^o V^o - [Y^-](V_M + V^o)$$

$$V_M(C_M^o - [M^+] + [X^-] + [Y^-]) = V^o([M^+] + C_X^o - [X^-] + C_Y^o - [Y^-])$$

which can be rearranged to give Equation 7-24.

28.
$$M^+ \quad + \quad X^- \quad \overset{K_{sp}}{\rightleftharpoons} \quad MX(s)$$
$$\text{Analyte} \qquad \text{Titrant}$$
$$C_M^o, V_M^o \qquad C_X^o, V_X^o$$

Mass balance for M: $C_M^o V_M = [M^+](V_M^o + V_X) + \text{mol } MX(s)$

Mass balance for X: $C_X^o V_X = [X^-](V_M^o + V_X) + \text{mol } MX(s)$

Equating mol MX(s) from both mass balances gives

$$C_M^O V_M - [M^+](V_M^O + V_X) = C_X^O V_X - [X^-](V_M^O + V_X)$$

which can be rearranged to $V_X = V_M \left(\dfrac{C_M^O - [M^+] + [X^-]}{C_X^O + [M^+] - [X^-]} \right)$

29. The spreadsheet is the same as the one in Figure 7-10, with different values of K_{sp} in cell A2.

30.

	A	B	C	D	E	F	G
1	Ksp(AgCl)=	pAg	[Ag+]	Cl-(Diluted)	[Cl-]	[Br-]	Vm
2	1.8E-10	11	1.00E-11	5.00E-02	5.00E-02	5.00E-02	0.000
3	Ksp(AgBr)=	10	1.00E-10	3.50E-02	3.50E-02	5.00E-03	21.429
4	5.E-13	9	1.00E-09	3.35E-02	3.35E-02	5.00E-04	24.627
5	Vo=	8	1.00E-08	2.95E-02	1.80E-02	5.00E-05	34.710
6	50	7	1.00E-07	2.55E-02	1.80E-03	5.00E-06	48.227
7	Co(Cl)=	5	1.00E-05	2.50E-02	1.80E-05	5.00E-08	49.992
8	0.05	4	1.00E-04	2.50E-02	1.80E-06	5.00E-09	50.098
9	Co(Br)=	3	1.00E-03	2.48E-02	1.80E-07	5.00E-10	51.010
10	0.05	2	1.00E-02	2.25E-02	1.80E-08	5.00E-11	61.111
11	Co(Ag)=						
12	0.1	Concentrations at first equivalence point:					
13		(Display more digits to answer question in problem)					
14		8.266402	5.41E-09	3.333E-02	3.324E-02	9.234E-05	25.000
15							
16	C2 = 10^-B2						
17	D2 = $A$8*($A$6/($A$6+G2))						
18	E2 = If((D2<($A$2/C2)),D2,$A$2/C2)						
19	F2 = $A$4/C2						
20	G2 = $A$6*($A$8+$A$10+C2-E2-F2)/($A$12-C2+E2+F2)						

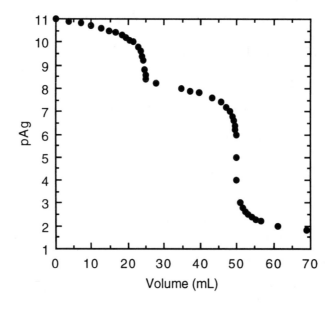

31. For anions X^-, Y^- and Z^-, the titration equation is

$$V_M = V^o \left(\frac{C_X^o + C_Y^o + C_Z^o + [M^+] - [X^-] - [Y^-] - [Z^-]}{C_M^o - [M^+] + [X^-] + [Y^-] + [Z^-]} \right)$$

	A	B	C	D	E	F	G	H	I
1	Ksp(AgCl)	pAg	[Ag+]	[Cl-]	[Cl-]	[Br-]	[Br-]	[I-]	Vm
2	1.8E-10			(Diluted)		(Diluted)			
3	Ksp(AgBr)	14.77	1.7E-15	5.0E-02	5.0E-02	5.0E-02	5.0E-02	4.9E-02	0.38
4	5.E-13	14	1.0E-14	3.6E-02	3.6E-02	3.6E-02	3.6E-02	8.3E-03	19.25
5	Ksp(AgI)	13	1.0E-13	3.4E-02	3.4E-02	3.4E-02	3.4E-02	8.3E-04	24.38
6	8.3E-17	12	1.0E-12	3.3E-02	3.3E-02	3.3E-02	3.3E-02	8.3E-05	24.94
7	Vo =	11	1.0E-11	3.3E-02	3.3E-02	3.3E-02	3.3E-02	8.3E-06	24.99
8	50	10	1.0E-10	2.6E-02	2.6E-02	2.6E-02	5.0E-03	8.3E-07	45.24
9	Co(Cl) =	9	1.0E-09	2.5E-02	2.5E-02	2.5E-02	5.0E-04	8.3E-08	49.50
10	0.05	8	1.0E-08	2.4E-02	1.8E-02	2.4E-02	5.0E-05	8.3E-09	55.89
11	Co(Br) =	7	1.0E-07	2.0E-02	1.8E-03	2.0E-02	5.0E-06	8.3E-10	72.78
12	0.05	6	1.0E-06	2.0E-02	1.8E-04	2.0E-02	5.0E-07	8.3E-11	74.78
13	Co(I) =	5	1.0E-05	2.0E-02	1.8E-05	2.0E-02	5.0E-08	8.3E-12	74.99
14	0.05	4	1.0E-04	2.0E-02	1.8E-06	2.0E-02	5.0E-09	8.3E-13	75.12
15	Co(Ag) =	3	1.0E-03	2.0E-02	1.8E-07	2.0E-02	5.0E-10	8.3E-14	76.26
16	0.1	2	1.0E-02	1.8E-02	1.8E-08	1.8E-02	5.0E-11	8.3E-15	88.89
17									
18	C3 = 10^-B3					F3 = $A$12*($A$8/($A$8+I3))			
19	D3 = $A$10*($A$8/($A$8+I3))					G3 = If((F3<($A$4/C3)),F3,$A$4/C3)			
20	E3 = If((D3<($A$2/C3)),D3,$A$2/C3)					H3 = $A$6/C3			
21	I3 = $A$8*($A$10+$A$12+$A$14+C3-E3-G3-H3)/($A$16-C3+E3+G3+H3)								

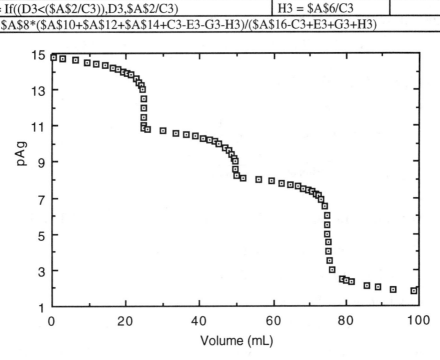

32. Consider the titration of C^+ (in a flask) by A^- (from a buret). Before the equivalence point there is excess C^+ in solution. Selective adsorption of C^+ on the CA crystal surface gives the crystal a positive charge. After the equivalence point there is excess A⁻ in solution. Selective adsorption of A⁻ on the CA crystal surface gives it a negative charge.

33. Beyond the equivalence point there is excess $Fe(CN)_6^{4-}$ in solution. Selective adsorption of this ion by the precipitate will give the particles a negative charge.

34. A known excess of Ag^+ is added to form AgI (s). In the presence of Fe^{3+}, the excess Ag^+ is titrated with standard SCN^- (to make AgSCN (s)). When Ag^+ is consumed, the SCN^- reacts with Fe^{3+} to form the red complex, $FeSCN^{2+}$.

35. AgCl is more soluble than AgSCN, and will slowly dissolve in the presence of excess SCN^-. This consumes the SCN^- and causes the red end point color to fade. Nitrobenzene coats the AgCl and inhibits reaction with SCN^-.

36. A certain amount of $Ag_2CrO_4(s)$ is needed before the red color is apparent. In the blank titration, we measure the amount of Ag^+ needed for $Ag_2CrO_4(s)$ to be observed. This is then subtracted from subsequent titrations, since it is in excess of the true equivalence point.

37. 50.00 mL of 0.365 0 M $AgNO_3$ = 18.25 mmol of Ag^+.
37.60 mL of 0.287 0 M KSCN = 10.79 mmol of SCN^-.
Difference = 18.25 - 10.79 = 7.46 mmol of I^- = 947 mg of I^-.

38. <u>Upper limit</u>: At 24.975 mL, 0.1% of Br^- is left in solution.
The Br^- concentration is
$$\underbrace{(0.001)}_{\substack{\text{fraction left}}} \cdot \underbrace{(0.100\ 0\ M)}_{\substack{\text{original} \\ [Br^-]}} \cdot \underbrace{\frac{20.00}{44.975}}_{\substack{\text{dilution} \\ \text{factor}}} = 4.45 \times 10^{-5}\ M$$

The Ag^+ concentration will be $K_{sp}/[Br^-] = \dfrac{5.0 \times 10^{-13}}{4.45 \times 10^{-5}} = 1.12 \times 10^{-8}\ M$

The $[CrO_4^{2-}]$ in equilibrium with this much Ag^+ is
$[CrO_4^{2-}] = K_{sp}/[Ag^+]^2 = 1.2 \times 10^{-12}/(1.12 \times 10^{-8})^2 = 9.6 \times 10^3\ M$.

This very high upper limit means that there is no attainable upper limit.
<u>Lower limit</u>: When the titration is 0.1% past the equivalence point, the
concentration of Ag^+ is $(0.080\ 0\ M)\left(\dfrac{0.025}{45.025}\right) = 4.4 \times 10^{-5}\ M$. The CrO_4^{2-} in
equilibrium with this much Ag^+ is $[CrO_4^{2-}] = K_{sp}/[Ag^+]^2 =$
$1.2 \times 10^{-12}/ (4.4 \times 10^{-5})^2 = 6.2 \times 10^{-4}\ M$.

ACTIVITY

1. As the ionic strength increases, the charges of the ionic atmospheres increase and the net ionic attractions decrease.

2. (a) True   (b) true   (c) true

3. (a) $\frac{1}{2}[0.008\ 7 \cdot 1^2 + 0.008\ 7 \cdot (-1)^2] = 0.008\ 7$ M

   (b) 0.001 2 M   (c) 0.08 M   (d) $0.03 + 0.15 + 0.12 = 0.30$ M

4. (a) Mg^{2+} has a smaller ionic radius than Ba^{2+}, so Mg^{2+} binds water molecules more strongly and has a larger hydrated radius.

   (b) As the charge on the ion decreases, water is bound less strongly and the hydrated radius decreases.

   (c) H^+ is unique among aqueous ions in that it forms H_3O^+ which is tightly bound to three more H_2O molecules and more loosely bound to another. OH^- forms a very strong hydrogen bond to just one H_2O molecule.

5. (a) 0.660   (b) 0.54   (c) 0.18 (Eu^{3+} is a lanthanide ion)   (d) 0.83

6. The ionic strength 0.030 M is halfway between the values 0.01 and 0.05 M. Therefore, the activity coefficient will be halfway between the tabulated values:
   $\mu = \frac{1}{2}(0.914 + 0.86) = 0.88_7$.

7. (a) $\log \gamma = \dfrac{-0.51 \cdot 2^2 \cdot \sqrt{0.083}}{1 + (600\sqrt{0.083}\ /\ 305)} = -0.375 \Rightarrow \gamma = 10^{-0.037\ 5} = 0.42_2$

   (b) $\gamma = (\dfrac{0.083 - 0.05}{0.1 - 0.05})(0.405 - 0.485) + 0.485 = 0.43_2$

8. $\gamma_\pm = (\gamma^1_{H^+}\ \gamma^1_{Cl^-})^{1/(1+1)} = [(0.933)(0.925)]^{1/2} = 0.929$

9. If [ether (aq)] becomes smaller, γ_{ether} must become larger, since K $(= [ether]\ \gamma_{ether})$ is a constant.

10. $\varepsilon = 79.755\ e^{-4.6 \times 10^{-3}(323.15 - 293.15)} = 69.474$

    $$\log \gamma = \dfrac{(-1.825 \times 10^6)[(69.474)(323.15)]^{-3/2}(-2)^2\sqrt{0.100}}{1 + \dfrac{400\sqrt{0.100}}{2.00\sqrt{(69.474)(323.15)}}}$$

    $= -0.482\ 6 \Rightarrow \gamma = 0.33$ (In the table, $\gamma = 0.355$ at 25°C)

11.   (a) To find the acitivity coefficient for $\mu = 0.001$, we write

$$\log \gamma = \frac{-3.23 \sqrt{0.001}}{1 + 2.57\sqrt{0.001}} + 0.198\,(0.001) = -0.094\,3.\quad \gamma = 10^{-0.094\,3} = 0.805$$

In a similar manner, we calculate the results below:

μ	0.001	0.01	0.1	0.5	1.0	1.5	2.0	3.0
γ	0.805	0.556	0.286	0.194	0.196	0.220	0.258	0.370
γ	0.870	0.675	0.405	($\leftarrow$ values from table)				

(b) $\log \gamma = \dfrac{-3.23(\pm0.32) \sqrt{0.1}}{1 + 2.57(\pm0.32) \sqrt{0.1}} + 0.198\,(\pm0.012) \cdot 0.1 = -0.543\,(\pm0.065\,2)$

$\gamma = 10^{-0.543(\pm0.065\,2)} = 0.286 \pm e.$ Table 3-1 tells us that for $y = 10^x$,

$e_y = y \cdot 2.303 \cdot e_x = (0.286)(2.303)(0.065\,2) = 0.043. \therefore \gamma = 0.286 \pm 0.043$

12.

	A	B	C	D	E
1	Ionic strength	Gamma (z=±1)	Gamma (z=±2)	Gamma (z=±3)	Gamma (z=±4)
2	0.0001	0.988	0.955	0.901	0.831
3	0.0003	0.980	0.924	0.836	0.727
4	0.001	0.965	0.867	0.725	0.565
5	0.003	0.942	0.787	0.583	0.383
6	0.01	0.901	0.660	0.393	0.190
7	0.03	0.847	0.515	0.225	0.071
8	0.1	0.769	0.350	0.094	0.015
9					
10	B2 = 10^(-0.51*Sqrt(A2)/(1+400*Sqrt(A2)/305))				
11	C2 = 10^(-0.51*4*Sqrt(A2)/(1+400*Sqrt(A2)/305))				
12	D2 = 10^(-0.51*9*Sqrt(A2)/(1+400*Sqrt(A2)/305))				
13	E2 = 10^(-0.51*16*Sqrt(A2)/(1+400*Sqrt(A2)/305))				

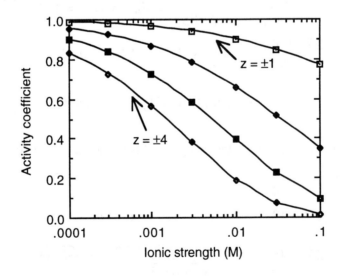

13.   (a) $\mu = 0.001$ M. $[Hg_2^{2+}]\,\gamma_{Hg_2^{2+}}[Br^-]^2\gamma_{Br^-}^2 = x\,(0.867)\,(2x)^2(0.964)^2$

$= 5.6 \times 10^{-23} \Rightarrow x = [Hg_2^{2+}] = 2.6 \times 10^{-8}$ M

(b) $\mu = 0.01$ M.   x $(0.660)(2x)^2(0.899)^2 = K_{sp} \Rightarrow x = 3.0 \times 10^{-8}$ M

(c) $\mu = 0.1$ M.   x $(0.335)(2x)^2(0.755)^2 = K_{sp} \Rightarrow x = 4.2 \times 10^{-8}$ M

14. $\mu = 0.1$ M. $[Ba^{2+}] \gamma_{Ba^{2+}} [IO_3^-]^2 \gamma_{IO_3^-}^2$

$= x (0.38) (2x + 0.066\,6)^2 (.775)^2 \approx \underbrace{x (0.38) (0.066\,6)^2}_{[IO_3^-]} (0.775)^2$

$= K_{sp} = 1.5 \times 10^{-9} \Rightarrow [Ba^{2+}] = 1.5 \times 10^{-6}$ M

15. First assume $\mu = 0$.

$[Pb^{2+}] \gamma_{Pb^{2+}} [F^-]^2 \gamma_{F^-}^2 = K_{sp}$

(x)   (1) $(2x)^2 (1)^2 = 3.6 \times 10^{-8} \Rightarrow x = 2.0_8 \times 10^{-3}$ M

With this value of x, the ionic strength is

$\mu = \frac{1}{2} [(2.08 \times 10^{-3}) (2)^2 + (4.16 \times 10^{-3}) (-1)^2] = 6.24 \times 10^{-3}$ M

Interpolation in Table 8-1 gives $\gamma_{Pb^{2+}} = 0.723$ and $\gamma_{F^-} = 0.920$

Now we repeat the cycle:

$(x)(0.723)(2x)^2(0.920)^2 = K_{sp} \Rightarrow x = 2.45 \times 10^{-3}$ M

$\Rightarrow \mu = 7.35 \times 10^{-3} \Rightarrow \gamma_{Pb^{2+}} = 0.706$ and $\gamma_{F^-} = 0.914$

Another cycle gives

$x (0.706)(2x)^2(0.914)^2 = K_{sp} \Rightarrow x = 2.48 \times 10^{-3}$ M

$\Rightarrow \mu = 7.44 \times 10^{-3} \Rightarrow \gamma_{Pb^{2+}} = 0.704$ and $\gamma_{F^-} = 0.913$

Finally a self-consistent answer is found:

$x (0.704)(2x)^2(0.913)^2 = K_{sp} \Rightarrow x = 2.48 \times 10^{-3}$ M

Answer : $[Pb^{2+}] = 2.5 \times 10^{-3}$ M

16. (a) Since we don't know the ionic strength, we will find it iteratively:

1. $[Ca^{2+}] [SO_4^{2-}] = x^2 = K_{sp} = 2.4 \times 10^{-5} \Rightarrow x = 4.90 \times 10^{-3}$ M

$\Rightarrow \mu = 0.019\,6$ M $\Rightarrow \gamma_{Ca^{2+}} = 0.629$ and $\gamma_{SO_4^{2-}} = 0.608$

2. $[Ca^{2+}] \gamma_{Ca^{2+}} [SO_4^{2-}] \gamma_{SO_4^{2-}} = 2.4 \times 10^{-5}$

(x) (0.629) (x) (0.608) $= 2.4 \times 10^{-5} \Rightarrow x = 7.92 \times 10^{-3}$ M

$\Rightarrow \mu = 0.031\,7$ M $\Rightarrow \gamma_{Ca^{2+}} = 0.572$ and $\gamma_{SO_4^{2-}} = 0.543$

3. (x) (0.572) (x) (0.543) $= 2.4 \times 10^{-5} \Rightarrow x = 8.79 \times 10^{-3}$ M

$\Rightarrow \mu = 0.035\,2$ M $\Rightarrow \gamma_{Ca^{2+}} = 0.555$ and $\gamma_{SO_4^{2-}} = 0.525$

4.   (x) (0.555) (x) (0.525) $= 2.4 \times 10^{-5} \Rightarrow x = 9.08 \times 10^{-3}$ M

$\Rightarrow \mu = 0.036\ 3$ M $\Rightarrow \gamma_{Ca^{2+}} = 0.550$ and $\gamma_{SO_4^{2-}} = 0.518$

5.   (x) (0.550) (x) (0.518) $= 2.4 \times 10^{-5} \Rightarrow x = 9.18 \times 10^{-3}$ M

(b) Since the concentration of Ca^{2+} and SO_4^{2-} are 9.2 mM, the remaining dissolved material is probably ion-paired $Ca^{2+}SO_4^{2-}$(aq).

17.

	A	B	C
1	Size (pm) of Li+ =	Ionic strength	x = [Li+] = [F-]
2	600	0.05015	0.05015
3	Size (pm) of F- =		
4	350		
5	Ksp =		
6	0.0017		
7	Activity coeff (Li+) =		
8	0.833		
9	Activity coeff (F-) =		
10	0.811		
11			
12	Formulas used:		
13	A8 = 10^((-0.51)*Sqrt(B2)/(1+(A2*Sqrt(B2)/305)))		
14	A10 = 10^((-0.51)*Sqrt(B2)/(1+(A4*Sqrt(B2)/305)))		
15	C2 = Sqrt(A6/(A8*A10))		
16	B2 = C2		

18.

	A	B	C
1	Size of Ca2+ =	Ionic strength	x = [Ca2+]
2	600	0.05199	0.01733
3	Size of OH- =		
4	350		
5	Ksp =		
6	6.5E-06		
7	Act coeff (Ca2+) =		
8	0.477		
9	Act coeff (OH-) =		
10	0.809		
11			
13	A8 = 10^(-0.51*4*Sqrt(B2)/(1+(A2*Sqrt(B2)/305)))		
14	A10 = 10^(-0.51*Sqrt(B2)/(1+(A4*Sqrt(B2)/305)))		
14	C2 = (0.25*A6/(A8*A10*A10))^0.333333333		
15	B2 = 3*C2		

19. The reaction is $3Hg_2^{2+} + 2Co(CN)_6^{3-} \rightarrow (Hg_2)_3[Co(CN)_6]_2$(s), with $V_e = 75.0$ mL. Excess $Hg_2(NO_3)_2 = 15.0$ mL. The first 75.0 mL was used up precipitating product. $[Co(CN)_6^{3-}]$ is negligible.

$$[Hg_2^{2+}] = \left(\frac{15.0}{140.0}\right)(0.010\ 0) = 1.071 \times 10^{-3}\ M$$

$$[Na^+] = \left(\frac{50.0}{140.0}\right)(0.030\ 0) = 1.071 \times 10^{-2}\ M$$

$$[NO_3^-] = \left(\frac{90.0}{140.0}\right)(0.020\ 0) = 1.286 \times 10^{-2}\ M$$

$$\mu = \frac{1}{2}[(1.071 \times 10^{-3}) \cdot 2^2 + (1.071 \times 10^{-2}) \cdot 1^2 + (1.286 \times 10^{-2}) \cdot 1^2] = 0.013\ 9\ M$$

$$\mathcal{A}^3_{Hg_2^{2+}} \cdot \mathcal{A}^2_{Co(CN)_6^{3-}} = K_{sp} \Rightarrow$$

$$\mathcal{A}_{Co(CN)_6^{3-}} = \sqrt{\frac{K_{sp}}{[Hg_2^{2+}]^3\ \gamma^3_{Hg_2^{2+}}}} = \sqrt{\frac{1.9 \times 10^{-37}}{(1.071 \times 10^{-3})^3(0.639)^3}}$$

$$= 2.4 \times 10^{-14} \Rightarrow pCo(CN)_6^{3-} = 13.61$$

20. Ionic strength $= 0.010\ M$ (from HCl) $+ 0.040\ M$ (from $KClO_4$ that gives $K^+ + ClO_4^-$) $= 0.050\ M$. Using Table 8-1, $\gamma_{H^+} = 0.86$.

$$pH = \log[H^+]\gamma_{H^+} = -\log(0.010)(0.86) = 2.07.$$

21. Ionic strength $= 0.010\ M$ from NaOH $+ 0.012\ M$ from $Ca(NO_3)_2 = 0.022\ M$.
Interpolating in Table 8-1 gives $\gamma_{OH^-} = 0.873$.

$$[H^+]\gamma_{H^+} = \frac{K_w}{[OH^-]\ \gamma_{OH^-}} = \frac{1.0 \times 10^{-14}}{(0.010)(0.873)} = 1.15 \times 10^{-12}$$

$$pH = -\log(1.15 \times 10^{-12}) = 11.94$$

If we had neglected activities, $pH \approx -\log[H^+] = -\log\frac{K_w}{[OH^-]} = 12.00$

CHAPTER 9
SYSTEMATIC TREATMENT OF EQUILIBRIUM

1. The sum of positive charges in a solution equals the magnitude of the sum of negative charges.

2. $$[H^+] + 2[Ca^{2+}] + [Ca(HCO_3)^+] + [Ca(OH)^+] + [K^+] =$$
$$[OH^-] + [HCO_3^-] + 2[CO_3^{2-}] + [ClO_4^-]$$

3. $$[H^+] = [OH^-] + [HSO_4^-] + 2[SO_4^{2-}]$$

4. $$[H^+] = [OH^-] + [H_2AsO_4^-] + 2[HAsO_4^{2-}] + 3[AsO_4^{3-}]$$

   structure : $H-O-\overset{\overset{\displaystyle O}{\|}}{\underset{\underset{\displaystyle O^-}{|}}{As}}-O^-$

5. (a) $2[Mg^{2+}] + [H^+] = [Br^-] + [OH^-]$

   (b) $2[Mg^{2+}] + [H^+] + [MgBr^+] = [Br^-] + [OH^-]$

6. 250 mL of 1.0×10^{-6} M charge $= 0.25 \times 10^{-6}$ moles of charge.

   $(0.25 \times 10^{-6}$ moles of charge$)\left(9.648 \times 10^4 \dfrac{\text{coulombs}}{\text{mole of charge}}\right) = 0.024\ 12$ C.

   Force $= -(8.988 \times 10^9)\dfrac{(0.024\ 12)\ (-0.024\ 12)}{1.5^2} = 2.3 \times 10^6$ N

   $(2.3 \times 10^6$ N$)\ (0.224\ 8$ pounds/N$) = 5.2 \times 10^5$ pounds

7. The sum of the amounts of all species containing a particular atom (or group of atoms) must equal the amount of that atom (or group) delivered to the solution.

8. (a) 0.20 M $= [Mg^{2+}]$ (c) 0.20 M $= [Mg^{2+}] + [MgBr^+]$

   (b) 0.40 M $= [Br^-]$ (d) 0.40 M $= [Br^-] + [MgBr^+]$

9. $[CH_3CO_2^-] + [CH_3CO_2H] = 0.1$ M

10. $Y_{total} = \dfrac{3}{2} X_{total}$

   $2[X_2Y_2^{2+}] + [X_2Y^{4+}] + 3[X_2Y_3] + [Y^{2-}] = \dfrac{3}{2}\{2[X_2Y_2^{2+}] + 2[X_2Y^{4+}] + 2[X_2Y_3]\}$

   $[Y^{2-}] = [X_2Y_2^{2+}] + 2[X_2Y^{4+}]$

11. Charge and mass are rigorously proportional to molarity.

12. 1. Pertinent reaction: $Zn_2[Fe(CN)_6](s) \rightleftarrows 2Zn^{2+} + Fe(CN)_6^{4-}$

   2. Charge balance: $2[Zn^{2+}] = 4[Fe(CN)_6^{4-}]$

   3. Mass balance: $[Zn^{2+}] = 2[Fe(CN)_6^{4-}]$

4. Equilibrium constant: $K_{sp} = 2.1 \times 10^{-16} = [Zn^{2+}]^2[Fe(CN)_6^{4-}]$

Now solve: If $[Zn^{2+}] = x$, then $[Fe(CN)_6^{4-}] = \frac{1}{2}x$ (from the charge or mass

balances). Putting these values into the K_{sp} equation gives:

$(x)^2 \, (\frac{1}{2}x) = 2.1 \times 10^{-16} \Rightarrow [Zn^{2+}] = x = 7.5 \times 10^{-6}$ M

13.   Charge balance: $2[Mg^{2+}] + [H^+] = [OH^-]$ (1)

Mass balance: $[Mg^{2+}] = 4.0 \times 10^{-8}$ M   (2)

Equilibrium: $K_w = [H^+][OH^-]$ (3)

For another mass balance we cannot write $[OH^-] = 2[Mg^{2+}]$ because OH^- comes
from both $Mg(OH)_2$ and H_2O ionization. Setting $[H^+] = x$ and
$[Mg^{2+}] = 4.0 \times 10^{-8}$ M in Equation 1 gives $[OH^-] = x + 8.0 \times 10^{-8}$. Putting this
into Equation 3 gives $K_w = (x)(x + 8.0 \times 10^{-8}) \Rightarrow x = [H^+] = 6.8 \times 10^{-8}$ M.
$[OH^-] = x + 8.0 \times 10^{-8} = 1.5 \times 10^{-7}$ M.

14.   (a) $\dfrac{[Pb^{2+}][\cancel{F^-}]^2}{[Sr^{2+}][\cancel{F^-}]^2} = \dfrac{K_{sp} \text{ (for } PbF_2)}{K_{sp} \text{ (for } SrF_2)} = \dfrac{3.6 \times 10^{-8}}{2.9 \times 10^{-9}} = 12.4$

(b) Mass balance: $[F^-] = 2[Pb^{2+}] + 2[Sr^{2+}]$

Part (a) gave $[Pb^{2+}] = 12.4 \, [Sr^{2+}]$, or
$$[F^-] = 2(12.4 \, [Sr^{2+}]) + 2 \, [Sr^{2+}] = 26.8 \, [Sr^{2+}]$$

The SrF_2 solubility product can then be used to find
$2.9 \times 10^{-9} = [Sr^{2+}][F^-]^2 = [Sr^{2+}](26.8 \, [Sr^{2+}])^2 \Rightarrow [Sr^{2+}] = 1.6 \times 10^{-4}$ M

$[F^-] = 26.8 \, [Sr^{2+}] = 4.3 \times 10^{-3}$ M $[Pb^{2+}] = 12.4 \, [Sr^{2+}] = 2.0 \times 10^{-3}$ M

15.   Mass balance: $0.10 = [M^{2+}] + [MX^-] + [MX_2(aq)]$

Charge balance: $2[M^{2+}] + [H^+] = [MX^-] + [X^-] + [OH^-]$

$2[M^{2+}] = [MX^-] + [X^-]$ (Since $[H^+] = [OH^-]$)

$K_1 = \dfrac{[MX^-]}{[M^{2+}][X^-]}$ $K_2 = \dfrac{[MX_2(aq)]}{[MX^-][X^-]}$

From the K_1 and K_2 equations, we can write

$[MX^-] = K_1[M^{2+}][X^-]$ (1)

$[MX_2(aq)] = K_2[MX^-][X^-] = K_2 K_1[M^{2+}][X^-]^2$ (2)

From the charge balance we can say $[X^-] = 2[M^{2+}] - [MX^-]$

Substituting the value of $[MX^-]$ from Equation (1) gives

$[X^-] = 2[M^{2+}] - K_1[M^{2+}][X^-]$

$[X^-] = \dfrac{2[M^{2+}]}{1 + K_1[M^{2+}]}$ (3)

Substituting values of $[MX^-]$, $[MX_2(aq)]$ and $[X^-]$ from Equations (1), (2), and

(3) into the mass balance gives

$$0.10 = [M^{2+}] + K_1[M^{2+}]\frac{2[M^{2+}]}{1 + K_1[M^{2+}]} + K_1K_2[M^{2+}]\left(\frac{2[M^{2+}]}{1 + K_1[M^{2+}]}\right)^2$$

16. (a) $K = \dfrac{[Zn^{2+}SO_4^{2-}]}{[Zn^{2+}]\,\gamma_{Zn^{2+}}\,[SO_4^{2-}]\,\gamma_{SO_4^{2-}}}$ (setting $\gamma_{Zn^{2+}SO_4^{2-}} = 1$)

Substituting $[Zn^{2+}SO_4^{2-}] = 0.010 - [Zn^{2+}]$ and $[SO_4^{2-}] = [Zn^{2+}]$

gives $K = \dfrac{0.010 - [Zn^{2+}]}{[Zn^{2+}]^2\,\gamma_{Zn^{2+}}\,\gamma_{SO_4^{2-}}}$ (1)

Setting $\gamma_{Zn^{2+}} = \gamma_{SO_4^{2-}} = 1$ and $K = 200$ allows us to calculate

$[Zn^{2+}] = 5.00 \times 10^{-3}$ M

(b) Ionic strength $= \frac{1}{2}([Zn^{2+}]\cdot 4 + [SO_4^{2-}]\cdot 4) = 0.020\,0$ M

Interpolation in Table 8-1 gives $\gamma_{Zn^{2+}} = 0.628$ and $\gamma_{SO_4^{2-}} = 0.606$

Putting these values of γ into Equation (1) gives $[Zn^{2+}] = 6.64 \times 10^{-3}$ M

(c) 3rd iteration: $\mu = 4\,(6.64 \times 10^{-3}) = 0.026\,6$ M

$\gamma_{Zn^{2+}} = 0.596$ $\gamma_{SO_4^{2-}} = 0.571$ $[Zn^{2+}] = 6.83 \times 10^{-3}$ M

4th iteration: $\mu = 0.027\,3$ M

$\gamma_{Zn^{2+}} = 0.593$ $\gamma_{SO_4^{2-}} = 0.567$ $[Zn^{2+}] = 6.85 \times 10^{-3}$ M

Ion-paired percent $= \dfrac{0.010 - 6.85 \times 10^{-3}}{0.010} \times 100 = 32\%$

(d) $\mu = 4\,(6.85 \times 10^{-3}) = 0.027\,4$ M

$\gamma_{Zn^{2+}} = 0.592$ $\gamma_{SO_4^{2-}} = 0.566$ $\gamma_\pm = \sqrt{(0.592)(0.566)} = 0.579$

17. As pH is lowered, the concentration of H$^+$ increases. Since H$^+$ reacts with basic anions, it increases the solubility of their salts. By promoting solubility of minerals such as galena, cerussite, kaolinite and bauxite, the concentrations of dissolved lead and aluminum in the environment are increased.

18. $[HF] = \dfrac{K_b}{[OH^-]}\,[F^-] = 15\,[F^-]$ at pH 2.00

$[F^-] + [HF] = 16[F^-] = 2[Ca^{2+}] \Rightarrow [F^-] = [Ca^{2+}]/8$

$[Ca^{2+}][F^-]^2 = [Ca^{2+}]\left(\dfrac{[Ca^{2+}]}{8}\right)^2 = K_{sp} \Rightarrow [Ca^{2+}] = 1.4 \times 10^{-3}$ M

$[F^-] = [Ca^{2+}]/8 = 1.7 \times 10^{-4}$ M $[HF] = 15[F^-] = 2.5 \times 10^{-3}$ M

19.   Call acrylic acid HA and acrylate anion A$^-$.

Charge balance: Invalid because pH is fixed.

Mass balance: $[A^-] + [HA] = 2[M^{2+}]$ (1)

Equilibria: $K_{sp} = [M^{2+}] [A^-]^2$ (2)

$$K_b = \frac{[HA] [OH^-]}{[A^-]}$$ (3)

$$K_w = [H^+] [OH^-]$$ (4)

Since $[OH^-] = 1.8 \times 10^{-10}$ M, we can use Equation 3 to find

$$[HA] = \frac{K_b}{[OH^-]} [A^-] = [A^-]$$

Putting $[HA] = [A^-]$ into Equation 1 gives $[A^-]+[A^-] = 2[M^{2+}] \Rightarrow [A^-] = [M^{2+}]$

Putting $[A^-] = [M^{2+}]$ into Equation 2 gives

$$K_{sp} = [M^{2+}] [M^{2+}]^2 \Rightarrow [M^{2+}] = 4.0 \times 10^{-5} \text{ M}$$

20.   Charge balance: Invalid because pH is fixed.

Mass balance: $[R_3NH^+] + [R_3N] = [Br^-]$ (1)

Equilibria: $K_{sp} = [R_3NH^+] [Br^-]$ (2)

$$K_a = \frac{[R_3N] [H^+]}{[R_3NH^+]}$$ (3)

$$K_w = [H^+] [OH^-]$$ (4)

If $[H^+] = 10^{-9.50}$, we can use Equation 3 to write

$[R_3N] = \dfrac{K_a}{[H^+]} [R_3NH^+] = 7.27 [R_3NH^+]$. Putting this into Equation 1 gives

$$[R_3NH^+] + 7.27 [R_3NH^+] = [Br^-]$$
$$8.27 [R_3NH^+] = [Br^-].$$

Putting this relation into Equation 2 gives

$$K_{sp} = [R_3NH^+] [Br^-] = \left(\frac{[Br^-]}{8.27}\right) [Br^-] \Rightarrow [Br^-] = 5.8 \times 10^{-4} \text{ M}$$

This must be equal to the solubility of $R_3NH^+Br^-$, since all Br^- originates from this salt.

21.   (a) Charge balance: Invalid because pH is fixed.

Mass balance: We are tempted to write $[OH^-] = 2[Pb^{2+}]$, but this is not true.

OH$^-$ also comes from water ionization and from the buffer used to fix the pH. We do not have enough information to write a mass balance.

Equilibria: $K = [Pb^{2+}] [OH^-]^2$ (1)

$K_w = [H^+] [OH^-]$ (2)

If pH = 10.50, $[OH^-] = 10^{-3.50}$ M. Putting this value into Equation 1 gives

$[Pb^{2+}] = K/(10^{-3.50})^2 = 5.0 \times 10^{-9}$ M

(b) Equilibria: $K = [Pb^{2+}][OH^-]^2$ (1')

$$K_a = \frac{[PbOH^+][H^+]}{[Pb^{2+}]}$$ (2')

As in part (a), we find $[Pb^{2+}] = 5.0 \times 10^{-9}$ M with Equation (1').

Putting this value into Equation (2') gives

$$[PbOH^+] = \frac{K_a[Pb^{2+}]}{[H^+]} = 2.1 \times 10^{-16}$ M$$

$[Pb]_{total} = [Pb^{2+}] + [PbOH^+] = 5 \times 10^{-9}$ M

(c) $K = [Pb^{2+}]\,\gamma_{Pb^{2+}}\,[OH^-]^2\,\gamma_{OH^-}^2 = [Pb^{2+}]\,(0.455)\,(10^{-3.50})^2$

$\underbrace{\qquad\qquad}$

$[OH^-]\,\gamma_{OH^-}$

$\Rightarrow [Pb^{2+}] = 1.1 \times 10^{-8}$ M

In this problem we used the relation

$$[OH^-]\,\gamma_{OH^-} = K_w/([H^+]\,\gamma_{H^+}) = K_w/10^{-10.50}.$$

22.   Charge balance: Invalid because pH is fixed.

Mass balance: $[Ag^+] = 3\{[PO_4^{3-}] + [HPO_4^{2-}] + [H_2PO_4^-] + [H_3PO_4]\}$ (1)

Equilibria: $K_{b1} = \dfrac{[HPO_4^{2-}][OH^-]}{[PO_4^{3-}]}$ (2)

$K_{b2} = \dfrac{[H_2PO_4^-][OH^-]}{[HPO_4^{2-}]}$ (3)

$K_{b3} = \dfrac{[H_3PO_4][OH^-]}{[H_2PO_4^-]}$ (4)

$K_w = [H^+][OH^-]$ (5)

$K_{sp} = [Ag^+]^3\,[PO_4^{3-}]$ (6)

From Equations (2), (3) (4), we can write

$$[HPO_4^{2-}] = \frac{K_{b1}[PO_4^{3-}]}{[OH^-]}$$

$$[H_2PO_4^-] = \frac{K_{b2}[HPO_4^-]}{[OH^-]} = \frac{K_{b2}K_{b1}[PO_4^{3-}]}{[OH^-]^2}$$

$$[H_3PO_4] = \frac{K_{b3}[H_2PO_4^-]}{[OH^-]} = \frac{K_{b3}K_{b2}K_{b1}[PO_4^{3-}]}{[OH^-]^3}$$

Using $[OH^-] = 10^{-8.00}$ M gives $[HPO_4^{2-}] = 2.3 \times 10^6\ [PO_4^{3-}]$

$[H_2PO_4^-] = 3.68 \times 10^7\ [PO_4^{3-}]$ $[H_3PO_4] = 5.15 \times 10^3\ [PO_4^{3-}]$

Putting these values into Equation (1) gives $[Ag^+] = [PO_4^{3-}]\,(1.17 \times 10^8)$

which can be used in Equation (6) to give

$$K_{sp} = [Ag^+]^3 \left(\frac{[Ag^+]}{1.17 \times 10^8}\right) \Rightarrow [Ag^+] = 4.3 \times 10^{-3} \text{ M}.$$

23. (a) $[NH_4^+] + [H^+] = 2[SO_4^{2-}] + [HSO_4^-] + [OH^-]$

(b) $[NH_3] + [NH_4^+] = 2\{[SO_4^{2-}] + [HSO_4^-]\}$

(c) $\dfrac{[NH_3][H^+]}{[NH_4^+]} = 5.70 \times 10^{-10}$.

Putting in $[H^+] = 10^{-9.25}$ M gives $[NH_3] = 1.014[NH_4^+]$.

$\dfrac{[HSO_4^-][OH^-]}{[SO_4^{2-}]} = 9.80 \times 10^{-13}$.

Putting in $[H^+] = 10^{-9.25}$ M gives $[HSO_4^-] = 5.51 \times 10^{-8}[SO_4^{2-}]$.

Putting these values of $[NH_3]$ and $[HSO_4^-]$ into the mass balance gives:

$$1.014[NH_4^+] + [NH_4^+] = 2\{[SO_4^{2-}] + 5.51 \times 10^{-8}[SO_4^{2-}]\}$$
$$[SO_4^{2-}] = 1.007[NH_4^+]$$

Now we use the K_{sp} equation:
$$K_{sp} = [NH_4^+]^2[SO_4^{2-}] = 1.007[NH_4^+]^3 \Rightarrow [NH_4^+] = 6.50 \text{ M}$$
$$[NH_3] = 1.014[NH_4^+] = 6.59 \text{ M}$$

24. (a) $[FeG^+] + [H^+] = [G^-] + [OH^-]$

(b) Total Fe $= 0.0500$ M $= [FeG_2] + [FeG^+]$ (1)

Total G $= 0.100$ M $= [FeG^+] + 2[FeG_2] + [G^-] + [HG]$ (2)

(c) Subtracting (1) from (2) gives $[FeG_2] = 0.0500 - [G^-] - [HG]$ (3)

Multiplying (1) by 2 and subtracting (2) gives

$[FeG^+] = [G^-] + [HG]$ (4)

$$K_b = \frac{[HG][OH^-]\gamma_{OH^-}}{[G^-]\gamma_{G^-}} \Rightarrow [HG] = \frac{K_b\gamma_{G^-}}{[OH^-]\gamma_{OH^-}}[G^-]$$

$$= \frac{(6.0 \times 10^{-5})(0.78)}{10^{-5.50}}[G^-] \Rightarrow [HG] = 14.80[G^-]$$

or $[G^-] = 0.06757[HG]$ (5)

Putting (5) into (3) gives $[FeG_2] = 0.0500 - 15.80[G^-]$ (6)

Putting (5) into (4) gives $[FeG^+] = 15.80[G^-]$ (7)

Using (6) and (7) in the K_2 equilibrium gives

$$K_2 = \frac{[FeG_2]}{[FeG^+]\gamma_{FeG^+}[G^-]\gamma_{G^-}} = \frac{0.0500 - 15.80[G^-]}{15.80[G^-](0.79)[G^-](0.78)}$$

$$\Rightarrow [G^-] = 1.038 \times 10^{-3} \text{ M}$$

$$[FeG^+] = 15.80[G^-] = 0.016_4 \text{ M}$$

25.

	A	B	C	D	E	F	G
1	Ksp=	pH	[H+]	[OH-]	[Ca2+]	[F-]	[HF]
2	3E-11	0	1E+00	1E-14	2.80E-02	3.73E-05	5.60E-02
3	Kb=	2	1E-02	1E-12	1.36E-03	1.70E-04	2.54E-03
4	1E-11	4	1E-04	1E-10	2.34E-04	4.08E-04	6.12E-05
5	Kw=	6	1E-06	1E-08	2.14E-04	4.27E-04	6.41E-07
6	1E-14	8	1E-08	1E-06	2.14E-04	4.27E-04	6.41E-09
7		10	1E-10	1E-04	2.14E-04	4.27E-04	6.41E-11
8		12	1E-12	1E-02	2.14E-04	4.27E-04	6.41E-13
9		14	1E-14	1E+00	2.14E-04	4.27E-04	6.41E-15
10	Formulas						
11	C2 = 10^-B2						
12	D2 = $A$6/C2						
13	E2 = (0.25*$A$2*(1+$A$4*C2/$A$6)^2)^0.333333333						
14	F2 = 2*E2/(1+$A$4*C2/$A$6)						
15	G2 = $A$4*F2/D2						

26.

	A	B	C	D	E	F	G	H
1	Ksp=	pH	[H+]	[OH-]	[Hg2+]	[S2-]	[HS-]	[H2S]
2	5E-54	0	1E+00	1E-14	6.6E-17	7.5E-38	6.0E-24	6.6E-17
3	Kb1=	2	1E-02	1E-12	6.6E-19	7.5E-36	6.0E-24	6.6E-19
4	0.8	4	1E-04	1E-10	6.6E-21	7.5E-34	6.0E-24	6.6E-21
5	Kb2=	6	1E-06	1E-08	6.9E-23	7.2E-32	5.8E-24	6.4E-23
6	1.1E-7	8	1E-08	1E-06	2.1E-24	2.4E-30	1.9E-24	2.1E-25
7	Kw=	10	1E-10	1E-04	2.0E-25	2.5E-29	2.0E-25	2.2E-28
8	1E-14	12	1E-12	1E-02	2.0E-26	2.5E-28	2.0E-26	2.2E-31
9		14	1E-14	1E+00	3.0E-27	1.7E-27	1.3E-27	1.5E-34
10	Formulas							
11	C2 = 10^-B2							
12	D2 = $A$8/C2							
13	E2 = Sqrt($A$2*(1+$A$4/D2+$A$4*$A$6/D2^2))							
14	F2 = $A$2/E2							
15	G2 = $A$4*C2*F2/$A$8							
16	H2 = $A$6*C2*G2/$A$8							

27. ZnC_2O_4 has the same stoichiometry as HgS. Therefore, we can use Equation 9-33 for $[Zn^{2+}]$:

$$[Zn^{2+}] = \sqrt{K_{sp}\left(1 + \frac{K_{b1}}{[OH^-]} + \frac{K_{b1}K_{b2}}{[OH^-]^2}\right)}$$

For $[C_2O_4^{2-}]$ we use the solubility product: $[C_2O_4^{2-}] = K_{sp}/[Zn^{2+}]$.

The pH must satisfy the charge balance: $[H^+] + 2[Zn^{2+}] - [OH^-] - 2[C_2O_4^{2-}] = 0$.

	A	B	C	D	E	F	G
1	Ksp =	pH	[H+]	[OH-]	[Zn2+]	[C2O4 2-]	[H+]+2[Zn2+]-
2	7.5E-9						[OH-]-2[C2O4 2-]
3	Kb1 =	7	1.0E-07	1.0E-07	8.7E-05	8.7E-05	3.115E-07
4	1E-10	8	1.0E-08	1.0E-06	8.7E-05	8.7E-05	-9.588E-07
5	Kb2 =	7.2	6.3E-08	1.6E-07	8.7E-05	8.7E-05	1.012E-07
6	1E-13	7.4	4.0E-08	2.5E-07	8.7E-05	8.7E-05	-8.730E-08
7	Kw =	7.31	4.9E-08	2.0E-07	8.7E-05	8.7E-05	-2.565E-09
8	1E-14	7.30725	4.9E-08	2.0E-07	8.7E-05	8.7E-05	3.877E-12
9	Formulas						
10	C3=10^-B3						
11	D3=$A$8/C3						
12	E3=Sqrt($A$2*(1 + $A$4/D3 + $A$4*$A$6/D3^2))						
13	F3=$A$2/E3						
14	G3=C3+2*E3-D3-2*F3						

28. By analogy to the HgS problem, we can use the mass balance and base equilibria to write

$$[PO_4^{3-}] + [HPO_4^{2-}] + [H_2PO_4^-] + [H_3PO_4] = \tfrac{1}{3}[Ag^+]$$

$$[PO_4^{3-}]\left(1 + \frac{K_{b1}}{[OH^-]} + \frac{K_{b1}K_{b2}}{[OH^-]^2} + \frac{K_{b1}K_{b2}K_{b3}}{[OH^-]^3}\right) = \tfrac{1}{3}[Ag^+]$$

Substituting this expression for $[PO_4^{3-}]$ into the solubility product gives

$$[Ag^+]^3[PO_4^{3-}] = K_{sp}$$

$$[Ag^+]^3 \frac{\tfrac{1}{3}[Ag^+]}{\left(1 + \dfrac{K_{b1}}{[OH^-]} + \dfrac{K_{b1}K_{b2}}{[OH^-]^2} + \dfrac{K_{b1}K_{b2}K_{b3}}{[OH^-]^3}\right)} = K_{sp}$$

$$[Ag^+] = (3K_{sp})^{1/4}\left(1 + \frac{K_{b1}}{[OH^-]} + \frac{K_{b1}K_{b2}}{[OH^-]^2} + \frac{K_{b1}K_{b2}K_{b3}}{[OH^-]^3}\right)^{1/4}$$

	A	B	C	D	E	F	G	H	I
1	Ksp=	pH	[H+]	[OH-]	[Ag+]	[PO4]	[HPO4]	[H2PO4]	[H3PO4]
2	2E-18	0	1E+00	1E-14	1E+01	9E-22	2E-09	3E-02	5E+00
3	Kb1=	2	1E-02	1E-12	5E-01	2E-17	5E-07	7E-02	1E-01
4	0.023	4	1E-04	1E-10	4E-02	4E-14	9E-06	1E-02	2E-04
5	Kb2=	6	1E-06	1E-08	4E-03	4E-11	8E-05	1E-03	2E-07
6	1.6E-7	8	1E-08	1E-06	7E-04	9E-09	2E-04	3E-05	4E-11
7	Kb3 =	10	1E-10	1E-04	2E-04	3E-07	7E-05	1E-07	2E-15
8	1E-12	12	1E-12	1E-02	7E-05	7E-06	2E-05	3E-10	4E-20
9	Kw=	14	1E-14	1E+00	5E-05	2E-05	4E-07	6E-14	9E-26
10	1E-14	9.87558	1E-10	8E-05	2E-04	2E-07	7E-05	2E-07	3E-15
11									
12			To find pH of saturated solution using pH in cell B10:						
13			[H+]+[Ag+]-[OH-]-3[PO4]-2[HPO4]-[H2PO4] =						
14			7E-11						
15									
16	C2 = 10^-B2			G2 = $A$4*F2/D2					
17	D2 = $A$8/C2			H2 = $A$6*G2/D2					
18	F2 = $A$2/(E2^3)			I2 = $A$8*H2/D2					
19	E2=(3*$A$2*(1+$A$4/D2+$A$4*$A$6/D2^2+$A$4*$A$6*$A$8/D2^3))^0.25								
20	C14 = C10+E10-D10-3*F10-2*G10-H10								

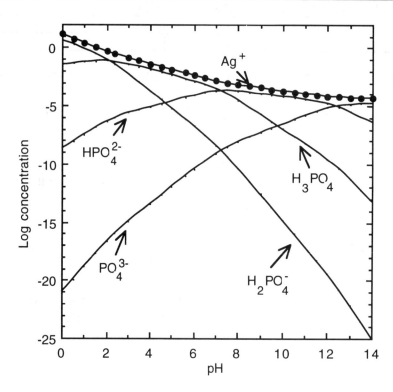

CHAPTER 10
MONOPROTIC ACID-BASE EQUILIBRIA

1. HBr (or any other acid or base) drives the reaction $H_2O \rightleftarrows H^+ + OH^-$ to the left, according to Le Châtelier's principle. If, for example, the solution contains 10^{-4} M HBr, the concentration of OH^- from H_2O is $K_w/[H^+] = 10^{-10}$ M. The concentration of H^+ from H_2O must also be 10^{-10} M, since H^+ and OH^- are created in equimolar quantities.

2. (a) $pH = -\log [H^+] = -\log (1.0 \times 10^{-3}) = 3.00$

(b) $[H^+] = K_w /[OH^-] = (1.0 \times 10^{-14})/(1.0 \times 10^{-2}) = 1.0 \times 10^{-12}$ M.
$pH = -\log [H^+] = 12.00$

3. $[H^+] = [OH^-] + [ClO_4^-] \Rightarrow [OH^-] = [H^+] - 5.0 \times 10^{-8}$

$[H^+]\,[OH^-] = K_w$

$[H^+]\,([H^+] - 5.0 \times 10^{-8}) = 1.0 \times 10^{-14} \Rightarrow [H^+] = 1.28 \times 10^{-7}$ M

$pH = -\log [H^+] = 6.89$

$[OH^-] = K_w/[H^+] = 7.8 \times 10^{-8}$ M $\Rightarrow$ $[H^+]$ from $H_2O = 7.8 \times 10^{-8}$ M

$$\text{Fraction of total } [H^+] \text{ from } H_2O = \frac{7.8 \times 10^{-8}\ M}{1.28 \times 10^{-7}\ M} = 0.61$$

4. (a) $pH = -\log [H^+]\gamma_{H^+}$

$1.092 = -\log (0.100)\,\gamma_{H^+} \Rightarrow \gamma_{H^+} = 0.809$

The tabulated activity coefficient is 0.83.

(b) $2.102 = -\log (0.010\,0)\gamma_{H^+} \Rightarrow \gamma_{H^+} = 0.791$

(c) The activity coefficient depends somewhat on the identity of the counterions in solution.

5. (a)
(b)
(c)
(d)

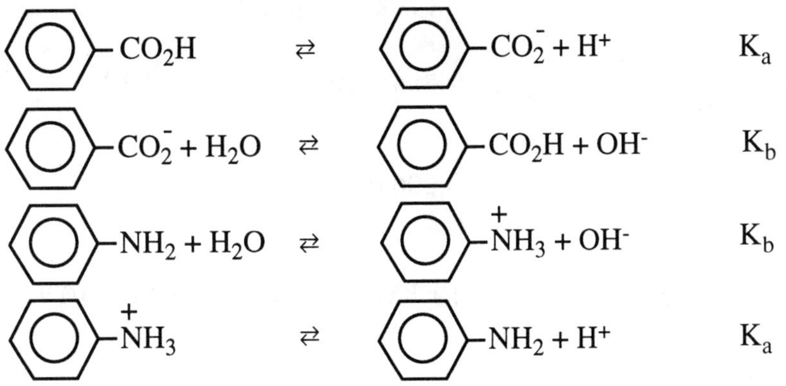

6. Let $x = [H^+] = [A^-]$ and $0.100 - x = [HA]$.

$$\frac{x^2}{0.100-x} = 1.00 \times 10^{-5} \Rightarrow x = 9.95 \times 10^{-4} \ M \Rightarrow pH = -\log x = 3.00$$

$$\alpha = \frac{[A^-]}{[A^-] + [HA]} = \frac{9.95 \times 10^{-4}}{0.100} = 9.95 \times 10^{-3}$$

7.
$$BH^+ \overset{K_a}{\rightleftarrows} B + H^+ \qquad\qquad K_a = K_w/K_b = 1.00 \times 10^{-10}$$

$$0.100-x \qquad\qquad x \quad\ x$$

$$\frac{x^2}{0.100-x} = 1.00 \times 10^{-10} \Rightarrow x = [B] = [H^+] = 3.16 \times 10^{-6} \ M \Rightarrow pH = 5.50$$

8.
$$(CH_3)_3NH^+ \rightleftarrows (CH_3)_3N + H^+ \qquad K_a = 1.58 \times 10^{-10}$$
$$ F-x \qquad\qquad\ x \qquad\ x$$

$$\frac{x^2}{0.060-x} = K_a \Rightarrow x = 3.0_8 \times 10^{-6} \Rightarrow pH = 5.51$$

$$[(CH_3)_3N] = x = 3.1 \times 10^{-6} \ M, \quad [(CH_3)_3NH^+] = F-x = 0.060 \ M$$

9.
$$HA \overset{K_a}{\rightleftarrows} H^+ + A^- \qquad\qquad K_a = \frac{[H^+][A^-]}{[HA]} = \frac{x^2}{F-x}$$
$$F-x \qquad\ x \quad\ x$$

$$\text{For } F = \frac{K_a}{10}, \quad \frac{x^2}{\frac{K_a}{10} - x} = K_a \Rightarrow x = 0.092 \ K_a; \quad \alpha = \frac{x}{F} = \frac{0.092 \ K_a}{0.100 \ K_a} = 92\%$$

$$\text{For } F = 10 \ K_a, \quad \frac{x^2}{10K_a - x} = K_a \Rightarrow x = 2.7 \ K_a; \quad \alpha = \frac{x}{F} = \frac{2.7 \ K_a}{10 \ K_a} = 27\%$$

$$\text{For } 99\% \text{ dissociation, } x = 0.99 \ F \Rightarrow K_a = \frac{(0.99 \ F)^2}{F - 0.99 \ F} \Rightarrow F = (0.010 \ 2)K_a$$

10.

$$ F - 10^{-2.78} \qquad\qquad 10^{-2.78} \quad\ 10^{-2.78}$$

$$K_a = \frac{(10^{-2.78})^2}{0.045 \ 0 - 10^{-2.78}} = 6.4 \times 10^{-5} = pK_a = 4.19$$

11. $\quad HA \rightleftarrows H^+ + A^- \quad \alpha = 0.006 \ 0 = \dfrac{x}{F}$
$$F-x \qquad x \qquad x$$

$$\text{Since } F = 0.045 \ 0 \ M, \ x = 2.7 \times 10^{-4} \ M \Rightarrow K_a = \frac{x^2}{F-x} = 1.6 \times 10^{-6} \Rightarrow pK_a = 5.79$$

12. (a) $HA \rightleftarrows H^+ + A^-$
$$ F-x \qquad x \qquad x$$

$$\frac{x^2}{F-x} = K_a \Rightarrow x = 9.4 \times 10^{-4} \Rightarrow pH = 3.03$$

$$\alpha = \frac{[A^-]}{[HA] + [A^-]} = \frac{x}{F} = 9.4\%$$

(b) pH = 7.00 because the acid is so dilute. From the K_a equilibrium we can

write $[A^-] = \dfrac{K_a}{[H^+]} [HA] = \dfrac{9.8 \times 10^{-5}}{1.0 \times 10^{-7}} [HA] = 980 [HA]$

$$\alpha = \frac{[A^-]}{[HA] + [A^-]} = \frac{980 [HA]}{[HA] + 980 [HA]} = \frac{980}{981} = 99.9\%$$

13. Phenol is a weak acid and will contribute negligible ionic strength. The ionic strength of the solution is 0.050 M.

$$HA \rightleftharpoons H^+ + A^- \qquad K_a = 1.05 \times 10^{-10}$$
$$ F-x \quad\quad x \quad\; x$$

$$\frac{[H^+] \gamma_{H^+} [A^-]\gamma_{A^-}}{[HA] \, \gamma_{HA}} = \frac{(x)(0.86)(x)(0.835)}{(0.050\ 0-x)(1.00)} = K_a \Rightarrow x = 2.7 \times 10^{-6}$$

$$\alpha = \frac{[A^-]}{[HA] + [A^-]} = \frac{2.7 \times 10^{-6}}{0.050\ 0} = 5.40 \times 10^{-5} = 0.005\ 4\%$$

14. $$Cr^{3+} + H_2O \overset{K_{a1}}{\rightleftharpoons} Cr(OH)^{2+} + H^+$$
$$0.010-x \phantom{ + H_2O \overset{K_{a1}}{\rightleftharpoons} Cr(OH)^{2+} } x x$$

$$\frac{x^2}{0.100-x} = 10^{-3.80} \Rightarrow x = 1.18 \times 10^{-3}\ M$$

$$pH = -\log x = 2.93 \qquad \alpha = \frac{x}{0.010} = 0.118$$

15. At 5°C, K_a increases with increasing temperature, so the reaction must be endothermic, by Le Châtelier's principle. At 45°C, K_a decreases with temperature, so the reaction must be exothermic.

16. For 10^{-5} F HA, $[H^+]$ and $[A^-]$ differ by more than 10% when $pK_a \geq 8$; for 10^{-4} F HA, $pK_a \geq 9$; for 10^{-3} F HA, $pK_a \geq 10$.

17. Let $x = [OH^-] = [BH^+]$ and $0.100-x = [B]$. $\dfrac{x^2}{0.100 - x} = 1.00 \times 10^{-5} \Rightarrow$

$x = 9.95 \times 10^{-4}\ M \Rightarrow [H^+] = \dfrac{K_w}{x} = 1.005 \times 10^{-11}$

$\Rightarrow pH = 11.00$

$$\alpha = \frac{[BH^+]}{[B] + [BH^+]} = \frac{9.95 \times 10^{-4}}{0.100} = 9.95 \times 10^{-3}$$

18. $(CH_3)_3N + H_2O \rightleftharpoons (CH_3)_3NH^+ + OH^- \qquad K_b = K_w/K_a = 6.33 \times 10^{-5}$
$$ F-x x x$$

$$\frac{x^2}{0.060 - x} = K_b \Rightarrow x = 1.9_2 \times 10^{-3} \Rightarrow pH = -\log \frac{K_w}{x} = 11.28$$

$$[(CH_3)_3NH^+] = x = 1.9_2 \times 10^{-3} \, M, \quad [(CH_3)_3N] = F - x = 0.058 \, M$$

19. $CN^- + H_2O \rightleftharpoons HCN + OH^- \quad K_b = K_w/K_a = 1.6 \times 10^{-5}$
 F-x $\qquad\qquad\qquad$ x $\qquad$ x

$$\frac{x^2}{0.050-x} = K_b \Rightarrow x = 8.9 \times 10^{-4} \Rightarrow pH = -\log \frac{K_w}{x} = 10.95$$

20. $CH_3CO_2^- + H_2O \rightleftharpoons CH_3CO_2H + OH^- \qquad K_b = K_w/K_a = 5.71 \times 10^{-10}$
 $\quad$ F-x $\qquad\qquad\qquad\qquad$ x $\qquad$ x

$$\frac{x^2}{(1.00 \times 10^{-1}) - x} = K_b \Rightarrow x = 7.56 \times 10^{-6} \Rightarrow \alpha = \frac{x}{F} = 0.007\,56\%$$

$$\frac{x^2}{(1.00 \times 10^{-2}) - x} = K_b \Rightarrow x = 2.39 \times 10^{-6} \Rightarrow \alpha = \frac{x}{F} = 0.023\,9\%$$

For 1.00×10^{-12} M sodium acetate, pH = 7.00 and we can say

$$[HA] = \frac{K_b[A^-]}{[OH^-]} = 5.71 \times 10^{-3} \, [A^-]$$

$$\alpha = \frac{[HA]}{[HA] + [A^-]} = \frac{5.71 \times 10^{-3} \, [A^-]}{(5.71 \times 10^{-3} + 1)[A^-]} = 0.568\%$$

The more dilute the solution, the greater is α.

21. $\qquad\quad B \qquad + \qquad H_2O \quad \rightleftharpoons \quad BH^+ \quad + \quad OH^-$
 $\;$ F - $(K_w/10^{-9.28})$ $\qquad\qquad\qquad\quad K_w/10^{-9.28} \quad K_w/10^{-9.28}$

$$K_b = \frac{(K_w/10^{-9.28})^2}{F - (K_w/10^{-9.28})} = 3.6 \times 10^{-9}$$

22. $\quad B + H_2O \rightleftharpoons BH^+ + OH^- \qquad\qquad \alpha = 0.020 = \frac{x}{F} \Rightarrow x = 2.0 \times 10^{-3} \, M$
 0.10 - x $\qquad$ x $\quad$ x

$$K_b = \frac{x^2}{0.10 - x} = \frac{(2.0 \times 10^{-3})^2}{0.10 - 2.0 \times 10^{-3}} = 4.0 \times 10^{-5}$$

23. As $[B] \rightarrow 0$, $pH \rightarrow 7$ and $[OH^-] \rightarrow 10^{-7}$ M.

$$K_b = \frac{[BH^+][OH^-]}{[B]} = 10^{-7} \frac{[BH^+]}{[B]} \Rightarrow [BH^+] = 10^7 \, K_b[B]$$

$$\alpha = \frac{[BH^+]}{[B] + [BH^+]} = \frac{10^7 \, K_b \, [B]}{[B] + 10^7 \, K_b \, [B]} = \frac{10^7 \, K_b}{1 + 10^7 \, K_b}$$

24. pH of 10^{-3} M B $(K_b = 10^{-5})$ is 9.978; pH of 10^{-5} M B $(K_b = 10^{-9})$ is 7.150.

	A	B	C	D
1	Kw=	Kb	OH-	pH
2	1.E-14	1.00E-05	9.512E-05	9.978
3	F=			
4	0.001			
5				
6	C2 = 0.5*(-B2+Sqrt(B2*B2+4*			
7	($A$2+B2*$A$4+B2*$A$2/C2)))			
8	D2 = -Log10($A$2/C2)			

25. I would weigh out 0.020 0 mol of acetic acid (= 1.201 g) and place it in a beaker with ~75 mL of water. While monitoring the pH with a pH electrode, I would add 3 M NaOH (~4 mL is required) until the pH is exactly 5.00. I would then pour the solution into a 100 mL volumetric flask and wash the beaker several times with a few milliliters of distilled water. Each washing would be added to the volumetric flask, to ensure quantitative transfer from the beaker to the flask. After swirling the volumetric flask to mix the solution, I would carefully add water up to the 100 mL mark, insert the cap, and invert 20 times to ensure complete mixing.

26. The pH of a buffer depends on the ratio of the concentrations of HA and A^- (pH = pK_a + log $[A^-]/[HA]$). When the volume of solution is changed, both concentrations are affected equally and their ratio does not change.

27. Buffer capacity is based on the ability of buffer to react with added acid or base, without making a large change in the ratio of concentrations $[A^-]/[HA]$. The greater the concentration of each component, the less relative change is brought about by reaction with a small increment of added acid or base.

28. When pH = pK_a, the ratio of concentrations $[A^-]/[HA]$ is unity. A given increment of added acid or base has the least effect on the ratio $[A^-]/[HA]$ when they start out being equal.

29. The Henderson-Hasselbalch is just a rearranged form of the K_a equilibrium expression, which is always true. When we make the approximation that [HA] and $[A^-]$ are unchanged from what we added, we are neglecting acid dissociation and base hydrolysis, which can change the concentrations in dilute solutions of moderately strong acids or bases.

30.

acid	pK_a
hydrogen peroxide	11.65
propanoic acid	4.874
cyanoacetic acid	2.472
4-aminobenenesulfonic acid	3.232 ← most suitable

31. $pH = pK_a + \log \frac{[A^-]}{[HA]} = 5.00 + \log \frac{0.050}{0.100} = 4.70$

32. $pH = 3.745 + \log \frac{[HCO_2^-]}{[HCO_2H]}$

pH:	3.000	3.745	4.000
$[HCO_2^-]/[HCO_2H]$:	0.180	1.00	1.80

33. $pH = pK_a + \log \frac{[NO_2^-]}{[HNO_2]}$, where $pK_a = 14.00 - pK_b = 3.15$

(a) If $pH = 2.00$, $[HNO_2]/[NO_2^-] = 14$

(b) If $pH = 10.00$, $[HNO_2]/[NO_2^-] = 1.4 \times 10^{-7}$

34.   1. Weigh out $(0.250 \text{ L})(0.050\ 0 \text{ M}) = 0.012\ 5$ mol of HEPES and dissolve in ~200 mL.

2. Adjust the pH to 7.45 with NaOH (HEPES is an acid).

3. Dilute to 250 mL.

35.

$$CH_3CH_2NH_2 \; + \; H^+ \; \rightarrow \; CH_3CH_2NH_3^+$$

Initial mmol:	1.41_9	x	—
Final mmol:	$1.41_9 - x$	—	x

$pH = pK + \log \frac{[CH_3CH_2NH_2]}{[CH_3CH_2NH_3^+]}$

$10.52 = 10.636 + \log \frac{1.41_9 - x}{x} \Rightarrow x = 0.804$ mmol

volume $= \frac{0.804 \text{ mmol}}{0.246 \text{ mmol/mL}} = 3.27$ mL

36.   (a)

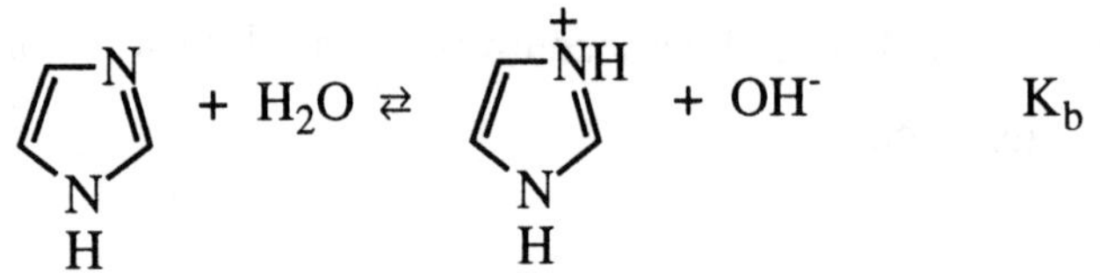

(b) FW of imidazole = 68.078.   FW of imidazole hydrochloride = 104.539.

$$pH = 6.993 + \log \frac{1.00/68.078}{1.00/104.539} = 7.18$$

(c)   B   +   H$^+$ →   BH

Initial mmol: 14.6$_9$ 2.46   9.57

Final mmol: 12.2$_3$ —   12.0$_3$

$$pH = 6.993 + \log \frac{12.2_3}{12.0_3} = 7.00$$

(d) The imidazole must be half neutralized to obtain pH = pK$_a$. Since there are 14.6$_9$ mmol of imidazole, this will require $\frac{1}{2}$(14.6$_9$) = 7.34 mmol of HClO$_4$ = 6.86 mL.

37.   (a) $pH = 2.865 + \log \dfrac{0.040\ 0}{0.080\ 0} = 2.56$

(b) Using Eqns. (10-20) and (10-21), and neglecting [OH$^-$], we can write

$$K_a = 1.36 \times 10^{-3} = \frac{[H^+](0.040 + [H^+])}{0.080\ 0 - [H^+]} \Rightarrow [H^+] = 2.48 \times 10^{-3}\ M \Rightarrow$$

pH = 2.61

(c) 0.080 mol of HNO$_3$ + 0.080 mol of Ca(OH)$_2$ react completely, leaving an excess of 0.080 mol of OH$^-$. This much OH$^-$ converts 0.080 mol of ClCH$_2$CO$_2$H into 0.080 mol of ClCH$_2$CO$_2^-$. The final concentrations are [ClCH$_2$CO$_2^-$] = 0.020 + 0.080 = 0.100 M and [ClCH$_2$CO$_2$H] = 0.180 − 0.080 = 0.100 M. So pH = pK$_a$ = 2.86.

38.   HA   +   OH$^-$ →   A$^-$ +   H$_2$O

Initial moles: 0.021 0   x   —

Final moles: 0.021 0 - x   —   x

$$pH = 7.40 = pK_a + \log \frac{[A^-]}{[HA]} = 7.56 + \log \frac{x}{0.021\ 0-x} \Rightarrow x = 8.59 \times 10^{-3}\ mol.$$

$$volume = \frac{8.59 \times 10^{-3}\ mol}{0.626\ M} = 13.7\ mL$$

39.   Since pK$_a$ for acetic acid is 4.757, we expect the solution to be acidic and will ignore [OH$^-$] in Eqns. (10-20) and (10-21).

[HA] = 0.002 0 - [H$^+$]   [A$^-$] = 0.004 00 + [H$^+$]

$$K_a = 1.75 \times 10^{-5} = \frac{[H^+](0.004\ 00 + [H^+])}{0.002\ 00 - [H^+]} \Rightarrow [H^+] = 8.69 \times 10^{-6}\ M$$

$$\Rightarrow [HA] = 0.001\ 99\ M \qquad [A^-] = 0.004\ 01\ M$$

40. Assuming $[B] = 0.100\ 0\ M$ and $[BH^+] = 0.020\ 0\ M$, we calculate

$$pH = pK_a + \log \frac{[B]}{[BH^+]} = 12.00 + \log \frac{0.010\ 0}{0.020\ 0} = 11.70$$

If we do not assume that $[B] = 0.010\ 0\ M$ and $[BH^+] = 0.020\ 0\ M$, we should use Eqns. (10-20) and (10-21) to find their concentrations. Since the solution is basic we can neglect $[H^+]$ in these equations and write

$[B] = 0.010\ 0 - x$ and $[BH^+] = 0.020\ 0 + x$ where $x = [OH^-]$.

Then we can say $K_b = 10^{-2.00} = \dfrac{[BH^+][OH^-]}{[B]} = \dfrac{(0.020\ 0 + x)\ (x)}{(0.010\ 0 - x)} \Rightarrow$

$x = 0.003\ 0_3\ M \qquad pH = -\log \dfrac{K_w}{x} = 11.48.$

41. (a) pH = 6.001; (b) pH = 6.371; (c) pH = 2.074; (d) pH = 11.617

Here is the spreadsheet for case (d):

	A	B	C
1	F(A-)	H+	pH
2	0.01	2.4142E-12	11.6172
3	F(HA)		
4	0.01		
5	Kw		
6	1.E-14		
7	Ka		
8	1.E-12		
9			
10	B2 = 0.5*(-A2-A8+Sqrt((A2+A8)^2+4*(A8*A4+A6+A8*A6/B2)))		
11	C2 = -Log10(B2)		

1. The K_a reaction, with a much greater equilibrium constant than K_b, releases H^+:

$$HA^- \rightleftarrows H^+ + A^{2-} \qquad K_a.$$

Each mole of H^+ reacts with one mole of OH^- from the K_b reaction:

$$HA^- + H_2O \rightleftarrows H_2A + OH^-.$$

The net result is that the K_b reaction is driven almost as far toward completion as the K_a reaction.

2.

$$\overset{R}{\underset{}{H_3\overset{+}{N}- \overset{|}{C}H - CO_2^-}}$$ pK values apply to $-\overset{+}{N}H_3$, $-CO_2H$ and, in some cases, R.

3.

$$K_{b1} = \frac{K_w}{K_2} = 4.37 \times 10^{-4}$$

$$K_{b2} = \frac{K_w}{K_1} = 8.93 \times 10^{-13}$$

4.   (a) $\dfrac{x^2}{0.100-x} = K_1 \Rightarrow x = 3.11 \times 10^{-3} = [H^+] = [HA^-] \Rightarrow pH = 2.51$

$[H_2A] = 0.100 - x = 0.096\,9\ M$ $\qquad [A^{2-}] = \dfrac{K_2[HA^-]}{[H^+]} = 1.00 \times 10^{-8}\ M$

(b) $[H^+] \approx \sqrt{\dfrac{K_1K_2F + K_1K_w}{K_1 + F}} = 1.00 \times 10^{-6} \Rightarrow pH = 6.00$

$[HA^-] \approx 0.100\ M$

$[H_2A] = \dfrac{[H^+][HA^-]}{K_1} = 1.00 \times 10^{-3}\ M$ $\quad [A^{2-}] = \dfrac{K_2[HA^-]}{[H^+]} = 1.00 \times 10^{-3}\ M$

(c) $\dfrac{x^2}{0.100-x} = \dfrac{K_w}{K_2} \Rightarrow x = [OH^-] = [HA^-] = 3.16 \times 10^{-4}\ M \Rightarrow pH = 10.50$

$[A^{2-}] = 0.100 - x = 9.97 \times 10^{-2}\ M$ $\qquad [H_2A] = \dfrac{[H^+][HA^-]}{K_1} = 1.00 \times 10^{-10}\ M$

	pH	$[H_2A]$	$[HA^-]$	$[A^{2-}]$
0.100 M H_2A	2.51	9.69×10^{-2}	3.11×10^{-3}	1.00×10^{-8}
0.100 M NaHA	6.00	1.00×10^{-3}	1.00×10^{-1}	1.00×10^{-3}
0.100 M Na_2A	10.50	1.00×10^{-10}	3.16×10^{-4}	9.97×10^{-2}

5.   (a) $H_2M = H^+ + HM^- \qquad K_1 = 1.42 \times 10^{-3}$
$\ \ F-x \quad\ \ x \qquad x$

$\dfrac{x^2}{0.100 - x} = K_1 \Rightarrow x = 1.12 \times 10^{-2} \Rightarrow pH = -\log x = 1.95$

$[H_2M] = 0.100 - x = 0.089\ M$

$$[HM^-] = x = 1.12 \times 10^{-2} \text{ M} \qquad\qquad [M^{2-}] = \frac{[HM^-] K_2}{[H^+]} = 2.01 \times 10^{-6} \text{ M}$$

(b) $\quad [H^+] = \sqrt{\dfrac{K_1 K_2 (0.100) + K_1 K_w}{K_1 + 0.100}} = 5.30 \times 10^{-5} \Rightarrow pH = 4.28$

$$[HM^-] \approx 0.100 \text{ M} \qquad\qquad [H_2M] = \frac{[HM^-][H^+]}{K_1} = 3.7 \times 10^{-3} \text{ M}$$

$$[M^{2-}] = \frac{K_2[HM^-]}{[H^+]} = 3.8 \times 10^{-3} \text{ M}$$

The method of Box 11-2 would give more accurate answers, since $[HM^-]$ is not that much greater than $[H_2M]$ or $[M^{2-}]$ in this case.

(c) $\quad M^{2-} + H_2O \rightleftarrows HM^- + OH^- \quad K_{b1} = K_w/K_{a2} = 4.98 \times 10^{-9}$
$ \underset{F-x}{} \underset{x}{} \underset{x}{}$

$$\frac{x^2}{0.100 - x} = K_{b1} \Rightarrow x = 2.23 \times 10^{-5} \Rightarrow pH = -\log \frac{K_w}{x} = 9.35$$

$$[M^{2-}] = 0.100 - x = 0.100 \text{ M} \qquad\qquad [HM^-] = x = 2.23 \times 10^{-5} \text{ M}$$

$$[H_2M] = \frac{[H^+][HM^-]}{K_1} = 7.04 \times 10^{-12} \text{ M}$$

6.

$$HN\!\!\bigcirc\!\!NH + H_2O \rightleftarrows HN\!\!\bigcirc\!\!\overset{+}{N}H_2 + OH^- \qquad K_{b1} = \frac{K_w}{K_2} = 5.38 \times 10^{-5}$$
$ \underset{F-x}{} \underset{x}{} \underset{x}{}$

$$\frac{x^2}{0.300 - x} = K_{b1} \Rightarrow x = 3.99 \times 10^{-3} \text{ M} \Rightarrow pH = -\log K_w/x = 11.60$$

$$[B] = 0.300 - x = 0.296 \text{ M} \qquad\qquad [BH^+] = x = 3.99 \times 10^{-3} \text{ M}$$

$$[BH_2^{2+}] = \frac{[BH^+][H^+]}{K_1} = 2.15 \times 10^{-9} \text{ M}$$

7.   For H_2A, $K_1 = 5.60 \times 10^{-2}$ and $K_2 = 5.42 \times 10^{-5}$

First approximation ($[HA^-]_1 \approx 0.001\,00$ M) :

$$[H^+]_1 = \sqrt{\frac{K_1 K_2 (0.001\,00) + K_1 K_w}{K_1 + 0.001\,00}} = 2.31 \times 10^{-4} \text{ M} \Rightarrow pH_1 = 3.64$$

$$[H_2A]_1 = \frac{[H^+]_1 [HA^-]_1}{K_1} = 4.13 \times 10^{-6} \text{ M}$$

$$[A^{2-}]_1 = \frac{K_2 [HA^-]_1}{[H^+]} = 2.35 \times 10^{-4} \text{ M}$$

Second approximation :

$[HA^-]_2 \approx 0.001\,00 - [H_2A]_1 - [A^{2-}]_1 = 0.000\,761$ M

$$[H^+]_2 = \sqrt{\frac{K_1 K_2 (0.000\,761) + K_1 K_w}{K_1 + 0.000\,761}} = 2.02 \times 10^{-4} \text{ M} \Rightarrow pH_2 = 3.70$$

$$[H_2A]_2 = \frac{[H^+]_2 [HA^-]_2}{K_1} = 2.75 \times 10^{-6} \text{ M}$$

$$[A^{2-}]_2 = \frac{K_2 [HA^-]_2}{[H^+]_2} = 2.04 \times 10^{-4} \text{ M}$$

Third approximation:

$$[HA^-]_3 \approx 0.001\,00 - [H_2A]_2 - [A^{2-}]_2 = 0.000\,793 \text{ M}$$

$$[H^+]_3 = \sqrt{\frac{K_1 K_2 (0.000\,793) + K_1 K_w}{K_1 + 0.000\,793}} = 2.05 \times 10^{-4} \text{ M} \Rightarrow \text{pH}_2 = 3.69$$

$$[H_2A]_3 = \frac{[H^+]_3 [HA^-]_3}{K_1} = 2.90 \times 10^{-6} \text{ M}$$

$$[A^{2-}]_3 = \frac{K_2 [HA^-]_3}{[H^+]_3} = 2.10 \times 10^{-4} \text{ M}$$

8. (a) Charge balance: $[K^+] + [H^+] = [OH^-] + [HP^-] + 2[P^{2-}]$ (1)

Mass balance : $[K^+] = [H_2P] + [HP^-] + [P^{2-}]$ (2)

Equilibria : $K_1 = \dfrac{[H^+]\gamma_{H^+} [HP^-] \gamma_{HP^-}}{[H_2P] \gamma_{H_2P}}$ (3)

$K_2 = \dfrac{[H^+]\gamma_{H^+} [P^{2-}] \gamma_{P^{2-}}}{[HP^-] \gamma_{HP^-}}$ (4)

$K_w = [H^+] \gamma_{H^+} [OH^-] \gamma_{OH^-}$ (5)

Solving for $[K^+]$ in Eqns. (1) and (2), and equating the results gives

$$[H_2P] + [H^+] - [P^{2-}] - [OH^-] = 0$$

Making substitutions from Eqns. (3), (4), and (5), we can write

$$\frac{[H^+]\gamma_{H^+} [HP^-] \gamma_{HP^-}}{K_1 \gamma_{H_2P}} + [H^+] - \frac{K_2[HP^-] \gamma_{HP^-}}{[H^+] \gamma_{H^+} \gamma_{P^{2-}}} - \frac{K_w}{[H^+] \gamma_{H^+} \gamma_{OH^-}} = 0$$

which can be rearranged to

$$[H^+] = \sqrt{\frac{\dfrac{K_1 K_2 [HP^-]\gamma_{HP^-}\gamma_{H_2P}}{\gamma_{H^+}\gamma_{P^{2-}}} + \dfrac{K_1 K_w \gamma_{H_2P}}{\gamma_{H^+}\gamma_{OH^-}}}{K_1\gamma_{H_2P} + [HP^-]\gamma_{H^+}\gamma_{HP^-}}}$$

(b) The ionic strength of 0.050 M KHP is 0.050 M, since the only major ions are K^+ and HP^-.

$[HP^-] \approx 0.050$ M, $\gamma_{HP^-} = 0.835$, $\gamma_{P^{2-}} = 0.485$, $\gamma_{H_2P} \approx 1.00$,

$\gamma_{H^+} = 0.86$, $\gamma_{OH^-} = 0.81$. Using these values in the equation above gives $[H^+] = 1.09 \times 10^{-4} \Rightarrow \text{pH} = 3.96$.

9. $["H_2CO_3"] = [CO_2 \text{ (aq)}] = KP_{CO_2} = 10^{-1.5} \cdot 10^{-3.5} = 10^{-5.0} \text{ M}$

$$H_2CO_3 \rightleftarrows HCO_3^- + H^+ \qquad\qquad K_{a1} = 4.45 \times 10^{-7}$$

$10^{-5.0} - x \qquad\qquad x \qquad\quad x$

$$\frac{x^2}{10^{-5.0} - x} = K_{a1} \Rightarrow x = 1.9 \times 10^{-6} \text{ M} \Rightarrow pH = 5.72$$

10. (a) $K = e^{(\Delta S^\circ/R) - (\Delta H^\circ/RT)} \Rightarrow \dfrac{\Delta S^\circ}{R} = -3.23 \ (\pm 0.53)$ and

$\dfrac{\Delta H^\circ}{RT} = \dfrac{-1.44 \ (\pm 0.15) \times 10^3}{T}$ Using the value R = 8.314 5 $\dfrac{J}{mol \cdot K}$ gives:

$\Delta S^\circ = [-3.23 \ (\pm 0.53)][8.314 \ 5] = -27 \ (\pm 4) \dfrac{J}{mol \cdot K}$

$\Delta H^\circ = [-1.44 \ (\pm 0.15) \times 10^3][8.314 \ 5] = -12 \ (\pm 1) \dfrac{kJ}{mol}$

(b) $K = \dfrac{[SO_3H^-]}{[HSO_3^-]} = e^{-[3.23 \ (\pm 0.53) + 1.44 \ (\pm 0.15) \times 10^3 \ (1/T)]}$

For T = 298 K, $K = e^{-[3.23 \ (\pm 0.53) + (1 \ 440 \ (\pm 150)) \ / \ 298]}$

$= e^{-3.23 \ (\pm 0.53) + 4.83 \ (\pm 0.50)}$ $K = e^{1.60 \ (\pm 0.73)} = 4.953 \ (\pm ?)$

For propagation of uncertainty, we consider $K = e^x$ in Table 3-1:

$\dfrac{e_K}{K} = e_x \Rightarrow e_K = Ke_x = (4.953)(\pm 0.73) = \pm 3.6 \Rightarrow K = 5.0 \ (\pm 3.6)$

11. Spreadsheet for case (b):

	A	B	C	D	E
1	K1 =	1st approx.	2nd approx.		20th approx.
2	0.0001	[HM-]	[HM-]		[HM-]
3	K2 =	1.000E-02	3.675E-03	......	6.125E-03
4	0.00001	[H+]	[H+]		[H+]
5	F =	3.147E-05	3.120E-05	......	3.137E-05
6	0.01	[H2M]	[H2M]		[H2M]
7	Kw =	3.147E-03	1.147E-03	......	1.921E-03
8	1.E-14	[M2-]	[M2-]		[M2-]
9		3.178E-03	1.178E-03	......	1.953E-03
10		pH	pH		pH
11		4.502	4.506	......	4.504
12	B3 = $A$6				
13	B5 = Sqrt(($A$2*$A$4*B3+$A$2*$A$8)/($A$2+B3))				
14	B7 = B5*B3/$A$2				
15	B9 = -Log10(B5)				
16	C3 = $A$6-B7-B9				
17	C5 = Sqrt(($A$2*$A$4*C3+$A$2*$A$8)/($A$2+C3))				
18	C7 = C5*C3/$A$2				
19	C9 = $A$4*C3/C5				
20	C11 = -Log10(C5)				

12. $pH = pK_a + \log \dfrac{[CO_3^{2-}]}{[HCO_3^-]}$

$10.00 = 10.329 + \log \dfrac{(\text{x g})/(105.99 \text{ g/mol})}{(5.00 \text{ g})/(84.01 \text{ g/mol})} \Rightarrow x = 2.96 \text{ g}$

13. We begin with $(25.0 \text{ mL})(0.023\ 3 \text{ M}) = 0.582\ 5$ mmol salicylic acid (H_2A, $pK_1 = 2.97$, $pK_2 = 13.74$). At pH 3.50, there will be a mixture of H_2A and HA^-.

$$H_2A \quad + \quad OH^- \quad \rightarrow \quad HA^- \quad + \quad H_2O$$

Initial mmol:	0.582 5	x	—
Final mmol:	0.582 5 - x	—	x

$3.50 = 2.97 + \log \dfrac{x}{0.582\ 5 - x} \Rightarrow x = 0.449\ 8$ mmol

$(0.449\ 8 \text{ mmol})/(0.202 \text{ M}) = 2.23 \text{ mL NaOH}$

14. Picolinic acid is HA, the intermediate form of a diprotic system with $pK_1 = 1.01$ and $pK_2 = 5.39$. To achieve pH 5.50, we need a mixture of $HA + A^-$.

$$HA \quad + \quad OH^- \quad \rightarrow \quad A^-$$

Initial mmol:	10.0	x	—
Final mmol:	10.0 - x	—	x

$5.50 = 5.39 + \log \dfrac{x}{10.0 - x} \Rightarrow x = 5.63 \text{ mmol} \approx 5.63 \text{ mL NaOH}$

Procedure: Dissolve 10.0 mmol (1.23 g) picolinic acid in ≈ 75 mL H_2O in a beaker. Add NaOH (≈ 5.63 mL) until the measured pH is 5.50. Transfer to a 100 mL volumetric flask and use small portions of H_2O to rinse the beaker into the flask. Dilute to 100.0 mL and mix well.

15. At pH 2.10, we have a mixture of SO_4^{2-} and HSO_4^-, since pK_a for HSO_4^- is 1.99.

$2.10 = 1.99 + \log \dfrac{[SO_4^{2-}]}{[HSO_4^-]} \Rightarrow HSO_4^- = 0.776\ 2\ [SO_4^{2-}]$

The reaction between H_2SO_4 and SO_4^{2-} produces 2 moles of HSO_4^-:

$$H_2SO_4 \quad + \quad SO_4^{2-} \quad \rightarrow \quad 2HSO_4^-$$

Initial mmol:	x	y	—
Final mmol:	—	y - x	2x

The Henderson-Hasselbalch equation told us that $[HSO_4^-] = 0.776\ 2\ [SO_4^{2-}] \Rightarrow$ $2x = 0.776\ 2\ (y - x)$. Since the total sulfur is 0.200 M, $x + y = 0.200$ mol. Substituting $x = 0.200 - y$ into the equation $2x = 0.776\ 2\ (y - x)$ gives $Na_2SO_4 = y = 0.156\ 3$ mol $= 22.20$ g and $H_2SO_4 = 0.043\ 7$ mol $= 4.29$ g.

16. pK_2 for phosphoric acid is 7.2, so it has a high buffer capacity at pH 7.45. At pH
8.5 the buffer capacity of phosphate would be low and it would not be very useful.

17.

$$\overset{+}{N}H_3 | CHCH_2CH_2CO_2H | CO_2H \underset{K_1}{\rightleftarrows} \overset{+}{N}H_3 | CHCH_2CH_2CO_2H | CO_2^- \underset{K_2}{\rightleftarrows} \overset{+}{N}H_3 | CHCH_2CH_2CO_2^- | CO_2^- \underset{K_3}{\rightleftarrows} NH_2 | CHCH_2CH_2CO_2^- | CO_2^-$$

glutamic acid

$$\overset{+}{N}H_3 | CHCH_2 - \bigcirc - OH | CO_2H \underset{K_1}{\rightleftarrows} \overset{+}{N}H_3 | CHCH_2 - \bigcirc - OH | CO_2^- \underset{K_2}{\rightleftarrows}$$

tyrosine

$$NH_2 | CHCH_2 - \bigcirc - OH | CO_2^- \underset{K_3}{\rightleftarrows} NH_2 | CHCH_2 - \bigcirc - O^- | CO_2^-$$

18.   (a)   For 0.050 0 M KH_2PO_4, $[H^+] = \sqrt{\dfrac{K_1K_2(0.050\ 0) + K_1K_w}{K_1 + 0.050\ 0}} = 1.98 \times 10^{-5}$

$\Rightarrow$ pH = 4.70

$4.70 = 2.148 + \log \dfrac{[H_2PO_4^-]}{[H_3PO_4]} \Rightarrow \dfrac{[H_3PO_4]}{[H_2PO_4^-]} = 2.8 \times 10^{-3}$

(b)   For 0.050 0 M K_2HPO_4, $[H^+] = \sqrt{\dfrac{K_2K_3(0.050\ 0) + K_2K_w}{K_2 + 0.050\ 0}} = 2.40 \times 10^{-10}$

$\Rightarrow$ pH = 9.62

$9.62 = 2.148 + \log \dfrac{[H_2PO_4^-]}{[H_3PO_4]} \Rightarrow \dfrac{[H_3PO_4]}{[H_2PO_4^-]} = 3.4 \times 10^{-8}$

19.   Lysine hydrochloride (H_2L^+) is

$$\overset{+}{N}H_3 | CHCH_2CH_2CH_2\overset{+}{N}H_3 | CO_2^-$$

for which $[H^+] = \sqrt{\dfrac{K_1K_2(0.010\ 0) + K_1K_w}{K_1 + 0.010\ 0}} = 1.99 \times 10^{-6}$ M $\Rightarrow$ pH = 5.70

$[H_2L^+] = 0.010\ 0$ M

$[H_3L^{2+}] = \dfrac{[H^+][H_2L^+]}{K_1} = 2.19 \times 10^{-6}$ M

$[HL] = \dfrac{K_2[H_2L^+]}{[H^+]} = 4.17 \times 10^{-6}$ M $[L^-] = \dfrac{K_3[HL]}{[H^+]} = 4.19 \times 10^{-11}$ M

20. (a) $pH = pK_3$ (citric acid) $+ \log \dfrac{[C^{3-}]\, \gamma_{C^{3-}}}{[HC^{2-}]\, \gamma_{HC^{2-}}}$

$pH = 6.396 + \log \dfrac{(1.00)(0.405)}{(2.00)(0.665)} = 5.88$

(b) If the ionic strength is raised to 0.10 M,

$pH = 6.396 + \log \dfrac{(1.00)(0.115)}{(2.00)(0.37)} = 5.59$

21. (a) HA   (b) A^-

(c) $pH = pK_a + \dfrac{[A^-]}{[HA]}$

$7.00 = 7.00 + \log \dfrac{[A^-]}{[HA]} \Rightarrow [A^-]/[HA] = 1.0$

$6.00 = 7.00 + \log \dfrac{[A^-]}{[HA]} \Rightarrow [A^-]/[HA] = 0.10$

22. (a) 4.00   (b) 8.00   (c) H_2A (d) HA^- (e) A^{2-}

23. (a) 9.00   (b) 9.00   (c) BH^+

(d) $12.00 = 9.00 + \log \dfrac{[B]}{[BH^+]} \Rightarrow [B]/[BH^+] = 1.0 \times 10^3$

24.

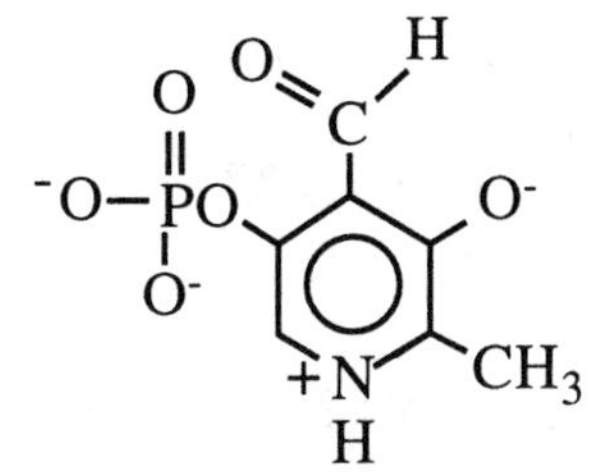

25. Fraction in form HA $= \alpha_{HA} = \dfrac{[H^+]}{[H^+] + K_a} = \dfrac{10^{-5}}{10^{-5} + 10^{-4}} = 0.091$

Fraction in form $A^- = \alpha_{A^-} = \dfrac{K_a}{[H^+] + K_a} = 0.909$.

$\dfrac{[A^-]}{[HA]} = \dfrac{\alpha_{A^-}}{\alpha_{HA}} = 10$, which makes sense.

26. $\alpha_{H_2A} = \dfrac{[H^+]^2}{[H^+]^2 + [H^+]K_1 + K_1 K_2}$, where $[H^+] = 10^{-7.00}$, $K_1 = 10^{-8.00}$ and $K_2 = 10^{-10.00} \Rightarrow \alpha_{H_2A} = 0.91$

27.

	pH 8.00	pH 10.00
α_{H_2A}	0.877	0.049 6
α_{HA^-}	0.123	0.694
$\alpha_{A^{2-}}$	4.54×10^{-4}	0.257

28.

pH :	1.00	1.91	6.00	6.33	10.00
α_{H_2A}	0.890	0.500	5.55×10^{-5}	1.89×10^{-5}	1.74×10^{-12}
α_{HA^-}	0.110	0.500	0.682	0.500	2.15×10^{-4}
$\alpha_{A^{2-}}$	5.10×10^{-7}	1.89×10^{-5}	0.318	0.500	1.00

29. (a) The derivation follows the outline of Equations 11-19 through 11-21. The results are

$$\alpha_{H_3A} = \frac{[H_3A]}{F} = \frac{[H^+]^3}{[H^+]^3 + [H^+]^2 K_1 + [H^+]K_1K_2 + K_1K_2K_3}$$

$$\alpha_{H_2A^-} = \frac{[H_2A^-]}{F} = \frac{[H^+]^2 K_1}{[H^+]^3 + [H^+]^2 K_1 + [H^+]K_1K_2 + K_1K_2K_3}$$

$$\alpha_{HA^{2-}} = \frac{[HA^{2-}]}{F} = \frac{[H^+] K_1 K_2}{[H^+]^3 + [H^+]^2 K_1 + [H^+]K_1K_2 + K_1K_2K_3}$$

$$\alpha_{A^{3-}} = \frac{[A^{3-}]}{F} = \frac{K_1 K_2 K_3}{[H^+]^3 + [H^+]^2 K_1 + [H^+]K_1K_2 + K_1K_2K_3}$$

(b) For phosphoric acid, $pK_1 = 2.148$, $pK_2 = 7.199$ and $pK_3 = 12.15$. At pH = 7.00, the expressions above give $\alpha_{H_3A} = 8.6 \times 10^{-6}$, $\alpha_{H_2A^-} = 0.61$, $\alpha_{HA^{2-}} = 0.39$ and $\alpha_{A^{3-}} = 2.7 \times 10^{-6}$.

30. $pH = pK_{NH_4^+} + \log \dfrac{[NH_3]}{[NH_4^+]} \Rightarrow 9.00 = 9.24 + \log \dfrac{[NH_3]}{[NH_4^+]} \Rightarrow \dfrac{[NH_3]}{[NH_4^+]} = 0.575$

Fraction unprotonated $= \dfrac{[NH_3]}{[NH_3] + [NH_4^+]} = \dfrac{0.575}{0.575 + 1} = 0.37$

31. The quantity of morphine in the solution is negligible compared to the quantity of cacodylic acid. The pH is determined by the reaction of cacodylic acid (HA) with NaOH :

	HA	+	OH⁻	→	A⁻	+	H₂O
Initial mmol :	1.000		0.800		—		—
Final mmol :	0.200		—		0.800		—

$$pH = pK_a + \log \frac{[A^-]}{[HA]} = 6.19 + \log \frac{0.800}{0.200} = 6.79$$

For morphine (B) at pH 6.79 we can write $pH = pK_{BH^+} + \log \dfrac{[B]}{[BH^+]} \Rightarrow$

$6.79 = 6.13 + \log \dfrac{[B]}{[BH^+]} \Rightarrow \log \dfrac{[B]}{[BH^+]} = 4.57 \Rightarrow [B] = 4.57\,[BH^+]$

Fraction in form $BH^+ = \dfrac{[BH^+]}{[B] + [BH^+]} = \dfrac{[BH^+]}{4.57\,[BH^+] + [BH^+]} = 18\%$

32.

	A	B	C	D	E	F	G
1	Ka1 =	pH	[H+]	Denominator	Alpha(H2A)	Alpha(HA-)	Alpha(A2-)
2	8.85E-04	1	1E-01	1.01E-02	9.91E-01	8.77E-03	2.82E-06
3	Ka2 =	3	1E-03	1.91E-06	5.23E-01	4.63E-01	1.48E-02
4	3.21E-05	5	1E-05	3.74E-08	2.68E-03	2.37E-01	7.60E-01
5		7	1E-07	2.85E-08	3.51E-07	3.11E-03	9.97E-01
6		9	1E-09	2.84E-08	3.52E-11	3.12E-05	1.00E+00
7		11	1E-11	2.84E-08	3.52E-15	3.12E-07	1.00E+00
8		13	1E-13	2.84E-08	3.52E-19	3.12E-09	1.00E+00
9							
10	C2 = 10^-B2				F2 = $A$2*C2/D2		
11	D2 = C2^2+$A$2*C2+$A$2*$A$4				G2 = $A$2*$A$4/D2		
12	E2 = C2^2/D2						

33.

	A	B	C	D	E	F	G	H
1	Ka1 =	pH	[H+]	Denominator	Alpha(H3A)	Alpha(H2A)	Alpha(HA)	Alpha(A)
2	6.76E-03	1	1E-01	1.07E-03	9.37E-01	6.33E-02	4.09E-10	1.39E-19
3	Ka2 =	3	1E-03	7.76E-09	1.29E-01	8.71E-01	5.63E-07	1.91E-14
4	6.46E-10	5	1E-05	6.77E-13	1.48E-03	9.98E-01	6.45E-05	2.19E-10
5	Ka3	7	1E-07	6.80E-17	1.47E-05	9.94E-01	6.42E-03	2.18E-06
6	3.39E-11	9	1E-09	1.13E-20	8.87E-08	6.00E-01	3.87E-01	1.31E-02
7		11	1E-11	1.92E-22	5.20E-12	3.51E-03	2.27E-01	7.69E-01
8		13	1E-13	1.48E-22	6.74E-18	4.55E-07	2.94E-03	9.97E-01
9								
10	C2 = 10^-B2							
11	D2 = C2^3+$A$2*C2^2+$A$2*$A$4*C2+$A$2*$A$4*$A$6							
14	E2 = C2^3/D2				G2 = $A$2*$A$4*C2/D2			
15	F2 = $A$2*C2^2/D2				H2 = $A$2*$A$4*$A$6/D2			

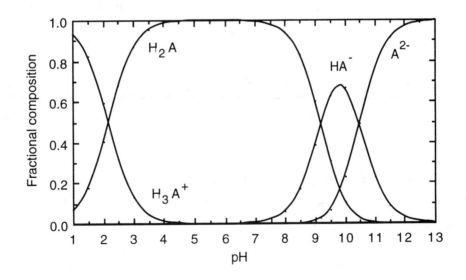

34.	A	B	C	D	E	F	G	H	I
1	Ka1 =	pH	[H+]	Denom.	Alph(H4A)	Alph(H3A)	Alph(H2A)	Alph(HA)	Alph(A)
2	1.58E-04	1	1E-01	1.0E-04	1.0E+00	1.6E-03	6.3E-09	2.5E-14	9.9E-25
3	Ka2 =	2	1E-02	1.0E-08	9.8E-01	1.6E-02	6.2E-07	2.5E-11	9.8E-21
4	3.98E-07	3	1E-03	1.2E-12	8.6E-01	1.4E-01	5.4E-05	2.2E-08	8.6E-17
5	Ka3 =	4	1E-04	2.6E-16	3.9E-01	6.1E-01	2.4E-03	9.7E-06	3.9E-13
6	3.98E-07	5	1E-05	1.7E-19	5.7E-02	9.1E-01	3.6E-02	1.4E-03	5.7E-10
7	Ka4 =	6	1E-06	2.5E-22	4.1E-03	6.4E-01	2.5E-01	1.0E-01	4.0E-07
8	3.98E-12	7	1E-07	3.3E-24	3.0E-05	4.8E-02	1.9E-01	7.6E-01	3.0E-05
9		8	1E-08	2.6E-25	3.9E-08	6.2E-04	2.4E-02	9.7E-01	3.9E-04
10		9	1E-09	2.5E-26	4.0E-11	6.3E-06	2.5E-03	9.9E-01	4.0E-03
11		10	1E-10	2.6E-27	3.8E-14	6.1E-08	2.4E-04	9.6E-01	3.8E-02
12		11	1E-11	3.5E-28	2.9E-17	4.5E-10	1.8E-05	7.2E-01	2.8E-01
13		12	1E-12	1.2E-28	8.0E-21	1.3E-12	5.0E-07	2.0E-01	8.0E-01
14		13	1E-13	1.0E-28	9.8E-25	1.5E-15	6.2E-09	2.5E-02	9.8E-01
15									
16	C2 = 10^-B2								
17	D2 = C2^4+$A$2*C2^3+$A$2*$A$4*C2^2+$A$2*$A$4*$A$6*C2								
18		+$A$2*$A$4*$A$6*$A$8							
19	E2 = C2^4/D2								
20	F2 = $A$2*C2^3/D2			H2 = $A$2*$A$4*$A$6*C2/D2					
21	G2 = $A$2*$A$4*C2^2/D2			I2 = $A$2*$A$4*$A$6*$A$8/D2					

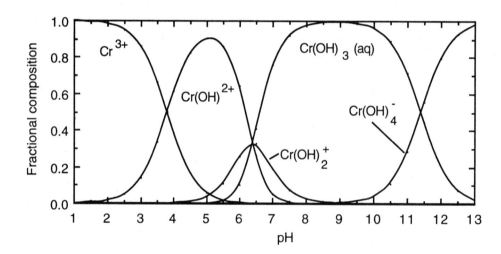

35. The isoelectric pH is the pH at which the protein has no net charge, even though it has many positive and negative sites. The isoionic pH is the pH of a solution containing only protein, H^+ and OH^-.

36. The <u>average</u> charge is zero. There is no pH at which <u>all</u> molecules have zero charge.

37. Isoionic pH $= \sqrt{\dfrac{K_1K_2(0.010) + K_1K_w}{K_1 + (0.010)}} \Rightarrow$ pH $= 5.72$

Isoelectric pH $= \dfrac{pK_1 + pK_2}{2} = 5.59$

ACID-BASE TITRATIONS

1. The equivalence point occurs when the quantity of titrant is exactly the stoichiometric amount needed for complete reaction with analyte. The end point occurs when there is an abrupt change in a physical property, such as pH or indicator color. Ideally, the end point is chosen to occur at the equivalence point.

2.
V_a	0	1	5	9	9.9	10	10.1	12
pH	13.00	12.95	12.68	11.96	10.96	7.00	3.04	1.75

Representative calculations:

$\underline{0\ mL}$: $pH = -\log \dfrac{K_w}{[OH^-]} = -\log \dfrac{10^{-14}}{0.100} = 13.00$

$\underline{1\ mL}$: $[OH^-] = \dfrac{9}{10}(0.100)\dfrac{100}{101} = 0.089\ 1\ M \Rightarrow pH = 12.95$

$\underline{10\ mL}$: $[OH^-] = [H^+] = 10^{-7}\ M$

$\underline{10.1\ mL}$: $[H^+] = \left(\dfrac{0.1}{110.1}\right)(1.00) = 9.08 \times 10^{-4}\ M \Rightarrow pH = 3.04$

3. If the analyte is too weak or too dilute, there is very little change in pH at the equivalence point.

4.
V_b	0	1	5	9	9.9	10	10.1	12
pH	3.00	4.05	5.00	5.95	7.00	8.98	10.96	12.25

Representative calculations:

$\underline{0\ mL}$: $\begin{array}{c} HA = H^+ + A^- \\ 0.100\text{-}x \quad x \quad x \end{array}$ $\dfrac{x^2}{0.100\text{-}x} = 10^{-5.00} \Rightarrow x = 9.95 \times 10^{-4}\ M$

$\Rightarrow pH = 3.00$

$\underline{1\ mL}$: $pH = pK_a + \log \dfrac{[A^-]}{[HA]} = 5.00 + \log \dfrac{1}{9} = 4.05$

$\underline{10\ mL}$: $\begin{array}{c} A^- + H_2O = HA + OH^- \\ \left(\frac{100}{110}\right)(0.100)\text{-}x \quad\quad x \quad\ x \end{array}$ $\dfrac{x^2}{0.090\ 9\text{-}x} = \dfrac{K_w}{K_a} \Rightarrow x = 9.53 \times 10^{-6}$

$\Rightarrow [H^+] = \dfrac{K_w}{x} \Rightarrow pH = 8.98$

$\underline{10.1\ mL}$: $[OH^-] = \left(\dfrac{0.1}{110.1}\right)(1.00) = 9.08 \times 10^{-4}\ M \Rightarrow pH = 10.96$

5. $pH = pK_a + \log \dfrac{[A^-]}{[HA]}$

$pK_a\text{-}1 = pK_a + \log \dfrac{[A^-]}{[HA]} \Rightarrow \dfrac{[A^-]}{[HA]} = \dfrac{1}{10}$

If the ratio $\dfrac{[A^-]}{[HA]}$ is to be $\dfrac{1}{10}$, then $\dfrac{1}{11}$ of the initial HA must remain as HA.

At this point, $[A^-]/[HA] = (1/11)/(10/11) = 1/10$. So $pH = pK_a\text{-}1$ when $V_b = V_e/11$.

In a similar manner, $pH = pK_a + 1$ when $V_b = 10 V_e/11$.

For anilinium ion, $pK_a = 4.601$. For the titration of 100 mL of 0.100 M anilinium ion with 0.100 M OH^-, the reaction is

$$\text{and } V_e = 100 \text{ mL.}$$

0 mL:

$$\frac{x^2}{0.100-x} = K_a \Rightarrow x = 1.57 \times 10^{-3} \Rightarrow pH = 2.80$$

$V_e/11 = 9.09$ mL: $pH = pK_a - 1 = 3.60$

$V_e/2 = 50.0$ mL: $pH = pK_a = 4.60$

$10V_e/11 = 90.91$ mL: $pH = pK_a + 1 = 5.60$

$V_e = 100.0$ mL: BH^+ has been converted to B.

$$B + H_2O \rightleftharpoons BH^+ + OH^-$$
$$F-x \qquad\qquad x \qquad x$$

$$K_b = \frac{K_w}{K_a} = \frac{x^2}{\left(\frac{100}{200}\right)(0.100)-x}$$

$$\Rightarrow x = 4.46 \times 10^{-6} \text{ M} \qquad pH = -\log\frac{K_w}{x} = 8.65$$

$1.2V_e = 120.0$ mL: There are 20.0 mL of excess NaOH.

$$[OH^-] = \left(\frac{20}{220}\right)(0.100) = 9.09 \times 10^{-3} \text{ M} \Rightarrow pH = 11.96$$

6.   The titration reaction is $HA + OH^- \rightarrow A^- + H_2O$. A volume of V mL of HA will require 2V mL of KOH to reach the equivalence point, because [HA] = 0.100 M and [KOH] = 0.050 0 M. The formal concentration of A^- at the equivalence point will be $\left(\frac{V}{V+2V}\right)(0.100) = 0.033\ 3$ M.   The pH is found by writing

$$A^- \rightleftharpoons H_2O = HA + OH^-$$
$$0.033\ 3-x \qquad\qquad x \qquad x$$

$$\frac{x^2}{0.033\ 3-x} = K_b = \frac{K_w}{K_a}$$

$$\Rightarrow x = 1.50 \times 10^{-6} \text{ M} \Rightarrow pH = 8.18$$

7.

$$K_a = 10^{-6.15}$$

$$H^+ + OH^- \rightleftharpoons H_2O \qquad\qquad 1/K_w = 10^{14.00}$$

$$K = \frac{K_a}{K_w} = 7.1 \times 10^7$$

8.   HA   +   OH$^-$ →   A$^-$ +   H$_2$O

Initial mmol: 5.857   x   —

Final mmol: 5.857-x   —   x

$$pH = 9.24 = pK_a + \log \frac{[A^-]}{[HA]} = 9.50 + \log \frac{x}{5.857-x} \Rightarrow x = 2.077 \text{ mmol}$$

$$[OH^-] = \frac{2.077 \text{ mmol}}{22.63 \text{ mL}} = 0.092 \text{ M}$$

9. $(CH_3)_3NH^+$ +   OH$^-$ → $(CH_3)_3N$ +   H$_2$O

Initial mmol: 1.00   0.40   —

Final mmol: 0.60   —   0.40

First find the ionic strength :

$$[(CH_3)_3NH^+] = 0.60 \text{ mmol}/14.0 \text{ mL} = 0.042\,86 \text{ M}$$

$$[Br^-] = 1.00 \text{ mmol}/14.0 \text{ mL} = 0.071\,43 \text{ M}$$

$$[Na^+] = 0.40 \text{ mmol}/14.0 \text{ mL} = 0.028\,57 \text{ M}$$

$$\mu = \tfrac{1}{2}\Sigma c_i z_i^2 = 0.071 \text{ M}$$

$$pH = pK_a + \log \frac{[B]\,\gamma_B}{[BH^+]\,\gamma_{BH^+}}$$

$$pH = 9.800 + \log \frac{(0.028\,6)(1.00)}{(0.042\,9)(0.80)} = 9.72$$

In the calculation above we used the size of $(CH_3)_3NH^+$ (400 pm) and the activity coefficient interpolated from Table 8-1.

10.   Titration of a weak base, B, produces the conjugate acid, BH$^+$, which is necessarily acidic.

11.
V_a	0	1	5	9	9.9	10	10.1	12
pH	11.00	9.95	9.00	8.05	7.00	5.02	3.04	1.75

Representative calculations:

<u>0 mL</u>: B + H$_2$O ⇌ BH + OH$^-$ $\dfrac{x^2}{0.100-x} = 10^{-5.00}$ M $\Rightarrow x = 9.95 \times 10^{-4}$ M

      0.100-x   x   x

$$[H^+] = \frac{K_w}{x} \Rightarrow pH = 11.00$$

<u>1 mL</u>: $pH = pK_{BH^+} + \log \dfrac{[B]}{[BH^+]} = 9.00 + \log \dfrac{9}{1} = 9.95$

<u>10 mL</u>: BH$^+$ ⇌   B + H$^+$ $\dfrac{x^2}{0.090\,9-x} = \dfrac{K_w}{K_b} \Rightarrow x = 9.53 \times 10^{-6}$

$\left(\dfrac{100}{110}\right)(0.100-x)$ x   x

$$\Rightarrow pH = 5.02$$

<u>10.1 mL</u>: $[H^+] = \left(\dfrac{0.1}{110.1}\right)(1.00) = 9.08 \times 10^{-4}$ M $\Rightarrow$ pH = 3.04

12. The maximum buffer capacity is reached when

$V = \frac{1}{2}V_e$, at which time $\dfrac{[B]}{[BH^+]} = 1$ and $pH = pK_a$ (for BH^+).

13.

$K = 1/K_a$ (for $C_6H_5CH_2NH_3^+$) $= 2.2 \times 10^9$

14. Titration reaction: $B + H^+ \rightarrow BH^+$. To find the equivalence point we write

$(50.0)(0.031\ 9) = (V_e)(0.050\ 0) \Rightarrow V_e = 31.9$ mL

$\underline{0\ mL}$: $\underset{0.031\ 9-x}{B} + H_2O \rightleftarrows \underset{x}{BH^+} + \underset{x}{OH^-}$ $\dfrac{x^2}{0.031\ 9-x} = K_b = \dfrac{K_w}{K_a} = 2.22 \times 10^{-5}$

$\Rightarrow x = 8.31 \times 10^{-4}$ M $\Rightarrow pH = 10.92$

$\underline{12.0\ mL}$: $B\quad +\quad H^+\quad \rightarrow\quad BH^+$

Initial :   31.9   12.0   —

Final: 19.9   —   12.0

$pH = pK_a + \log \dfrac{[B]}{[BH^+]} = 9.35 + \log \dfrac{19.9}{12.0} = 9.57$

$\underline{1/2V_e}$: $pH = pK_a = 9.35$

$\underline{30\ mL}$: $pH = pK_a + \log \dfrac{1.9}{30.0} = 8.15$

$\underline{V_e}$:   B has been converted to BH^+ at a concentration of $\left(\dfrac{50.0}{81.9}\right)(0.031\ 9)$

$= 0.019\ 5$ M

$\underset{0.0195-x}{BH^+} \rightleftarrows \underset{x}{B} + \underset{x}{H^+}$ $\dfrac{x^2}{0.019\ 5-x} = K_a \Rightarrow x = 2.96 \times 10^{-6}$ M

$\Rightarrow pH = 5.53$

$\underline{35.0\ mL}$: $[H^+] = \left(\dfrac{3.1}{85.0}\right)(0.050\ 0) = 1.82 \times 10^{-3}$ M $\Rightarrow pH = 2.74$

15. Titration reaction : $CN^- + H^+ \rightarrow HCN$

At the equivalence point, moles of $CN^- =$ moles of H^+

$(0.100\ M)\ (50.00\ mL) = (0.438\ M)\ (V_e) \Rightarrow V_e = 11.42$ mL

(a) $CN^-\quad +\quad H^+\quad \rightarrow\quad HCN$

Initial :   11.42   4.20   —

Final :   7.22   —   4.20

$pH = pK_a + \log \dfrac{7.22}{4.20} = 9.45$

(b) 11.82 mL is 0.40 mL past the equivalence point.

$[H^+] = \left(\dfrac{0.40}{61.82}\right)(0.438\ M) = 2.83 \times 10^{-3}$ M $\Rightarrow pH = 2.55$

(c) At the equivalence point we have made HCN at a formal concentration of

$$\left(\frac{50.00}{61.42}\right)(0.100) = 0.081\ 4\ M.$$

$$HCN \rightleftharpoons H^+ + CN^-$$
$$0.0814-x \qquad x \qquad x$$

$$\frac{x^2}{0.081\ 4-x} = K_a \Rightarrow x = 7.10 \times 10^{-6}$$
$$\Rightarrow pH = 5.15$$

16. The molecule is neutral at the isoelectric point. Since the isoionic point occurs at a lower pH, the protein must be positively charged at the isoionic point.

17. The equivalence point could be attained by mixing pure HA plus NaCl. This is equivalent to an isoionic solution of HA.

18. (a) $HA \rightleftharpoons A^- + H^+$ K_a
                    $\underline{B + H^+ \rightleftharpoons BH^+}$ $\underline{K = K_b/K_w}$
                    $B + HA \rightleftharpoons BH^+ + A^-$ $K = K_aK_b/K_w$
                    $= 10^{-2.86}\ 10^{-3.36} / 10^{-14.00} = 10^{7.78}$

(b) In the upper curve, $\frac{3}{2}V_e$ is half way between the first and second equivalence points. The pH is simply pK_2, since there is a 1:1 mixture of HA^- and A^{2-}.

In the lower curve, pK_2 $(= pK_{BH+})$ occurs when there is a 1:1 mixture of B and BH^+. To achieve this condition, all of B is first transformed into BH^+ by reaction with HA until V_e is reached. Then, at $2V_e$ one more equivalent of B has been added, giving a 1:1 mole ratio $B:BH^+$, so $pH = pK_{BH+}$.

19.

V_a	0	1	5	9	10	11	15	19	20	22
pH	11.49	10.95	10.00	9.05	8.00	6.95	6.00	5.05	3.54	1.79

Representative calculations:

$\underline{0\ mL}$: $B + H_2O \overset{K_{b1}}{\rightleftharpoons} BH^+ + OH^-$ $\frac{x^2}{0.100-x} = 10^{-4.00} \Rightarrow x = 3.11 \times 10^{-3}\ M$
                   $0.100-x \qquad\quad x \qquad x$
$$pH = -\log\frac{K_w}{x} = 11.49$$

$\underline{1\ mL}$: $pH = pK_{BH+} + \log\frac{[B]}{[BH^+]} = 10.00 + \log\frac{9}{1} = 10.95\ M$

$\underline{10\ mL}$: Predominant form is BH^+ with formal concentration $\frac{100}{110}(0.100) = 0.090\ 9\ M$

$$[H^+] \approx \sqrt{\frac{10^{-6.00}\ 10^{-10.00}\ (0.090\ 9) + 10^{-6.00}\ 10^{-14.00}}{10^{-6.00} + 0.090\ 9}}$$

$$= 1.00 \times 10^{-8} \Rightarrow pH = 8.00$$

$\underline{11\ mL}$: $pH = pK_{BH_2^{2+}} + \log\frac{[BH^+]}{[BH_2^{2+}]} = 6.00 + \log\frac{9}{1} = 6.95$

$\underline{20\ mL}$: $BH_2^{2+} \rightleftharpoons BH^+ + H^+$ $\dfrac{x^2}{0.083\ 3\text{-}x} = 10^{-6.00} \Rightarrow x = 2.88 \times 10^{-4}$

   $\dfrac{100}{120}(0.100)\text{-}x \quad\quad x \quad\quad x$ $\Rightarrow\ pH = 3.54$

$\underline{22\ mL}$: $[H^+] = \left(\dfrac{2}{122}\right)(1.00) = 1.64 \times 10^{-2}\ M \Rightarrow pH = 1.79$

20.

V_b	0	1	5	9	10	11	15	19	20	22
pH	2.51	3.05	4.00	4.95	6.00	7.05	8.00	8.95	10.46	12.21

Representative calculations:

$\underline{0\ mL}$: $H_2A \rightleftharpoons HA^- + H^+$ $\dfrac{x^2}{0.100\text{-}x} = 10^{-4.00} \Rightarrow x = 3.11 \times 10^{-3}\ M$

   $\quad\quad 0.100\text{-}x \quad\ x \quad\ x$ $\Rightarrow\ pH = 2.51$

$\underline{1\ mL}$: $pH = pK_1 + \log\dfrac{[HA^-]}{[H_2A]} = 4.00 + \log\dfrac{1}{9} = 3.05$

$\underline{10\ mL}$: Predominant form is HA^- with formal concentration $\left(\dfrac{100}{110}\right)(0.100) =$

   0.090 9 M.

$$[H^+] \approx \sqrt{\dfrac{10^{-4.00}\ 10^{-8.00}\ (0.090\ 9) + 10^{-4.00}\ 10^{-14.00}}{10^{-4.00} + 0.090\ 9}}$$

   $= 9.99 \times 10^{-7} \Rightarrow pH = 6.00$

$\underline{11\ mL}$: $pH = pK_2 + \log\dfrac{[A^{2-}]}{[HA^-]} = 8.00 + \log\dfrac{1}{9} = 7.05$

$\underline{20\ mL}$: $A^{2-} + H_2O \rightleftharpoons HA^- + OH^-$ $\dfrac{x^2}{0.083\ 3\text{-}x} = \dfrac{K_w}{K_2} \Rightarrow x = 2.88 \times 10^{-4}\ M$

   $\left(\dfrac{100}{120}\right)(0.100)\text{-}x \quad\quad x \quad\quad x$ $pH = \text{-}\log\dfrac{K_w}{x} = 10.46$

$\underline{22\ mL}$: $[OH^-] = \left(\dfrac{2}{122}\right)(1.00) = 1.64 \times 10^{-2}\ M \Rightarrow pH = 12.21$

21.   Titration reactions:

HN⟮⟯NH + H$^+$ → H$_2$N$^+$⟮⟯NH $V_e = 40.0\ mL$

H$_2$N$^+$⟮⟯NH + H$^+$ → H$_2$N$^+$⟮⟯N$^+$H$_2$ $V_e = 80.0\ mL$

$\underline{0\ mL}$: $B + H_2O \rightleftharpoons BH^+ + OH^-$ $\dfrac{x^2}{0.100\text{-}x} = K_{b1} = \dfrac{K_w}{K_{a2}}$

   $\quad 0.100\text{-}x \quad\quad x \quad\ x$ $\Rightarrow x = 2.29 \times 10^{-3}\ M \Rightarrow pH = 11.36$

$\underline{10.0\ mL}$: $pH = pK_2 + \log\dfrac{[B]}{[BH^+]} = 9.731 + \log\dfrac{3}{1} = 10.21$

$\underline{20.0\ mL}$: $pH = pK_2 = 9.73$

30.0 mL: $pH = pK_2 + \log \frac{1}{3} = 9.25$

40.0 mL: B has been converted to BH^+ at a formal concentration of $\left(\frac{40.0}{80.0}\right)(0.100)$

$= 0.050\ 0\ M.$ $[H^+] = \sqrt{\dfrac{K_1 K_2 F + K_1 K_w}{K_1 + F}} \Rightarrow pH = 7.53$

50.0 mL: $pH = pK_1 + \log \dfrac{[BH^+]}{[BH_2^{2+}]} = 5.333 + \log \dfrac{3}{1} = 5.81$

60.0 mL: $pH = pK_1 = 5.33$

70.0 mL: $pH = pK_1 + \log \frac{1}{3} = 4.85$

80.0 mL: B has been converted to BH_2^{2+} at a formal concentration of

$\left(\dfrac{40.0}{120.0}\right)(0.100) = 0.033\ 3\ M$

$$BH_2^{2+} \rightleftharpoons BH^+ + H^+$$
$$0.0333-x \qquad x \qquad x$$

$\dfrac{x^2}{0.033\ 3-x} = K_1 \Rightarrow$

$x = 3.91 \times 10^{-4}\ M \Rightarrow pH = 3.41$

90.0 mL: $[H^+] = \left(\dfrac{10.0}{130.0}\right)(0.100) \Rightarrow pH = 2.11$

100.0 mL: $[H^+] = \left(\dfrac{20.0}{140.0}\right)(0.100) \Rightarrow pH = 1.85$

22.

Initial mmol: 0.500   0.164   —

Final mmol: 0.336   —   0.164

$pH = pK_1 + \log \dfrac{0.336}{0.164} = 5.09$

23. (a) Titration reactions:

$H_2NCH_2CO_2^- + H^+ \rightarrow H_3^+NCH_2CO_2^- \qquad V_e = 50.0\ mL$

$H_3^+NCH_2CO_2^- + H^+ \rightarrow H_3^+NCH_2CO_2H \qquad V_e = 100.0\ mL$

At the second equivalence point the formal concentration of

$H_3^+NCH_2CO_2H$ is $\left(\dfrac{50.0}{150.0}\right)(0.100) = 0.333\ M$

$H_3^+NCH_2CO_2H \rightleftharpoons H_3^+NCH_2CO_2^- + H^+$ $\dfrac{x^2}{0.033\ 3-x} = K_1 \Rightarrow$
$\quad 0.0333-x \qquad\qquad x \qquad\quad x$

$x = 1.02 \times 10^{-2}\ M \Rightarrow pH = 1.99$

(b) At $V_a = 90.0$ mL, the approximation gives $pH = pK_1 + \log \frac{1}{4} = 1.75$, which

is <u>lower</u> than the correct value at 100.0 mL. At $V_a = 101.0$ mL, the

approximation gives $[H^+] = \left(\dfrac{1.0}{151.0}\right)(0.100) = 6.62 \times 10^{-4}\ M \Rightarrow pH = 3.18,$

which is <u>higher</u> than the correct value at 100.0 mL.

24. (a)

$$
\overset{+}{\underset{|}{N}}H_3
$$
$$
\underset{|}{C}HCH_2CH_2CO_2H + OH^- \quad \rightarrow
$$
$$
CO_2^-
$$

$$
\overset{+}{\underset{|}{N}}H_3
$$
$$
\underset{|}{C}HCH_2CH_2CO_2^- + H_2O
$$
$$
CO_2^-
$$

(b) V mL of glutamic acid will require $\dfrac{0.100}{0.025}$ V $= 4.00$ V mL of RbOH to reach

the equivalence point. The formal concentration of product will be

$$\left(\frac{V}{V + 4.00\,V}\right)(0.100) = 0.020\ 0\ M.$$

$$[H^+] = \sqrt{\frac{K_2K_3(0.020\ 0) + K_2K_w}{K_2 + 0.020\ 0}} = 6.53 \times 10^{-8}\ M \Rightarrow pH = 7.18$$

25.

$$
\overset{+}{\underset{|}{N}}H_3
$$
$$
\underset{|}{C}HCH_2-\!\!\bigcirc\!\!-OH + H^+ \quad \rightarrow
$$
$$
CO_2^-
$$
$$
H_2T
$$

$$
\overset{+}{\underset{|}{N}}H_3
$$
$$
\underset{|}{C}HCH_2-\!\!\bigcirc\!\!-OH
$$
$$
CO_2H
$$
$$
H_3T^+
$$

One volume of tyrosine (0.010 0 M) requires 2.5 volumes of $HClO_4$ (0.004 00 M), so the formal concentration of tyrosine at the equivalence point is

$$\left(\frac{1}{1+2.5}\right)(0.010\ 0\ M) = 0.002\ 86\ M.$$ The pH is calculated from the acid dissociation

of H_3T^+.

$$
\begin{array}{ccc}
H_3T^+ & \rightleftarrows & H_2T + H^+ \\
0.002\ 86\text{-}x & & x \quad\quad x
\end{array}
\qquad
\frac{x^2}{0.002\ 86\text{-}x} = K_1 \Rightarrow
$$

$$x = 0.002\ 16\ M \Rightarrow pH = 2.66$$

26. (a) $C^{2-} + H^+ \rightarrow HC^-$. $V_e = 20.0$ mL. At the equivalence point the formal concentration of HC^- is $\left(\dfrac{40.0}{60.0}\right)(0.030\ 0) = 0.020\ 0\ M.$

$$[H^+] = \sqrt{\frac{K_2K_3(0.020\ 0) + K_2K_w}{K_2 + 0.020\ 0}} = 2.776 \times 10^{-10}\ M \Rightarrow pH = 9.56$$

(b)
$$
\begin{array}{ccc}
H_3C^+ & \rightleftarrows & H_2C + H^+ \\
0.0500\text{-}x & & x \quad\quad x
\end{array}
\qquad
\frac{x^2}{0.050\ 0\text{-}x} = K_1
$$

$$\Rightarrow x = 2.29 \times 10^{-2}\ M \Rightarrow pH = 1.64$$

$$pH = pK_3 + \log\frac{[C^{2-}]}{[HC^-]}$$

$$1.64 = 10.77 + \log\frac{[C^{2-}]}{[HC^-]} \Rightarrow \frac{[C^{2-}]}{[HC^-]} = 7.4 \times 10^{-10}$$

27. The two values of pK_a for oxalic acid are 1.252 and 4.266. At a pH of 4.40 the $C_2O_4^{2-}$ has not yet been half neutralized.

$$C_2O_4^{2-} \quad + \quad H^+ \quad \rightarrow \quad HC_2O_4^-$$

Initial mmol:	x	16.0	—
Final mmol:	x-16.0	—	16.0

$$pH = 4.40 = pK_2 + \log\frac{[C_2O_4^{2-}]}{[HC_2O_4^-]} = 4.266 + \log\frac{x-16.0}{16.0}$$

$$\Rightarrow x = 37.8 \text{ mmol of } K_2C_2O_4 = 6.28 \text{ g}$$

28. Neutral alanine is designated HA.

$$HA \quad + \quad OH^- \quad \rightarrow \quad A^- \quad + \quad H_2O$$

Initial mmol:	1.260 5	0.516	—	
Final mmol:	0.744 5	—	0.516	

$$pH = pK_2 + \log\frac{[A^-]\,\gamma_{A^-}}{[HA]\,\gamma_{HA}}$$

$$9.57 = pK_2 + \log\frac{(0.516)(0.77)}{(0.744\ 5)(1)} \Rightarrow pK_2 = 9.84$$

29. A Gran plot allows us to find the equivalence point by extrapolating from points measured prior to the equivalence point.

30. It is evident that the end point is near 23.4 mL, at which point the derivative dpH/dV_b is greatest. A graph of $V_b 10^{-pH}$ versus V_b can be extrapolated to the x axis to give $V_e = 23.40$ mL.

V_b (mL)	$V_b\, 10^{-pH}$	V_b (mL)	$V_b\, 10^{-pH}$	V_b (mL)	$V_b\, 10^{-pH}$
21.01	15.22×10^{-6}	22.10	8.40×10^{-6}	22.97	2.76×10^{-6}
21.10	14.94	22.27	7.37	23.01	2.41
21.13	14.62	22.37	6.60	23.11	1.79
21.20	14.33	22.48	5.91	23.17	1.46
21.30	13.75	22.57	5.29	23.21	1.16
21.41	12.90	22.70	4.53	23.30	0.75
21.51	12.10	22.76	4.14	23.32	0.42
21.61	11.61	22.80	3.78	23.40	0.12
21.77	10.42	22.85	3.46	23.46	0.01
21.93	9.35	22.91	3.16	23.55	0.003

31.

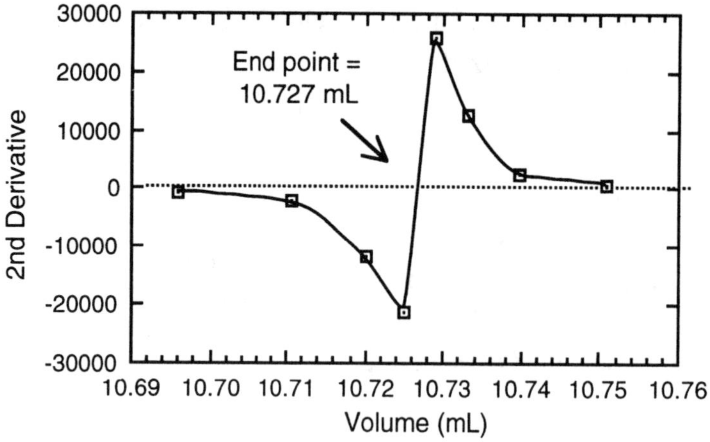

32. The quotient $[HIn]/[In^-]$ changes from 10:1 when $pH = pK_{HIn} - 1$ to 1:10 when $pH = pK_{HIn} + 1$. This change is generally sufficient to cause a complete color change.

33. Strong acids, such as H_2SO_4, HCl, HNO_3, and $HClO_4$, have $pK_a < 0$.

34. Yellow, green, blue

35. (a) red (b) orange (c) yellow

36. (a) red (b) orange (c) yellow (d) red

37. No. When a weak acid is titrated with a strong base, the solution contains A^- at the equivalence point. A solution of A^- must have a pH above 7.

38. (a) The titration reaction is $F^- + H^+ \rightarrow HF$

If V mL of NaF are used, $V_e = \frac{1}{2}$ V, since the concentration of $HClO_4$ is twice as great as the concentration of NaF. The formal concentration of HF at the equivalence point is $\left(\dfrac{V}{V + \frac{1}{2}V}\right)(0.030\ 0) = 0.020\ 0$ M.

The pH is determined by the acid dissociation of HF.

$$\begin{array}{cccc} HF & \rightleftarrows & H^+ & + & F^- \\ 0.0200-x & & x & & x \end{array} \qquad \dfrac{x^2}{0.020\ 0-x} = K_a \implies x = 3.36 \times 10^{-3}$$

$$\implies pH = 2.47$$

(b) The pH is so low that there would not be much (if any) break in the titration curve at the equivalence point. A sharp change in indicator color will not be seen.

39. (a) violet (red + blue) (b) blue (c) yellow

40. (a) $NH_4^+ \rightleftarrows NH_3 + H^+$ $\qquad \dfrac{x^2}{0.010-x} = K_a \Rightarrow x = 2.39 \times 10^{-6} \text{ M}$

$\qquad\qquad$ 0.010-x $\quad$ x $\quad$ x

$\qquad\qquad\qquad\qquad\qquad\qquad\qquad\qquad\qquad\qquad\qquad \Rightarrow \text{pH} = 5.62$

$\qquad$ (b) One possible indicator is methyl red, using the yellow end point.

41. Grams of cleaner titrated $= \left(\dfrac{4.373}{10.231 + 39.466}\right)(10.231 \text{ g}) = 0.900\,3 \text{ g}$

mol HCl used = mol NH_3 present $= (0.014\,22 \text{ L})(0.106\,3 \text{ M}) = 1.512 \text{ mmol}$

1.512 mmol NH_3 = 25.74 mg NH_3

wt% $NH_3 = \dfrac{2.574 \times 10^{-2} \text{ g}}{0.900\,3 \text{ g}} \times 100 = 2.859\%$

42. $A_{604} = \varepsilon_{In^-} [In^-] (1.00) \Rightarrow [In^-] = \dfrac{0.118}{4.97 \times 10^4} = 2.37 \times 10^{-6} \text{ M}$

Since the indicator was diluted with KOH solution, the formal concentration of indicator is 0.700×10^{-5} M.

$\qquad [HIn] = 7.00 \times 10^{-6} - 2.37 \times 10^{-6} = 4.63 \times 10^{-6} \text{ M}$

$\qquad \text{pH} = pK_{In} + \log \dfrac{[In^-]}{[HIn]} = 7.95 + \log \dfrac{2.37}{4.63} = 7.66$

Call benzene-1,2,3-tricarboxylic acid H_3A, with $pK_1 = 2.88$, $pK_2 = 4.75$ and $pK_3 = 7.13$. Since the pH is 7.66, the main species is A^{3-} and the second main species is HA^{2-}. Enough KOH to react with H_3A and H_2A^- must have been added, and there is enough KOH to react with part of the HA^{2-}.

$\qquad\qquad\qquad HA^{2-} + OH^- \rightarrow A^{3-} + H_2O$

Initial mmol: $\quad$ 1.000 $\qquad$ x $\qquad\qquad$ —

Final mmol: $\quad$ 1.000-x $\qquad$ — $\qquad\qquad$ x

$\qquad\qquad \text{pH} = pK_3 + \log \dfrac{[A^{3-}]}{[HA^{2-}]}$

$\qquad\qquad 7.66 = 7.13 + \log \dfrac{x}{1.00-x} \Rightarrow x = 0.772 \text{ mmol of } OH^-$

The total KOH added is 2.772 mmol. The molarity is $\dfrac{2.772 \text{ mmol}}{20.0 \text{ mL}} = 0.139 \text{ M}$

43. The pH of the solution is 7.50, and the total concentration of indicator is 5.00×10^{-5} M. At pH 7.50, there is a negligible amount of H_2In, since $pK_1 = 1.00$. We can write

$\qquad [HIn^-] + [In^{2-}] = 5.0 \times 10^{-5}$

$\qquad \text{pH} = pK_2 + \log \dfrac{[In^{2-}]}{[HIn^-]}$

$$7.50 = 7.95 + \log \frac{[In^{2-}]}{5.00 \times 10^{-5} - [In^{2-}]} \Rightarrow [In^{2-}] = 1.31 \times 10^{-5} \text{ M}$$

$$[HIn] = 3.69 \times 10^{-5} \text{ M}$$

$$A_{435} = \varepsilon_{435}[HIn^-] + \varepsilon_{435}[In^{2-}]$$

$$= (1.80 \times 10^4)(3.69 \times 10^{-5}) + (1.15 \times 10^4)(1.31 \times 10^{-5}) = 0.815$$

44. The quantity of HIn is small compared to aniline and sulfanilic acid. Calling aniline B and sulfanilic acid HA, we can write

	B	+	HA	$\overset{K}{\rightleftharpoons}$	BH$^+$	+	A$^-$
Initial mmol :	2.00		1.500		—		—
Final mmol :	2.00-x		1.500-x		x		x

$$K = \frac{K_a K_b}{K_w} = \frac{(10^{-3.232})(K_w/10^{-4.601})}{K_w} = 23.39$$

$$\frac{x^2}{(2.00-x)(1.500-x)} = 23.39 \Rightarrow x = 1.372 \text{ mmol}$$

$$pH = pK_{BH+} + \log \frac{[B]}{[BH^+]} = 4.601 + \log \frac{2.00-1.372}{1.372} = 4.26$$

For HIn we can write:

absorbance $= 0.110 = (2.26 \times 10^4)(5.00)[HIn] + (1.53 \times 10^4)(5.00)[In^-]$.
Substituting $[HIn] = 1.23 \times 10^{-6} - [In^-]$ gives $[In^-] = 7.94 \times 10^{-7}$ and $[HIn] = 4.36 \times 10^{-7}$. The Henderson-Hasselbalch equation for HIn is therefore

$$pH = pK_{HIn} + \log \frac{[In^-]}{[HIn]} \Rightarrow 4.26 = pK_{HIn} + \log \frac{7.94 \times 10^{-7}}{4.36 \times 10^{-7}} \Rightarrow pK_{HIn} = 4.00$$

45. Tris(hydroxymethyl) aminomethane ($H_2NC(CH_2OH)_3$), mercuric oxide (HgO), sodium carbonate (Na_2CO_3) and borax ($NaB_4O_7 \cdot 10H_2O$) can be used to standardize HCl. Potassium acid phthalate ($HO_2C\text{-}C_6H_4\text{-}CO_2^-K^+$), HCl azeotrope and potassium hydrogen iodate ($KH(IO_3)_2$) can be used to standardize NaOH.

46. The greater the equivalent mass, the more primary standard is required. There is less relative error in weighing a large mass of reagent than a small mass.

47. Potassium acid phthalate is dried at 105° and weighed accurately into a flask. It is titrated with NaOH, using a pH electrode or phenolphthalein to observe the end point.

48. Grams of tris titrated $= \dfrac{4.963}{(1.023+99.367)}(1.023) = 0.050\ 57 = 0.417\ 5$ mmol

Concentration of HNO$_3$ = $\dfrac{0.417\ 5 \text{ mmol}}{5.262 \text{ g solution}}$ = 0.079 34 mol/kg solution

49.   The mmoles of HgO in 0.194 7 g = 0.898 9, which will make 1.798 mmol of OH$^-$ by reaction with Br$^-$ plus H$_2$O. HCl molarity = 1.798 mmol/17.98 mL = 0.100 0 M

50.   30 mL of 0.05 M OH$^-$ = 1.5 mmol = 0.30 g of potassium acid phthalate.

51.   (a) From a graph of weight percent vs. pressure, HCl = 20.254% when p = 746 torr.

   (b) We need 0.100 00 mole of HCl = 3.646 1 g

   $$\frac{3.646\ 1\ \text{g HCl}}{0.202\ 54\ \text{HCl/g solution}} = 18.001_8\ \text{g of solution.}$$

   The mass required (weighed in air) is given by Equation 2-1.

   $$m' = \frac{(18.001\ 8)\left(1 - \dfrac{0.001\ 2}{1.096}\right)}{\left(1 - \dfrac{0.001\ 2}{8.0}\right)} = 17.985\ \text{g}$$

52.   When an acid that is stronger than H$_3$O$^+$ is added to H$_2$O, it reacts to give H$_3$O$^+$ and is "leveled" to the strength of H$_3$O$^+$. Similarly, bases stronger than OH$^-$ are leveled to the strength of OH$^-$.

53.   Methanol and ethanol have nearly the same acidity as water. Both equilibria below are driven to the right because of the high concentration of H$_2$O.

   $$CH_3O^- + H_2O \rightarrow CH_3OH + OH^-$$
   $$CH_3CH_2O^- + H_2O \rightarrow CH_3CH_2OH + OH^-$$

54.   (a) In acetic acid, strong acids are not leveled to the strength of CH$_3$CO$_2$H$_2^+$. Therefore, very weak bases can be titrated in acetic acid.

   (b) If tetrabutylammonium hydroxide were added to an acetic acid solution, most of the hydroxide would react with acetic acid instead of analyte. However, OH$^-$ will not react with pyridine, so this solvent would be suitable.

55.   Each reacts with H$_2$O to give OH$^-$:

   $$NH_2^- + H_2O \rightarrow NH_3 + OH^-$$
   $$C_6H_5^- + H_2O \rightarrow C_6H_6 + OH^-$$

56.   The surfactant forms micelles whose organic interior dissolves the product of dissociation of the ammonium cation:

   $$R_3NH^+ \rightleftarrows R_3N + H^+$$

   This removes the product from the aqueous phase and drives the equilibrium constant to the right, effectively increasing the strength of the weak acid, R$_3$NH$^+$.

57.   Titration reaction: $K^+HP^- + Na^+OH^- \rightarrow K^+Na^+P^{2-} + H_2O$

Begin with C_aV_a moles of K^+HP^- and add C_bV_b moles of NaOH

Fraction of titration $= \phi = \dfrac{C_bV_b}{C_aV_a}$

Charge balance: $[H^+] + [Na^+] + [K^+] = [HP^-] + 2[P^{2-}] + [OH^-]$

Substitutions: $[K^+] = \dfrac{C_aV_a}{V_a + V_b}$ $[Na^+] = \dfrac{C_bV_b}{V_a + V_b}$

$[HP^-] = \alpha_{HP^-}\dfrac{C_aV_a}{V_a + V_b}$ $[P^{2-}] = \alpha_{P^{2-}}\dfrac{C_aV_a}{V_a + V_b}$

Putting these expressions into the charge balance gives

$$[H^+] + \dfrac{C_bV_b}{V_a + V_b} + \dfrac{C_aV_a}{V_a + V_b} = \alpha_{HP^-}\dfrac{C_aV_a}{V_a + V_b} + 2\alpha_{P^{2-}}\dfrac{C_aV_a}{V_a + V_b} + [OH^-]$$

Multiply by $V_a + V_b$ and collect terms:

$$[H^+]V_a + [H^+]V_b + C_bV_b + C_aV_a = \alpha_{HP^-}C_aV_a + 2\alpha_{P^{2-}}C_aV_a + [OH^-]V_a + [OH^-]V_b$$

$$V_a([H^+] + C_a - \alpha_{HP^-}C_a - 2\alpha_{P^{2-}}C_a - [OH^-]) = V_b([OH^-] - [H^+] - C_b)$$

$$\dfrac{V_b}{V_a} = \dfrac{\alpha_{HP^-}C_a + 2\alpha_{P^{2-}}C_a - C_a - [H^+] + [OH^-]}{C_b + [H^+] - [OH^-]}$$

Multiply both sides by $\dfrac{1/C_a}{1/C_b}$:

$$\phi = \dfrac{C_bV_b}{C_aV_a} = \dfrac{\alpha_{HP^-} + 2\alpha_{P^{2-}} - 1 - \dfrac{[H^+] - [OH^-]}{C_a}}{1 + \dfrac{[H^+] - [OH^-]}{C_b}}$$

58.

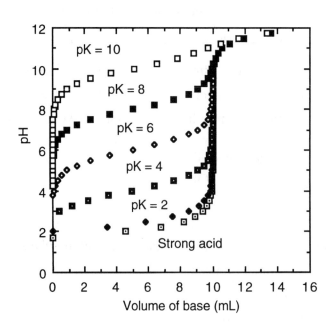

59.

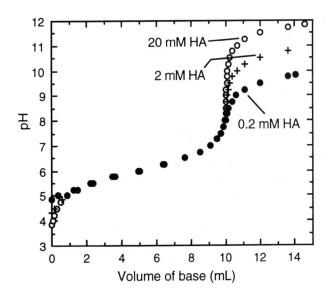

60.

	A	B	C	D	E	F	G
1	Ca =	pH	[H+]	[OH-]	Alpha(BH+)	Phi	Va (mL)
2	0.1	2.00	1.00E-02	1.00E-12	9.90E-01	1.66E+00	16.557
3	Cb =	2.90	1.26E-03	7.94E-12	9.26E-01	1.00E+00	10.020
4	0.02	3.50	3.16E-04	3.16E-11	7.60E-01	7.78E-01	7.780
5	Vb =	4.00	1.00E-04	1.00E-10	5.00E-01	5.06E-01	5.055
6	50	4.50	3.16E-05	3.16E-10	2.40E-01	2.42E-01	2.419
7	K(BH+) =	6.00	1.00E-06	1.00E-08	9.90E-03	9.95E-03	0.100
8	1E-04	8.15	7.08E-09	1.41E-06	7.08E-05	5.17E-07	0.000
9	Kw =						
10	1E-14			E2 = C2/(C2+$A$8)			
11		C2 = 10^-B2		F2 = (E2+(C2-D2)/$A$4)/(1-(C2-D2)/$A$2)			
12		D2 = $A$10/C2		G2 = F2*$A$4*$A$6/$A$2			

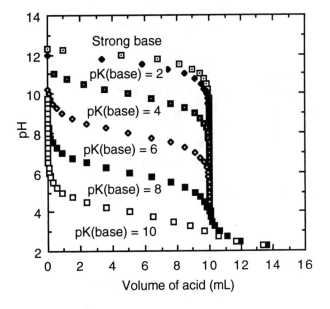

61.   (a)

	A	B	C	D	E	F	G	H
1	Cb =	pH	[H+]	[OH-]	Alpha(A-)	Alpha(BH+)	Phi	Vb (mL)
2	0.1	2.86	1.4E-03	7.2E-12	6.76E-02	1.00E+00	-1.4E-03	-0.01
3	Ca =	3.00	1.0E-03	1.0E-11	9.09E-02	1.00E+00	4.1E-02	0.41
4	0.02	4.00	1.0E-04	1.0E-10	5.00E-01	1.00E+00	4.9E-01	4.95
5	Va =	5.00	1.0E-05	1.0E-09	9.09E-01	9.99E-01	9.1E-01	9.09
6	50	6.00	1.0E-06	1.0E-08	9.90E-01	9.90E-01	1.0E+00	10.00
7	Ka =	7.00	1.0E-07	1.0E-07	9.99E-01	9.09E-01	1.1E+00	10.99
8	1E-04	8.00	1.0E-08	1.0E-06	1.00E+00	5.00E-01	2.0E+00	20.00
9	Kw =							
10	1E-14		A14 = $A$10/$A$12			D2 = $A$10/C2		
11	Kb =		C2 = 10^-B2			E2 = $A$8/(C2+$A$8)		
12	1E-06			F2 = C2/(C2+$A$14)				
13	K(BH+) =			G2 = (E2-(C2-D2)/$A$4)/(F2+(C2-D2)/$A$2)				
14	1E-08			H2 = F2*$A$4*$A$6/$A$2				

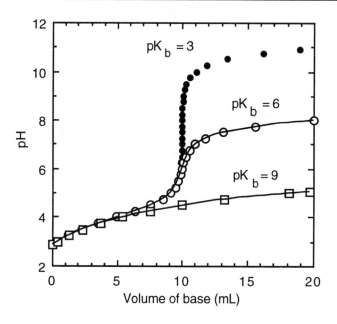

(b)   HA   +   B $\rightleftarrows$ A$^-$ + BH$^+$

$K_a = 1.75 \times 10^{-5}$ $K_b = 1.59 \times 10^{-10}$

$V_a = 212$ mL $K_{BH^+} = 6.28 \times 10^{-5}$

$C_a = 0.200$ M $V_b = 325$ mL

$C_b = 0.050\,0$ M

To find the equilibrium constant we write

HA $\rightleftarrows$ A$^-$ + H$^+$ K_a

H$^+$ + B $\rightleftarrows$ BH$^+$ $1/K_b$

_____

HA   +   B $\rightleftarrows$ A$^-$ + BH$^+$ $K = K_a/K_b = 1.10 \times 10^5$

A <u>pH of 4.16</u> gives $V_b = 325.0$ mL in the spreadsheet below:

	A	B	C	D	E	F	G	H
1	Cb =	pH	[H+]	[OH-]	Alpha(A-)	Alpha(BH+)	Phi	Vb (mL)
2	0.05	4.00	1.0E-04	1.0E-10	1.49E-01	6.14E-01	2.4E-01	204.28
3	Ca =	4.2	6.3E-05	1.6E-10	2.17E-01	5.01E-01	4.3E-01	365.98
4	0.2	4.1	7.9E-05	1.3E-10	1.81E-01	5.58E-01	3.2E-01	272.78
5	Va =	4.15	7.1E-05	1.4E-10	1.98E-01	5.30E-01	3.7E-01	315.79
6	212	4.1598	6.9E-05	1.4E-10	2.02E-01	5.24E-01	3.8E-01	325.03
7	Ka =							
8	1.750E-05		A14 = $A$10/$A$12					
9	Kw =		C2 = 10^-B2					
10	1.E-14		D2 = $A$10/C2					
11	Kb =		E2 = $A$8/(C2+$A$8)					
12	1.592E-10		F2 = C2/(C2+$A$14)					
13	K(BH+) =		G2 = (E2-(C2-D2)/$A$4)/(F2+(C2-D2)/$A$2)					
14	6.281E-05		H2 = F2*$A$4*$A$6/$A$2					

62.

	A	B	C	D	E	F	G	H
1	Cb =	pH	[H+]	[OH-]	Alpha(HA-)	Alpha(A2-)	Phi	Vb (mL)
2	0.1	2.86	1.4E-03	7.3E-12	6.83E-02	5.00E-07	5.0E-05	0.000
3	Ca =	4.00	1.0E-04	1.0E-10	5.00E-01	5.00E-05	4.9E-01	4.946
4	0.02	6.00	1.0E-06	1.0E-08	9.80E-01	9.80E-03	1.0E+00	9.999
5	Va =	8.00	1.0E-08	1.0E-06	5.00E-01	5.00E-01	1.5E+00	15.000
6	50	10.0	1.0E-10	1.0E-04	9.90E-03	9.90E-01	2.0E+00	19.971
7	Kw =	12.0	1.0E-12	1.0E-02	1.00E-04	1.00E+00	2.8E+00	27.777
8	1E-14							
9	K1 =		C2 = 10^-B2			D2 = $A$10/C2		
10	1E-4		E2 = C2*$A$10/(C2^2+C2*$A$10+$A$10*$A$12)					
11	K2 =		F2 = $A$10*$A$12/(C2^2+C2*$A$10+$A$10*$A$12)					
12	1.E-08		G2 = (E2+2*F2-(C2-D2)/$A$4)/(1+(C2-D2)/$A$2)					
13			H2 = F2*$A$4*$A$6/$A$2					

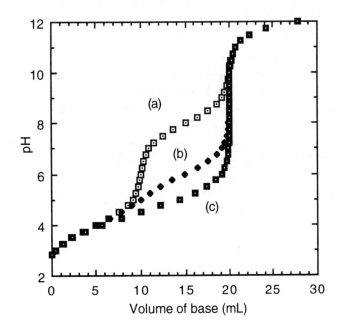

63.

	A	B	C	D	E	F	G	H
1	Cb =	pH	[H+]	[OH-]	Alpha(BH2)	Alpha(BH)	Phi	Va (mL)
2	0.1	1.75	1.8E-02	5.6E-13	9.62E-01	3.83E-02	2.6E+00	26.023
3	Ca =	2.00	1.0E-02	1.0E-12	9.34E-01	6.61E-02	2.3E+00	22.599
4	0.1	3.00	1.0E-03	1.0E-11	5.86E-01	4.14E-01	1.6E+00	16.117
5	Vb =	4.00	1.0E-04	1.0E-10	1.24E-01	8.76E-01	1.1E+00	11.258
6	10	5.00	1.0E-05	1.0E-09	1.39E-02	9.85E-01	1.0E+00	10.127
7	Kw =	6.00	1.0E-06	1.0E-08	1.39E-03	9.85E-01	9.9E-01	9.875
8	1.E-14	7.00	1.0E-07	1.0E-07	1.24E-04	8.76E-01	8.8E-01	8.764
9	KB1 =	8.00	1.0E-08	1.0E-06	5.86E-06	4.14E-01	4.1E-01	4.145
10	7.079E-7	9.00	1.0E-09	1.0E-05	9.34E-08	6.61E-02	6.6E-02	0.660
11	KB2 =	10.00	1.0E-10	1.0E-04	9.93E-10	7.03E-03	6.0E-03	0.060
12	1.41E-11	10.42	3.8E-11	2.6E-04	1.44E-10	2.68E-03	5.4E-05	0.001
13	KA1 =							
14	7.077E-4		C2 = 10^-B2				D2 = $A$10/C2	
15	KA2 =			E2 = C2*C2/(C2^2+C2*$A$14+$A$14*$A$16)				
16	1.413E-8			F2 = C2*$A$14/(C2^2+C2*$A$14+$A$14*$A$16)				
17				G2 = (F2+2*E2+(C2-D2)/$A$2)/(1-(C2-D2)/$A$4)				
18				H2 = G2*$A$2*$A$6/$A$4				

64.

	A	B	C	D	E	F	G	H	I
1	Ca =	pH	[H+]	[OH-]	Alpha(H2A-)	Alpha(HA2-)	Alpha(A3-)	Phi	Vb (mL)
2	0.02	1.91	1.2E-0.2	8.1E-13	6.19E-01	4.78E-05	3.22E-12	3.63E-03	0.036
3	Cb =	2.28	5.2E-03	1.9E-12	7.92E-01	1.43E-04	1.43E-04	5.03E-01	5.035
4	0.1	4.03	9.3E-05	1.1E-10	9.85E-01	1.00E-02	8.921E-8	1.00E+00	9.998
5	Va =	6.02	9.5E-07	1.0E-08	5.01E-01	4.98E-01	4.332E-4	1.50E+00	14.992
6	50	7.55	2.8E-08	3.5E-07	2.80E-02	9.44E-01	0.027805	2.00E+00	19.998
7	Kw =	9.08	8.3E-10	1.2E-05	4.38E-04	5.00E-01	0.499250	2.50E+00	24.997
8	1.E-14	10.58	2.6E-11	3.8E-04	8.50E-07	3.07E-02	0.969282	3.00E+00	29.997
9	K1 =	12.00	1.0E-12	1.0E-02	1.27E-09	1.20E-03	0.998796	3.89E+00	38.876
10	0.02								
11	K2 =		C2 = 10^-B2			I2 = H2*$A$2*$A$6/$A$4			
12	9.5E-07		D2 = $A$8/C2						
13	K3 =								
14	8.3E-10								
15									
16	E2 = C2*C2*$A$10/(C2^3+C2^2*$A$10+C2*$A$10*$A$12+$A$10*$A$12*$A$14)								
17	F2 =C2*$A$10*$A$12/(C2^3+C2^2*$A$10+C2*$A$10*$A$12+$A$10*$A$12*$A$14)								
18	G2 = $A$10*$A$12*$A$14/(C2^3+C2^2*$A$10+C2*$A$10*$A$12+$A$10*$A$12*$A$14)								
19	H2 = (E2+2*F2+3*G2-(C2-D2)/$A$2)/(1+(C2-D2)/$A$4)								

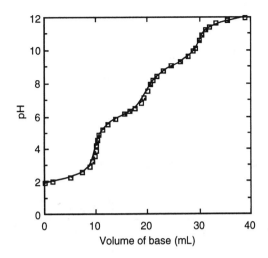

65.
$$\phi = \frac{C_aV_a}{C_bV_b} = \frac{\alpha_{BH^+} + 2\alpha_{BH_2^{2+}} + 3\alpha_{BH_3^{3+}} + 4\alpha_{BH_4^{4+}} + \dfrac{[H^+] - [OH^-]}{C_b}}{1 - \dfrac{[H^+] - [OH^-]}{C_a}}$$

	A	B	C	D	E	F	G	H	I	J
1	Cb =	pH	[H+]	[OH-]	Alph(BH)	Alph(BH2)	Alph(BH3)	Alph(BH4)	Phi	Va (mL)
2	0.02	10.84	1.4E-11	6.9E-04	3.5E-02	2.5E-06	6.1E-15	5.5E-25	0.00	0.003
3	Ca =	9.00	1.0E-09	1.0E-05	7.1E-01	3.6E-03	5.9E-10	3.7E-18	0.72	7.183
4	0.1	6.00	1.0E-06	1.0E-08	1.7E-01	8.3E-01	1.4E-04	8.7E-10	1.83	18.334
5	Vb =	3.00	1.0E-03	1.0E-11	1.7E-04	8.6E-01	1.4E-01	8.9E-04	2.22	22.165
6	50	1.70	2.0E-02	5.0E-13	2.1E-06	2.1E-01	7.0E-01	8.7E-02	4.84	48.398
7	Kw =									
8	1E-14	Denominator = (C2^4+C2^3*$A$10+C2^2*$A$10*$A$12+								
9	Ka1 =	C2*$A$10*$A$12*$A$14+$A$10*$A$12*$A$14*$A$16)								
10	0.16	C2 = 10^-B2								
11	Ka2 =	D2 = $A$8/C2								
12	0.006	E2 = C2*$A$10*$A$12*$A$14/Denominator								
13	Ka3 =	F2 =C2^2*$A$10*$A$12/Denominator								
14	2.E-07	G2 = C2^3*$A$10/Denominator								
15	Ka4	H2 = C2^4/Denominator								
16	4E-10	I2 = =(E2+2*F2+3*G2+4*H2+(C2-D2)/$A$2)/(1-(C2-D2)/$A$4)								
17		J2 = I2*$A$2*$A$6/A								

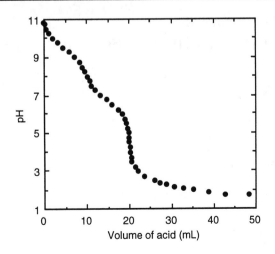

CHAPTER 13
EDTA TITRATIONS

1. The chelate effect is the observation that multidentate ligands form more stable metal complexes than do similar, monodentate ligands. This happens because the entropy of binding by one multidentate ligand is less unfavorable than the entropy of binding by many smaller ligands.

2. $\alpha_{Y^{4-}}$ gives the fraction of all free EDTA in the form Y^{4-}.

    (a) At pH 3.50:
    $$\alpha_{Y^{4-}} = \frac{10^{-0.0}10^{-1.5}...10^{-10.24}}{(10^{-3.50})^6+(10^{-3.50})^510^{-0.0}+...+10^{-0.0}10^{-1.5}...10^{-10.24}} = 3.4 \times 10^{-10}$$

    (b) At pH 10.50, $\alpha_{Y^{4-}} = 0.64$

3. (a) $K_f' = \alpha_{Y^{4-}} K_f = 0.054 \times 10^{8.79} = 3.3 \times 10^7$

    (b)
    $$Mg^{2+} + EDTA \rightleftarrows MgY^{2-}$$
    $$x \qquad x \qquad 0.050-x$$

    $$\frac{0.050-x}{x^2} = 3.3 \times 10^7 \Rightarrow [Mg^{2+}] = 3.9 \times 10^{-5}\ M$$

4. $[Ca^{2+}] = 10^{-9.00}\ M$, so essentially all calcium in solution is CaY^{2-}.

    $$[CaY^{2-}] = \frac{1.95\ g}{(200.12\ g/mol)\ (0.500\ L)} = 1.949 \times 10^{-2}\ M$$

    $$K_f' = (0.054)\ (4.9 \times 10^{10}) = \frac{[CaY^{2-}]}{[EDTA]\ [Ca^{2+}]} = \frac{(1.949 \times 10^{-2})}{[EDTA]\ (10^{-9.00})}$$

    $$\Rightarrow [EDTA] = 7.37 \times 10^{-3}\ M$$

    Total EDTA needed = mol CaY^{2-} + mol free EDTA

    $= (1.949 \times 10^{-2}\ M)\ (0.500\ L) + (7.37 \times 10^{-3}\ M)\ (0.500\ L) = 0.013\ 4$ mol

    $= 5.00\ g\ Na_2EDTA \cdot 2\ H_2O$

5. (a) mmol EDTA = mmol M^{n+}

    $(V_e)(0.050\ 0\ M) = (100.0\ mL)(0.050\ 0\ M) \Rightarrow V_e = 100.0\ mL$

    (b) $[M^{n+}] = \left(\frac{1}{2}\right)\ (0.050\ 0\ M)\ \left(\frac{100}{150}\right) = 0.016\ 7\ M$

                fraction   original   dilution
            remaining concentration factor

    (c) 0.054 (Table 13-1)

    (d) $K_f' = (0.054)(10^{12.00}) = 5.4 \times 10^{10}$

    (e) $[MY^{n-4}] = (0.050\ 0\ M)\left(\frac{100}{200}\right) = 0.025\ 0\ M$

$$\frac{[MY^{n-4}]}{[M^{n+}][EDTA]} = \frac{0.025\ 0-x}{x^2} = 5.4 \times 10^{10} \Rightarrow x = [M^{n+}] = 6.8 \times 10^{-7}\ M$$

(f) $[EDTA] = (0.050\ 0\ M)\left(\frac{10.0}{210.0}\right) = 2.38 \times 10^{-3}\ M$

$[MY^{n-4}] = (0.050\ 0\ M)\left(\frac{100.0}{210.0}\right) = 2.38 \times 10^{-2}\ M$

$$\frac{[MY^{n-4}]}{[M^{n+}][EDTA]} = \frac{(2.38 \times 10^{-2})}{[M^{n+}](2.38 \times 10^{-3})} = 5.4 \times 10^{10} \Rightarrow [M^{n+}] = 1.9 \times 10^{-10}\ M$$

6. $Co^{2+} + EDTA \rightleftarrows CoY^{2-}$ $\alpha_{Y^{4-}} K_f = 4.7 \times 10^{11}$

$V_e = (25.00)\left(\frac{0.020\ 26\ M}{0.038\ 55\ M}\right) = 13.14\ mL$

(a) <u>12.00 mL</u> : $[Co^{2+}] = \left(\frac{13.14 - 12.00}{13.14}\right)(0.020\ 26\ M)\left(\frac{25.00}{37.00}\right)$

$= 1.19 \times 10^{-3}\ M \Rightarrow pCo^{2+} = 2.93$

(b) <u>V_e</u>: Formal concentration of CoY^{2-} is $\left(\frac{25.00}{38.14}\right)(0.020\ 26\ M) = 1.33 \times 10^{-2}\ M$

$$\begin{array}{ccccc} Co^{2+} & + & EDTA & \rightleftarrows & CoY^{2-} \\ x & & x & & 1.33 \times 10^{-2} - x \end{array}$$

$\frac{1.33 \times 10^{-2}-x}{x^2} = \alpha_{Y^{4-}} K_f \Rightarrow x = 1.68 \times 10^{-7}\ M \Rightarrow pCo^{2+} = 6.77$

(c) <u>14.00 mL</u>: Formal concentration of CoY^{2-} is $\left(\frac{25.00}{39.00}\right)(0.020\ 26\ M)$

$= 1.30 \times 10^{-2}\ M$

Formal concentration of EDTA is $\left(\frac{14.0 - 13.14}{39.00}\right)(0.038\ 55\ M) = 8.50 \times 10^{-4}\ M$

$[Co^{2+}] = \frac{[CoY^{2-}]}{[EDTA]\ K_f'} = 3.3 \times 10^{-11}\ M \Rightarrow pCo^{2+} = 10.49$

7. Titration reaction : $Mn^{2+} + EDTA \rightleftarrows MnY^{2-}$ $K' = \alpha_{Y^{4-}} K_f = 4.1 \times 10^{11}$

The equivalence point is 50.0 mL.

Sample calculations :

<u>20.0 mL</u> : The fraction of Mn^{2+} that has reacted is 2/5 and the fraction remaining
is 3/5.

$[Mn^{2+}] = \left(\frac{30.0}{50.0}\right)(0.020\ 0\ M)\left(\frac{25.0}{45.0}\right) = 6.67 \times 10^{-3}\ M \Rightarrow pMn^{2+} = 2.18$

<u>50.0 mL</u> : The formal concentration of MnY^{2-} is

$[MnY^{2-}] = \left(\frac{25.0}{75.0}\right)(0.020\ 0\ M) = 0.006\ 67\ M$

$$\begin{array}{ccccc} Mn^{2+} & + & EDTA & \rightleftarrows & MnY^{2-} \\ x & & x & & 0.006\ 67-x \end{array}$$

$$\frac{0.006\ 67\text{-}x}{x^2} = \alpha_{Y^{4-}}\ K_f \Rightarrow 1.28 \times 10^{-7} \Rightarrow pMn^{2+} = 6.89$$

<u>60.0 mL</u> : There are 10.0 mL of excess EDTA.

$$[EDTA] = \left(\frac{10.0}{85.0}\right)(0.010\ 0\ M) = 1.176 \times 10^{-3}\ M$$

$$[MnY^{2-}] = \left(\frac{25.0}{85.0}\right)(0.020\ 0\ M) = 5.88 \times 10^{-3}\ M$$

$$[Mn^{2+}] = \frac{[MnY^{2-}]}{[EDTA]K_f'} = 1.20 \times 10^{-11} \Rightarrow pMn^{2+} = 10.92$$

Volume (mL)	pMn^{2+}	Volume	pMn^{2+}	Volume	pMn^{2+}
0	1.70	49.0	3.87	50.1	8.92
20.0	2.18	49.9	4.87	55.0	10.62
40.0	2.81	50.0	6.90	60.0	10.92

8.   Titration reaction :

$$Ca^{2+} + EDTA \rightleftarrows CaY^{2-} \quad K_f' = \alpha_{Y^{4-}}\ K_f = 1.76 \times 10^{10}$$

The equivalence point is 50.0 mL.

Sample calculations :

<u>20.0 mL</u> : The fraction EDTA consumed is 2/5.

$$[EDTA] = \left(\frac{30.0}{50.0}\right)(0.020\ 0\ M)\left(\frac{25.0}{45.0}\right) = 0.006\ 67\ M$$

$$[CaY^{2-}] = \left(\frac{20.0}{50.0}\right)(0.020\ 0\ M)\left(\frac{25.0}{45.0}\right) = 0.004\ 44\ M$$

$$[Ca^{2+}] = \frac{[CaY^{2-}]}{[EDTA]K_f'} = 3.79 \times 10^{-11} \Rightarrow pCa^{2+} = 10.42$$

<u>50.0 mL</u> : The formal concentration of CaY^{2-} is

$$[CaY^{2-}] = \left(\frac{25.0}{75.0}\right)(0.020\ 0\ M) = 0.006\ 67\ M$$

$$\begin{array}{cccc} Ca^{2+} & + & EDTA & \rightleftarrows & CaY^{2-} \\ x & & x & & 0.006\ 67\text{-}x \end{array}$$

$$\frac{0.006\ 67\text{-}x}{x^2} = \alpha_{Y^{4-}}\ K_f \Rightarrow x = 6.16 \times 10^{-7}\ M \Rightarrow pCa^{2+} = 6.21$$

<u>50.1 mL</u> : There is an excess of 0.1 mL of Ca^{2+}.

$$[Ca^{2+}] = \left(\frac{0.1}{75.1}\right)(0.010\ 0\ M) = 1.33 \times 10^{-5}\ M \Rightarrow pCa^{2+} = 4.88$$

Volume (mL)	pCa^{2+}	Volume	pCa^{2+}	Volume	pCa^{2+}
0	(∞)	49.0	8.56	50.1	4.88
20.0	10.42	49.9	7.55	55.0	3.20
40.0	9.64	50.0	6.21	60.0	2.93

9. There is more VO^{2+} than EDTA in this solution.

$$[VO^{2+}] = \left(\frac{0.10}{29.9}\right)(0.010\ 0\ M) = 3.34 \times 10^{-5}\ M$$

$$[VOY^{2-}] = \left(\frac{9.90}{29.90}\right)(0.010\ 0\ M) = 3.31 \times 10^{-3}\ M$$

$$[Y^{4-}] = \frac{[VOY^{2-}]}{[VO^{2+}]\ K_f} = 1.57 \times 10^{-17}\ M \quad [HY^{3-}] = \frac{[H^+]\ [Y^{4-}]}{K_6} = 2.7 \times 10^{-11}\ M$$

10.

	A	B	C	D	E
1	Cm =	pM	M	Phi	V(ligand)
2	0.001	3.00	1.00E-03	0.000	0.000
3	Vm =	3.50	3.16E-04	0.663	0.663
4	10	4.50	3.16E-05	0.965	0.965
5	C(ligand) =	5.00	1.00E-05	0.989	0.989
6	0.01	6.54	2.91E-07	1.000	1.000
7	Kf' =	8.00	1.00E-08	1.009	1.009
8	1.07E+10	8.50	3.16E-09	1.030	1.030
9		9.50	3.16E-10	1.296	1.296
10					
11	C2 = 10^-B2				
12	D2 = (1+$A$8*C2-(C2+C2*C2*$A$8)/$A$2)/				
13		(C2*$A$8+(C2+C2*C2*$A$8)/$A$6)			
14	E2 = D2*$A$2*$A$4/$A$6				

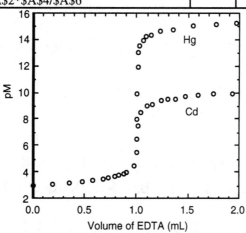

11. The spreadsheet is similar to that of the previous problem.

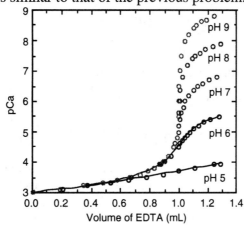

12.

	A	B	C	D	E
1	Cm =	pM	M	Phi	Vm
2	0.08	14.28	5.25E-15	0.004	0.099
3	V(ligand) =	12.48	3.31E-13	0.201	5.016
4	50	12.06	8.71E-13	0.398	9.941
5	C(ligand) =	11.70	2.00E-12	0.602	15.049
6	0.04	11.28	5.25E-12	0.799	19.978
7	Kf =	10.50	3.16E-11	0.960	23.999
8	7.58E+11	6.73	1.86E-07	1.000	25.000
9		2.98	1.05E-03	1.040	25.995
10		2.30	5.01E-03	1.201	30.013
11					
12	C2 = 10^-B2				
13	D2 = (C2*$A$8+(C2+C2*C2*$A$8)/$A$6)/				
14		(1+$A$8*C2-(C2+C2*C2*$A$8)/$A$2)			
15	E2 = D2*$A$6*$A$4/$A$2				

13. An auxiliary complexing agent forms a weak complex with analyte ion, thereby keeping it in solution without interfering with the EDTA titration. For example, NH_3 keeps Zn^{2+} in solution at high pH.

14. (a) $\beta_2 = K_1K_2 = 10^{2.23}\, 10^{1.40} = 4.3 \times 10^3$

(b) $\alpha_{Cu^{2+}} = \dfrac{1}{1+\beta_1[L]+\beta_2[L]^2} = \dfrac{1}{1+10^{2.23}(0.100) + 10^{3.63}(0.100)^2} = 0.017$

15. $Cu^{2+} + Y^{4-} \rightleftarrows CuY^{2-}$ $K_f = 10^{18.80} = 6.3 \times 10^{18}$

$\alpha_{Y^{4-}} = 0.85$ at pH 11.00 (Table 13-1)

For Cu^{2+} and NH_3, Appendix I gives $K_1 = 10^{3.99}$, $K_2 = 10^{3.34}$, $K_3 = 10^{2.73}$ and $K_4 = 10^{1.97}$. This means $\beta_1 = K_1 = 9.8 \times 10^3$, $\beta_2 = K_1K_2 = 2.1 \times 10^7$, $\beta_3 = K_1K_2K_3 = 1.15 \times 10^{10}$ and $\beta_4 = K_1K_2K_3K_4 = 1.07 \times 10^{12}$.

$\alpha_{Cu^{2+}} = \dfrac{1}{1+\beta_1(0.100)+\beta_2(0.100)^2+\beta_3(0.100)^3+\beta_4(0.100)^4} = 8.4 \times 10^{-9}$

$K_f' = \alpha_{Y^{4-}}\, K_f = 5.4 \times 10^{18}$

$K_f'' = \alpha_{Y^{4-}}\, \alpha_{Cu^{2+}}\, K_f = 4.5 \times 10^{10}$

Equivalence point = 50.00 mL

(a) At 0 mL, the total concentration of copper is Cu^{2+} = 0.001 00 M and

$[Cu^{2+}] = \alpha_{Cu^{2+}} C_{Cu^{2+}} = 8.4 \times 10^{-12}$ M $\Rightarrow$ $pCu^{2+} = 11.08$

(b) At 1.00 mL, $C_{Cu^{2+}} = \left(\dfrac{49.00}{50.00}\right)(0.001\,00\ M)\left(\dfrac{50.00}{51.00}\right) = 9.61 \times 10^{-4}$ M

$\quad\quad\quad\quad\quad\quad$ fraction $\quad$ original $\quad$ dilution
$\quad\quad\quad\quad\quad$ remaining $\quad$ concentration $\quad$ factor

$[Cu^{2+}] = \alpha_{Cu^{2+}} C_{Cu^{2+}} = 8.1 \times 10^{-12}$ M $\Rightarrow$ $pCu^{2+} = 11.09$

(c) At 45.00 mL, $C_{Cu^{2+}} = \left(\dfrac{5.00}{50.00}\right)(0.001\ 00)\left(\dfrac{50.00}{95.00}\right) = 5.26 \times 10^{-5}$ M

$[Cu^{2+}] = \alpha_{Cu^{2+}}\,C_{Cu^{2+}} = 4.4 \times 10^{-13}$ M $\Rightarrow$ $pCu^{2+} = 12.35$

(d) At the equivalence point, we can write

$$C_{Cu^{2+}} \;+\; EDTA \;\rightleftarrows\; CuY^{2-}$$

$$x \qquad\qquad x \qquad\qquad \left(\dfrac{50.00}{100.00}\right)(0.001\ 00)\text{-}x$$

$\dfrac{0.000\ 500\text{-}x}{x^2} = 4.5 \times 10^{10} \Rightarrow x = C_{Cu^{2+}} = 1.05 \times 10^{-7}$ M

$[Cu^{2+}] = \alpha_{Cu^{2+}}\,C_{Cu^{2+}} = 8.9 \times 10^{-16}$ M $\Rightarrow$ $pCu^{2+} = 15.06$

(e) Past the equivalence point at 55.00 mL, we can say

$[EDTA] = \left(\dfrac{5.00}{105.00}\right)(0.001\ 00\ \text{M}) = 4.76 \times 10^{-5}$ M

$[CuY^{2-}] = \left(\dfrac{50.00}{105.00}\right)(0.001\ 00\ \text{M}) = 4.76 \times 10^{-4}$ M

$K_f' = \dfrac{[CuY^{2-}]}{[Cu^{2+}][EDTA]} = \dfrac{(4.76 \times 10^{-4})}{[Cu^{2+}]\,(4.76 \times 10^{-5})}$

$\Rightarrow [Cu^{2+}] = 1.85 \times 10^{-18}$ M $\Rightarrow$ $pCu^{2+} = 17.73$

16. (a) $\alpha_{ML} = \dfrac{[ML]}{C_M} = \dfrac{\beta_1[M][L]}{[M]\{1+\beta_1[L] + \beta_2[L]^2\}} = \dfrac{\beta_1[L]}{1 + \beta_1[L] + \beta_2[L]^2}$

$\alpha_{ML_2} = \dfrac{[ML_2]}{C_M} = \dfrac{\beta_2[M][L]^2}{[M]\{1+\beta_1[L] + \beta_2[L]^2\}} = \dfrac{\beta_2[L]^2}{1 + \beta_1[L] + \beta_2[L]^2}$

(b) For $[L] = 0.100$ M, $\beta_1 = 1.7 \times 10^2$ and $\beta_2 = 4.3 \times 10^3$, we get

$\alpha_{ML} = 0.28$ and $\alpha_{ML_2} = 0.70$

17. Let T = transferrin

(a) $Fe^{3+} + T \overset{K_1}{\rightleftarrows} FeT$ $\qquad\qquad\qquad K_1 = \dfrac{[FeT]}{[Fe^{3+}][T]}$

$Fe^{3+} + FeT \overset{K_2}{\rightleftarrows} Fe_2T$ $\qquad\qquad K_2 = \dfrac{[Fe_2T]}{[Fe^{3+}][FeT]}$

(b) $K_1 = \dfrac{[Fe_aT] + [Fe_bT]}{[Fe^{3+}][T]} = \dfrac{[Fe_aT]}{[Fe^{3+}][T]} + \dfrac{[Fe_bT]}{[Fe^{3+}][T]} = k_{1a} + k_{1b}$

$\dfrac{1}{K_2} = \dfrac{[Fe^{3+}]([Fe_aT] + [Fe_bT])}{[Fe_2T]} = \dfrac{[Fe^{3+}][Fe_aT]}{[Fe_2T]} + \dfrac{[Fe^{3+}][Fe_bT]}{[Fe_2T]} = \dfrac{1}{k_{2b}} + \dfrac{1}{k_{2a}}$

(c) $k_{1a}k_{2b} = \dfrac{[Fe_aT]}{[Fe^{3+}][T]}\,\dfrac{[Fe_2T]}{[Fe^{3+}][Fe_aT]} = \dfrac{[Fe_bT]}{[Fe^{3+}][T]}\,\dfrac{[Fe_2T]}{[Fe^{3+}][Fe_bT]} = k_{1b}k_{2a}$

(d) Substituting from Eqn. (a) into Eqn. (c) gives

$19.44 = \dfrac{[FeT]^2}{(1 - [FeT] - [Fe_2T])\,[Fe_2T]}$ $\qquad\qquad$ (d)

Substituting from Eqn. (b) into Eqn.(d) gives

$$19.44 = \frac{(0.8 - 2[Fe_2T])^2}{\{1 - (0.8 - 2[Fe_2T]) - [Fe_2T]\}\,[Fe_2T]} \quad \underset{\substack{\text{quadratic}\\\text{equation}}}{\overset{\text{solve}}{\Rightarrow}} \quad [Fe_2T] = 0.077\ 3$$

Using this value for $[Fe_2T]$ in Eqns. (a) and (b) gives $[FeT] = 0.645$ and

$[T] = 0.277\ 3$. Now we also know that $\dfrac{k_{1a}}{k_{1b}} = \dfrac{[Fe_aT]}{[Fe_bT]} = 6.0$, which tells us

that $[Fe_aT] = \left(\dfrac{6.0}{7.0}\right)[FeT] = 0.553\ 2$ and $[Fe_bT] = \left(\dfrac{1.0}{7.0}\right)[FeT] = 0.092\ 2$.

The final result is $[T] = 0.277\ 3$; $[Fe_aT] = 0.553\ 2$; $[Fe_bT] = 0.092\ 2$;

$[Fe_2T] = 0.077\ 3$.

18. In place of Equation 13-8 we write

$$M_{tot} + EDTA \rightleftarrows M(EDTA) \qquad K_f'' = \frac{[M(EDTA)]}{[M]_{tot}\,[EDTA]}$$

where $[M]_{tot}$ is the concentration of all metal not bound to EDTA. $[EDTA]$ is the concentration of all EDTA not bound to metal. The mass balances are

$$\text{Metal:} \quad [M]_{tot} + [M(EDTA)] = \frac{C_m V_m}{C_m + V_{EDTA}}$$

$$\text{EDTA:} \quad [EDTA] + [M(EDTA)] = \frac{C_{EDTA} V_{EDTA}}{C_m + V_{EDTA}}$$

These equations have the same form as the first three equations in Section 13-4, with K_f replaced by K_f'', $[M]$ replaced by $[M]_{tot}$ and $[L]$ replaced by $[EDTA]$. The derivation therefore leads to Equation 13-11, with K_f replaced by K_f'', $[M]$ replaced by $[M]_{tot}$ and C_l replaced by C_{EDTA}.

19. (a)

	A	B	C	D	E	F
1	Cm =	pM	M	[M]t	Phi	V(ligand)
2	0.001	8.11	7.76E-09	4.31E-04	0.397	19.869
3	Vm =	12.05	8.91E-13	4.95E-08	1.000	50.000
4	50	15.36	4.37E-16	2.43E-11	1.201	60.057
5	C(ligand) =					
6	0.001	C2 = 10^-B2				
7	Kf' =	D2 = C2/$A$10				
8	2.0E+11	E2 = (1+$A$8*D2-(D2+D2*D2*$A$8)/$A$2)/				
9	Alpha(M) =		(D2*$A$8+(D2+D2*D2*$A$8)/$A$6)			
10	0.000018	F2 = E2*$A$2*$A$4/$A$6				

(b) The spreadsheet is similar to the one given in part (a).

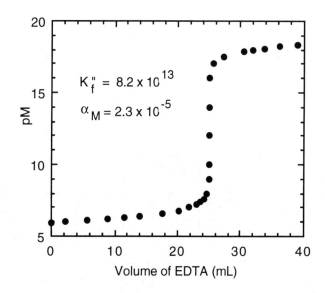

20. $[L] + [ML] + 2[ML_2] = \dfrac{C_lV_l}{V_m + V_l}$

$[L] + \alpha_{ML}\dfrac{C_mV_m}{V_m + V_l} + 2\alpha_{ML_2}\dfrac{C_mV_m}{V_m + V_l} = \dfrac{C_lV_l}{V_m + V_l}$

Multiply both sides by $V_m + V_l$:

$[L]V_m + [L]V_l + \alpha_{ML}C_mV_m + 2\alpha_{ML_2}C_mV_m = C_lV_l$

Collect terms $\quad V_l([L] - C_l) = V_m(-[L] - \alpha_{ML}C_m - 2\alpha_{ML_2}C_m)$

$\dfrac{V_l}{V_m} = \dfrac{[L] + \alpha_{ML}C_m + 2\alpha_{ML_2}C_m}{C_l - [L]} \qquad \phi = \dfrac{C_lV_l}{C_mV_m} = \dfrac{\dfrac{[L]}{C_m} + \alpha_{ML} + 2\alpha_{ML_2}}{1 - \dfrac{[L]}{C_l}}$

21.

	A	B	C	D	E	F	G	H	I	J	K
1		pL	[L]	Alpha(M)	Alpha(ML)	Alpha(ML2)	Phi	V(ligand)	[M]	[ML]	[ML2]
2	Cm = 0.05	4	0.0001	0.983	0.017	0.000	0.019	0.019	0.0491	0.0008	0.0000
3	Vm = 10	3	0.001	0.852	0.145	0.004	0.172	0.172	0.0419	0.0071	0.0002
4		2.8	0.00158	0.781	0.210	0.008	0.260	0.260	0.0381	0.0103	0.0004
5	C(ligand) =	2.6	0.00251	0.688	0.294	0.019	0.383	0.383	0.0331	0.0141	0.0009
6	0.5	2.4	0.00398	0.573	0.388	0.039	0.550	0.550	0.0272	0.0184	0.0019
7	B1 =	2.2	0.00630	0.446	0.478	0.076	0.766	0.766	0.0207	0.0222	0.0035
8	170	2	0.01	0.319	0.543	0.137	1.039	1.039	0.0145	0.0246	0.0062
9	B2 =	1.9	0.01258	0.262	0.560	0.178	1.199	1.199	0.0117	0.0250	0.0080
10	4300	1.8	0.01584	0.209	0.564	0.226	1.377	1.377	0.0092	0.0248	0.0099
11		1.7	0.01995	0.164	0.556	0.280	1.579	1.579	0.0071	0.0240	0.0121
12		1.6	0.02511	0.125	0.535	0.340	1.808	1.808	0.0053	0.0226	0.0144
13		1.5	0.03162	0.094	0.504	0.403	2.073	2.073	0.0039	0.0209	0.0167
14		1.4	0.03981	0.069	0.464	0.467	2.385	2.385	0.0028	0.0187	0.0189
15		1.3	0.05011	0.049	0.419	0.532	2.761	2.761	0.0019	0.0164	0.0208
16		1.25	0.05623	0.041	0.396	0.563	2.981	2.981	0.0016	0.0152	0.0217
17											
18	C2 = 10^-B2										
19	D2 = 1/(1+$A$8*C2+$A$10*C2*C2)						H2 = G2*$A$2*$A$4/$A$6				
20	E2 = $A$8*C2/(1+$A$8*C2+$A$10*C2*C2)						I2 = D2*$A$2*$A$4/($A$4+H2)				
21	F2 = $A$10*C2*C2/(1+$A$8*C2+$A$10*C2*C2)						J2 = E2*$A$2*$A$4/($A$4+H2)				
22	G2 = (C2/$A$2+E2+2*F2)/(1-C2/$A$6)						K2 = F2*$A$2*$A$4/($A$4+H2)				

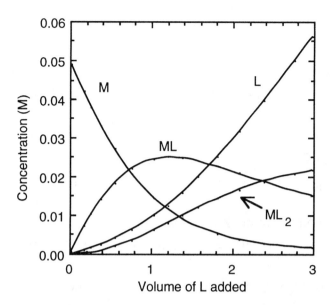

22. Only a small amount of indicator is employed. Most of the Mg^{2+} is not bound to indicator. The free Mg^{2+} reacts with EDTA before MgIn reacts. Therefore the concentration of MgIn is constant until all of the Mg^{2+} has been consumed. Only when MgIn begins to react does the color change.

23.   1.   With metal ion indicators.
    2.   With a mercury electrode (described in Exercise B in Chapter 15).
    3.   With a glass electrode.
    4.   By polarography (described in Chapter 18).

24. HIn^{2-}, wine red, blue

25. Buffer (a) (pH 6-7) will give a yellow $\rightarrow$ blue color change that will be easier to observe than the violet $\rightarrow$ blue change expected with the other buffers.

26. A back titration is necessary if the analyte precipitates in the absence of EDTA, if it reacts too slowly with EDTA, or if it blocks the indicator.

27. In a displacement titration, analyte displaces a metal ion from a complex. The displaced metal ion is then titrated. An example is the liberation of Ni^{2+} from $Ni(CN)_4^{2-}$ by the analyte Ag^+. The liberated Ni^{2+} is then titrated by EDTA to find out how much Ag^+ was present.

28. The Mg^{2+} in a solution of Mg^{2+} and Fe^{3+} can be titrated by EDTA if the Fe^{3+} is masked with CN^- to form $Fe(CN)_6^{3-}$, which does not react with EDTA.

29. Hardness refers to the total concentration of alkaline earth cations in water, which normally means $[Ca^{2+}] + [Mg^{2+}]$. Hardness gets its name from the reaction of these cations with soap to form insoluble curds. Temporary hardness, due to $Ca(HCO_3)_2$, is lost by precipitation of $CaCO_3(s)$ upon heating. Permanent hardness derived from other salts, such as $CaSO_4$, is not affected by heat.

30. $(50.0 \text{ mL})(0.010\ 0 \text{ mmol/mL}) = 0.500 \text{ mmol } Ca^{2+}$, which requires 0.500 mmol EDTA = 10.0 mL EDTA.

0.500 mmol Al^{3+} requires the same amount of EDTA, 10.0 mL.

31.   mmol EDTA = mmol Ni^{2+} + mmol Zn^{2+}

$\quad$ 1.250 $\quad$ = $\quad$ x $\quad$ + $\quad$ 0.250 $\Rightarrow$ 1.000 mmol Ni^{2+} in 50.0 mL = 0.020 0 M

32.   The formula weight of $MgSO_4$ is 120.36. The 50.0 mL aliquot contains

$$\left(\frac{50.0 \text{ mL}}{500 \text{ mL}}\right)\left(\frac{0.450 \text{ g}}{120.36 \text{ g/mol}}\right) = 0.373\ 9 \text{ mmol of } Mg^{2+}$$

Since 37.6 mL of EDTA reacts with this much Mg^{2+}, the EDTA solution contains 0.373 9 mmol / 37.6 mL = 9.944×10^{-3} mmol/mL. The formula weight of $CaCO_3$ is 100.09. 1.00 mL of EDTA will react with 9.944×10^{-3} mmol of $CaCO_3$ = 0.995 mg.

33.   For 1.00 mL of unknown :

$\quad$ 25.00 mL of EDTA = 0.968 0 mmol

$\quad$ -23.54 mL of Zn^{2+} = 0.500 7 mmol

$\quad$ $\overline{\quad Co^{2+} + \quad Ni^{2+} \quad\quad = 0.467\ 3 \text{ mmol}}$

For 2.000 mL of unknown :

$\quad$ 25.00 mL of EDTA = 0.968 0 mmol

$\quad$ -25.63 mL of Zn^{2+} = 0.545 2 mmol

$\quad$ $\overline{Ni^{2+} \text{ in 2.000 mL}\ \ = 0.422\ 8 \text{ mmol}}$

Co^{2+} in 2.000 mL of unknown = 2 (0.467 3) - 0.422 8 = 0.511 8 mmol. The Co^{2+} will react with 0.511 8 mmol of EDTA, leaving 0.968 0 - 0.511 8 = 0.456 2 mmol EDTA. $\quad$ mL Zn needed = $\dfrac{0.456\ 2 \text{ mmol}}{0.021\ 27 \text{ mmol/mL}}$ = 21.45 mL

34.   Total EDTA $\quad$ = (25.0 mL) (0.045 2 M) $\quad$ = $\quad$ 1.130 mmol

$\quad$ - Mg^{2+} required $\;$ = (12.4 mL) (0.012 3 M) $\quad$ = $\quad$ 0.153 mmol

$\quad\overline{Ni^{2+} + Zn^{2+} \quad\quad\quad\quad\quad\quad\quad\quad = \quad 0.977 \text{ mmol}}$

Zn^{2+} = EDTA displaced by 2,3-dimercapto-1-propanol

$\quad$ = (29.2 mL) (0.012 3 M) = 0.359 mmol

$\quad \Rightarrow Ni^{2+}$ = 0.977 - 0.359 = 0.618 mmol; $[Ni^{2+}] = \dfrac{0.618 \text{ mmol}}{50.0 \text{ mL}} = 0.012\ 4$ M

$$[Zn^{2+}] = \frac{0.359 \text{ mmol}}{50.0 \text{ mL}} = 0.007\ 18 \text{ M}$$

35. The precipitation reaction is $Cu^{2+} + S^{2-} \rightarrow CuS$ (s).

Total Cu^{2+} used $= (25.00 \text{ mL}) (0.043\ 32 \text{ M}) = 1.083\ 0 \text{ mmol}$

- Excess Cu^{2+} $= (12.11 \text{ mL}) (0.039\ 27 \text{ M}) = 0.475\ 6 \text{ mmol}$

mmol of S^{2-} $= 0.607\ 4 \text{ mmol}$

$$[S^{2-}] = 0.607\ 4 \text{ mmol}/25.00 \text{ mL} = 0.024\ 30 \text{ M}$$

36. mmol Bi in reaction $= (25.00 \text{ mL}) (0.086\ 40 \text{ M}) = 2.160 \text{ mmol}$

EDTA required $= (14.24 \text{ mL}) (0.043\ 7 \text{ M}) = 0.622 \text{ mmol}$

mmol Bi that reacted with Cs $= 2.160 - 0.622 = 1.538 \text{ mmol}$

Since 2 mol Bi react with 3 mol Cs to give $Cs_3Bi_2I_9$,

mmol Cs^+ in unknown $= \frac{3}{2} (1.538) = 2.307 \text{ mmol}$

$$[Cs^+] = \frac{2.307 \text{ mmol}}{25.00 \text{ mL}} = 0.092\ 28 \text{ M}.$$

CHAPTER 14
FUNDAMENTALS OF ELECTROCHEMISTY

1. Electric charge (coulombs) refers to the quantity of positive or negative particles. Current (amperes) is the quantity of charge moving past a point in a circuit each second. Electric potential (volts) measures the work that can be done by (or must be done to) each coulomb of charge as it moves from one point to another.

2. (a) $1/1.602\ 177\ 33 \times 10^{-19}$ C/electron $= 6.241\ 506\ 3 \times 10^{18}$ electrons/C.

   (b) $F = 96\ 485.309$ C/mol

3. (a) I = coulombs/s. Every mol of O_2 accepts 4 mol of e^-. 16 mol O_2/day
   $= 64$ mol e^-/day $= 7.41 \times 10^{-4}$ mol e^-/s $= 71.5$ C/s $= 71._5$ A.

   (b) $I =$ Power/$E = 500$ W/115 V $= 4.35$ A. The resting human uses 16 times as much current as the refrigerator.

   (c) Power $= E \cdot I = (1.1$ V$) (71._5$ A$) = 79$ W

4. (a) $I = \dfrac{6.00\ V}{2.0 \times 10^3\ \Omega} = 3.00$ mA $= 3.00 \times 10^{-3}$ C/s

   $\left(\dfrac{3.00 \times 10^{-3}\ C/s}{9.649 \times 10^4\ C/mol}\right) (6.022 \times 10^{23}\ e^-/mole) = 1.87 \times 10^{16}\ e^-/s$

   (b) $P = E \cdot I = (6.00$ V$) (3.00 \times 10^{-3}$ A$) = 1.80 \times 10^{-2}$ W

   $\Rightarrow \dfrac{1.80 \times 10^{-2}\ J/s}{1.87 \times 10^{16}\ e^-/s} = 9.63 \times 10^{-19}$ J/e^-

   (c) 30.0 min $= 1\ 800$ s $= 3.37 \times 10^{19}$ electrons $= 5.60 \times 10^{-5}$ mol

   (d) $E = \sqrt{PR} + \sqrt{(100\ W)(2.00 \times 10^3\ \Omega)} = 447$ V

5. (a) $I_2 + 2e^- \rightleftarrows 2I^-$
   Oxidant

   (b) $2S_2O_3^{2-} \rightleftarrows S_4O_6^{2-} + 2e^-$
   Reducant

   (c) 1.00 g $S_2O_3^{2-}$ /$(112.12$ g/mol$) = 8.92$ mmol $S_2O_3^{2-} = 8.92$ mmol e^-
   $(8.92 \times 10^{-3}$ mol$) (9.649 \times 10^4$ C/mol$) = 861$ C

   (d) Current (A) $=$ coulombs/s $= 861$ C/60 s $= 14.3$ A.

6. (a) Oxidation numbers of reactants:

	N (in NH_4^+)	Cl (in ClO_4^-)	Al
	-3	+7	0
Oxidation numbers of products:	N	Cl	Al
	0	-1	+3

   (b) Formula weight of reactants $= 6($FW $NH_4ClO_4) + 10($FW Al$) = 974.75$
   Heat released per gram $= 9\ 334$ kJ/974.75 g $= 9.576$ kJ/g

7. (a) $Fe(s) \mid FeO(s) \mid KOH(aq) \mid Ag_2O(s) \mid Ag(s)$

oxidation : $Fe(s) + 2OH^- \rightleftarrows FeO(s) + H_2O + 2e^-$

reduction : $Ag_2O + H_2O + 2e^- \rightleftarrows 2Ag(s) + 2OH^-$

(b) $Pb(s) \mid PbSO_4(s) \mid K_2SO_4(aq) \parallel H_2SO_4(aq) \mid PbSO_4(s) \mid PbO_2(s) \mid Pb(s)$

oxidation: $Pb(s) + SO_4^{2-} \rightleftarrows PbSO_4(s) + 2e^-$

reduction: $PbO_2 + 4H^+ + SO_4^{2-} + 2e^- \rightleftarrows PbSO_4(s) + 2H_2O$

8. (a)

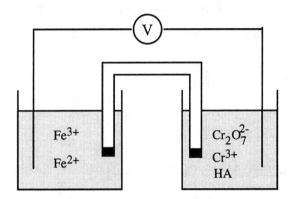

(b) oxidation: $Fe^{2+} \rightleftarrows Fe^{3+} + e^-$

reduction: $Cr_2O_7^{2-} + 14H^+ + 6e^- \rightleftarrows 2Cr^{3+} + 7H_2O$

(c) $Cr_2O_7^{2-} + 6Fe^{2+} + 14H^+ \rightleftarrows 2Cr^{3+} + 6Fe^{3+} + 7H_2O$

9. (a) oxidation: $Zn(s) \rightleftarrows Zn^{2+} + 2e^-$

reduction: $Cl_2(l) + 2e^- \rightleftarrows 2Cl^-$

(b) One mol of Cl_2 requires 2 mol of e^-.

Moles of Cl_2 consumed in 1.00 hr $= \frac{1}{2}$ (mol of e^-/hr) $=$

$\left[\frac{1}{2} \left(1.00 \times 10^3 \frac{C}{s} \right) / (9.64 \times 10^4 \ C/mol) \right] (3\ 600\ s/hr) = 18.7$ mol of $Cl_2 = 1.32$ kg

10. Cl_2 is strongest because it has the most positive reduction potential.

11. Become stronger : $Cr_2O_7^{2-}$, MnO_4^-, IO_3^-

Unchanged : Cl_2, Fe^{3+}

12. (a) Since it is harder to reduce Fe(III) to Fe(II) in the presence of CN^-, Fe(III) is stabilized more than Fe(II).

(b) Since it is easier to reduce Fe(III) to Fe(II) in the presence of phenanthroline, Fe(II) is stabilized more than Fe(III).

13. (a) $Zn(s) \mid Zn^{2+}(0.1\ M) \parallel Cu^{2+}(0.1\ M) \mid Cu(s)$

right half-cell: $Cu^{2+} + 2e^- \rightleftarrows Cu(s)$ $E_+^{\circ} = 0.339$ V left half-cell: $Zn^{2+} + 2e^- \rightleftarrows Zn(s)$ $E_-^{\circ} = -0.762$ V

$$E = \left\{ 0.339 - \frac{0.059\ 16}{2} \log \frac{1}{0.1} \right\} - \left\{ -0.762 - \frac{0.059\ 16}{2} \log \frac{1}{0.1} \right\} = 1.101 \text{ V}$$

Since the voltage is positive, electrons are transferred from Zn to Cu. The net reaction is $Cu^{2+} + Zn(s) \rightleftarrows Cu(s) + Zn^{2+}$.

(b) Since Cu^{2+} ions are consumed in the right half-cell, Zn^{2+} ions must migrate from the left half-cell into the salt bridge to help balance charge. I hope you like our Zn^{2+}, because that is what your body will take up.

14. (a) $E = -0.238 - \dfrac{0.059\ 16}{3} \log \dfrac{P_{AsH_3}}{[H^+]^3}$

(b) $E = -0.238 - \dfrac{0.059\ 16}{3} \log \dfrac{1.00/760}{(10^{-3.00})^3} = -0.359$ V

15. (a) $Pt(s) \mid Br_2(l) \mid HBr(aq, 0.10\ M) \parallel Al(NO_3)_3(aq, 0.010\ M) \mid Al(s)$

(b) right half-cell: $Al^{3+} + 3e^- \rightleftarrows Al(s)$ $E_+^{\circ} = -1.677$ V
   left half-cell: $Br_2(l) + 2e^- \rightleftarrows 2Br^-$ $E_-^{\circ} = 1.078$ V

right half-cell: $E_+ = \left\{ -1.677 - \dfrac{0.059\ 16}{3} \log \dfrac{1}{[0.010]} \right\} = -1.716_4$ V

left half-cell: $E_- = \left\{ 1.078 - \dfrac{0.059\ 16}{2} \log [0.10]^2 \right\} = 1.137_2$ V

$E = E_+ - E_- = -1.716_4 - 1.137_2 = -2.854$ V. Since the voltage is negative, electrons flow from the right-hand electrode to the left-hand electrode. Reduction occurs at the left-hand electrode. The spontaneous reaction is

$$\tfrac{3}{2} Br_2(l) + Al(s) \rightleftarrows 3Br^- + Al^{3+}$$

(c) 14.3 mL of Br_2 = 44.6 g = 0.279 mol of Br_2. 12.0 g of Al = 0.445 mol of Al. The reaction requires 3/2 mol of Br_2 for every mol of Al. The Br_2 will be used up first.

(d) 0.231 mL of Br_2 = 0.721 g of Br_2 = 4.51×10^{-3} mol Br_2 = 9.02×10^{-3} mol e^- = 870 C. Work = $E \cdot q$ = (1.50)(870) = 1.31 kJ.

(e) $I = \sqrt{P/R} = \sqrt{(1.00 \times 10^{-4})/(1.20 \times 10^3)}$ = 2.89×10^{-4} A
   = 2.99×10^{-9} mol e^-/s = 9.97×10^{-10} mol Al/s = 2.69×10^{-8} g/s

16. (a) right half-cell: $E_+ = \left\{ 0.222 - 0.059\ 16 \log [Cl^-] \right\}$ = 0.281_2 V

left half-cell: $E_- = \left\{ -0.350 - \dfrac{0.059\ 16}{2} \log [F^-]^2 \right\} = -0.290_8$ V

$$E = E_+ - E_- = 0.281_2 - (-0.290_8) = 0.572 \text{ V}$$

(b) $[Pb^{2+}] = K_{sp} \text{ (for } PbF_2) / [F^-]^2 = (3.6 \times 10^{-8}) / (0.10)^2 = 3.6 \times 10^{-6} \text{ M}$

$[Ag^+] = K_{sp} \text{ (for } AgCl) / [Cl^-] = (1.8 \times 10^{-10}) / (0.10) = 1.8 \times 10^{-9} \text{ M}$

right half-cell: $E_+ = \left\{ 0.799 - 0.059\ 16 \log \dfrac{1}{[Ag^+]} \right\} = 0.281_2 \text{ V}$

left half-cell: $E_- = \left\{ -0.126 - \dfrac{0.059\ 16}{2} \log \dfrac{1}{[Pb^{2+}]} \right\} = -0.287_0 \text{ V}$

$$E = E_+ - E_- = 0.281_2 - (-0.287_0) = 0.568 \text{ V}$$

The agreement between the two calculations is reasonable.

17. $\dfrac{RT}{nF} \ln Q = \dfrac{(\ln 10)(8.314\ 510 \text{ J mol}^{-1} \text{ K}^{-1})(298.15 \text{ K})}{n\ (9.648\ 530\ 9 \times 10^4 \text{ C mol}^{-1})} \log Q = \dfrac{0.059\ 16}{n} \log Q$

If $T = 0° \text{ C} = 273.15 \text{ K}$, $\dfrac{RT}{nF} \ln Q = \dfrac{0.054\ 20}{n} \log Q.$

When $T = 37° \text{ C} = 310.15 \text{ K}$, $\dfrac{RT}{nF} \ln Q = \dfrac{0.061\ 54}{n} \log Q.$

18. $0.798\ 3 = E°_{Ag^+ | Ag} - 0.059\ 16 \log \dfrac{[0.010\ 00]\ (0.914)}{(727.2/760)^{1/2}\ [0.010\ 00](0.898)}$

$\Rightarrow E°_{Ag^+ | Ag} = 0.799\ 3 \text{ V}$

19. Balanced reaction: $HOBr + 2e^- + H^+ \rightleftarrows Br^- + H_2O$

$HOBr \rightarrow \frac{1}{2} Br_2$ $\qquad\qquad\qquad \Delta G°_1 = -1F(1.584)$

$\frac{1}{2} Br_2 \rightarrow Br^-$ $\qquad\qquad\qquad \Delta G°_2 = -1F(1.098)$

$\overline{}$ $\qquad\qquad \overline{}$

$HOBr \rightarrow Br^-$ $\qquad\qquad\qquad \Delta G°_3 = \Delta G°_1 + \Delta G°_2 = -2FE°_3$

$E°_3 = \dfrac{-1F(1.584) - 1F(1.098)}{-2F} = 1.341 \text{ V}$

20. $2X^+ + 2e^- \rightleftarrows 2X(s)$ $\qquad\qquad\qquad E°_+ = E°_2$

$X^{3+} + 2e^- \rightleftarrows X^+$ $\qquad\qquad\qquad E°_- = E°_1$

$\overline{3X^+ \rightleftarrows X^{3+} + 2X(s)}$ $\qquad\qquad \overline{E°_3 = E°_2 - E°_1}$

Whenever $E°_2 > E°_1$, then $E°_3$ will be greater than 0 and disproportionation will be spontaneous.

21. right half-cell : $Cu^{2+} + 2e^- \rightleftarrows Cu(s)$ $\qquad\qquad E°_+ = 0.339 \text{ V}$

left half-cell : $Ni^{2+} + 2e^- \rightleftarrows Ni(s)$ $\qquad\qquad E°_- = -0.236 \text{ V}$

The ionic strength of the right half-cell is 0.090 M and the ionic strength of the left half-cell is 0.080 M. At $\mu = 0.090$ M, $\gamma_{Cu^{2+}} = 0.421$.

At $\mu = 0.080$ M , $\gamma_{Ni^{2+}} = 0.437$.

$$E_+ = E_+^\circ - \frac{0.059\ 16}{2} \log \frac{1}{[Cu^{2+}]\gamma_{Cu^{2+}}}$$

$$= 0.339 - \frac{0.059\ 16}{2} \log \frac{1}{(0.030)(0.421)} = 0.282_8\ V$$

$$E_- = E_-^\circ - \frac{0.059\ 16}{2} \log \frac{1}{[Ni^{2+}]\gamma_{Ni^{2+}}}$$

$$= -0.236 - \frac{0.059\ 16}{2} \log \frac{1}{(0.020)(0.437)} = -0.296_9\ V$$

$$E = E_+ - E_- = 0.580\ V$$

22. (a) $E^\circ = \dfrac{-\Delta G^\circ}{nF} = \dfrac{(+257 \times 10^3\ J/mol)}{(2)(9.648\ 5 \times 10^4\ C/mol)} = 1.33\ V$

(b) $K = 10^{nE^\circ/0.059\ 16} = 9 \times 10^{44}$

23. (a)

$$4[Co^{3+} + e^- \rightleftarrows Co^{2+}] \qquad\qquad E_+^\circ = 1.92\ V$$
$$-\ 2[\tfrac{1}{2}O_2 + 2H^+ + 2e^- \rightleftarrows H_2O] \qquad\qquad E_-^\circ = 1.229\ V$$
$$\overline{4Co^{3+} + 2H_2O \rightleftarrows 4Co^{2+} + O_2 + 4H^+ \qquad\qquad E^\circ = 0.69_1\ V}$$

$$\Delta G^\circ = -4FE^\circ = -2.7 \times 10^5\ J \qquad\qquad K = 10^{4E^\circ/0.059\ 16} = 10^{47}$$

(b)

$$Ag(S_2O_3)_2^{3-} + e^- \rightleftarrows Ag(s) + 2S_2O_3^{2-} \qquad E_+^\circ = 0.017\ V$$
$$-\ Fe(CN)_6^{3-} + e^- \rightleftarrows Fe(CN)_6^{4-} \qquad E_-^\circ = 0.356\ V$$
$$\overline{Ag(S_2O_3)_2^{3-} + Fe(CN)_6^{4-} \rightleftarrows Ag(s) + 2S_2O_3^{2-} + Fe(CN)_6^{3-} \qquad E^\circ = -0.339\ V}$$

$$\Delta G^\circ = -1FE^\circ = 32.7\ kJ$$
$$K = 10^{1E^\circ/0.059\ 16} = 1.9 \times 10^{-6}$$

24. (a)

$$5Ce^{4+} + 5e^- \rightleftarrows 5Ce^{3+} \qquad\qquad E_+^\circ = 1.70\ V$$
$$-\ MnO_4^- + 8H^+ + 5e^- \rightleftarrows Mn^{2+} + 4H_2O \qquad E_-^\circ = 1.507\ V$$
$$\overline{5Ce^{4+} + Mn^{2+} + 4H_2O \rightleftarrows 5Ce^{3+} + MnO_4^- + 8H^+ \qquad E^\circ = 0.19_3\ V}$$

(b) $\Delta G^\circ = -5FE^\circ = -93._1\ kJ. \qquad K = 10^{5E^\circ/0.059\ 16} = 2 \times 10^{16}$

(c) $E = \left\{ 1.70 - \dfrac{0.059\ 16}{5} \log \dfrac{[Ce^{3+}]^5}{[Ce^{4+}]^5} \right\} - \left\{ 1.507 - \dfrac{0.059\ 16}{5} \log \dfrac{[Mn^{2+}]}{[MnO_4^-][H^+]^8} \right\}$

$$= -0.02_0\ V$$

(d) $\Delta G = -5FE = +10\ kJ$

(e) At equilibrium, $E = 0 \Rightarrow E^\circ = \dfrac{0.059\ 16}{5} \log \dfrac{[Ce^{3+}]^5[MnO_4^-][H^+]^8}{[Ce^{4+}]^5[Mn^{2+}]}$

$$\Rightarrow [H^+] = 0.62 \Rightarrow pH = 0.21$$

25. right half-cell : $Sn^{4+} + 2e^- \rightleftarrows Sn^{2+}$

left half-cell : $-2VO^{2+} + 4H^+ + 2e^- \rightleftarrows 2V^{3+} + 2H_2O$

net reaction : $Sn^{4+} + 2V^{3+} + 2H_2O \rightleftarrows Sn^{2+} + 2VO^{2+} + 4H^+$

$$E = E° - \frac{0.059\,16}{2} \log \frac{[VO^{2+}]^2[H^+]^4[Sn^{2+}]}{[V^{3+}]^2[Sn^{4+}]}$$

$$-0.289 = E° - \frac{0.059\,16}{2} \log \frac{(0.116)^2(1.57)^4(0.031\,8)}{(0.116)^2(0.031\,8)} \Rightarrow E° = -0.266 \text{ V}$$

$$\Rightarrow K = 10^{2E°/0.059\,16} = 1.0 \times 10^{-9}$$

26.
$Pd(OH)_2(s) + 2e^- \rightleftarrows Pd(s) + 2OH^-$ $\qquad E_+°$

$- Pd^{2+} + 2e^- \rightleftarrows Pd(s)$ $\qquad E_-° = 0.915 \text{ V}$

$Pd(OH)_2 \overset{K_{sp}}{\rightleftarrows} Pd^{2+} + 2OH^-$ $\qquad E° = E_+° - 0.915$

But $K_{sp} = 3 \times 10^{-28} \Rightarrow E° = \frac{0.059\,16}{2} \log K_{sp} = -0.814$

$-0.814 = E_1° - 0.915 \Rightarrow E_1° = 0.101 \text{ V}$

27.
$Br_2(l) + 2e^- \rightleftarrows 2Br^-$ $\qquad E_+° = 1.078 \text{ V}$

$- Br_2(aq) + 2e^- \rightleftarrows 2Br^-$ $\qquad E_-° = 1.098 \text{ V}$

$Br_2(l) \overset{K}{\rightleftarrows} Br_2(aq)$ $\qquad E° = -0.020 \text{ V}$

At equilibrium, $E° = 0$. Therefore $0 = -0.020 - \frac{0.059\,16}{2} \log \frac{[Br_2(aq)]}{[Br_2(l)]}$

$$\Rightarrow K = \frac{[Br_2(aq)]}{[Br_2(l)]} = 0.21_1 \text{ M} = 34 \text{ g/L}$$

28.
$FeY^- + e^- \rightleftarrows FeY^{2-}$ $\qquad E_+°$

$- FeY^- + e^- \rightleftarrows Fe^{2+} + Y^{4-}$ $\qquad E_-° = -0.730 \text{ V}$

$Fe^{2+} + Y^{4-} \rightleftarrows FeY^{2-}$ $\qquad E° = E_+° + 0.730$

But $E° = 0.059\,16 \log [K_f \text{ (for FeY}^{2-})] = 0.847 \text{ V} \Rightarrow E_+° = E° - E_-° = 0.117 \text{ V}$

29. $E°(T) = E° + \frac{dE°}{dT} \Delta T$ (according to Appendix H footnote)

$E°(50° \text{ C}) = -1.677 \text{ V} + (0.533 \times 10^{-3} \frac{mV}{K})(25 \text{ K}) = -1.664 \text{ V}$

30.
$H_2PO_4^- + H^+ + 2e^- \rightleftarrows HPO_3^{2-} + H_2O$ $\qquad E_+°$

$- HPO_4^{2-} + 2H^+ + 2e^- \rightleftarrows HPO_3^{2-} + H_2O$ $\qquad E_-° = -0.234 \text{ V}$

$H_2PO_4^- \rightleftarrows HPO_4^{2-} + H^+$ $\qquad E° = E_+° - E_-°$

$E° = \frac{0.059\,16}{2} \log K_{a2} \text{ (for H}_3PO_4) = -0.213 \text{ V}$

$$E_+^\circ = E^\circ - 0.234 \text{ V} = -0.447 \text{ V}$$

31. 1. $2Cu(s) + 2I^- \rightleftarrows 2CuI(s) + 2e^-$ $\qquad \Delta G_1^\circ = +2F(-0.185) = -35.7_0 \text{ kJ}$

2. $2Cu^{2+} + 4e^- \rightleftarrows 2Cu(s)$ $\qquad\qquad \Delta G_2^\circ = -4F(0.339) = -130._{83} \text{ kJ}$

3. hydro $\rightleftarrows$ quinone $+ 2H^+ + 2e^-$ $\qquad \Delta G_3^\circ = +2F(0.700) = 135._{08} \text{ kJ}$

4. (1)+(2)+(3): $2Cu^{2+} + 2I^- +$ hydro $\rightleftarrows$ $\qquad \Delta G_4^\circ = \Delta G_1^\circ + \Delta G_2^\circ + \Delta G_3^\circ$

$\qquad\qquad 2CuI(s) +$ quinone $+ 2H^+$ $\qquad\qquad\quad = -31.4 \text{ kJ}$

Since $2e^-$ are transferred in the net reaction, $E_4^\circ = \dfrac{-\Delta G_4^\circ}{2F} = +0.163 \text{ V}$

$K = 10^{2(0.163)/0.059\,16} = 3.2 \times 10^5$

32. In the right half-cell, the reaction $Hg^{2+} + Y^{4-} \rightleftarrows HgY^{2-}$ is at equilibrium, even though the net cell reaction $Hg^{2+} + H_2 \rightleftarrows Hg(l) + 2H^+$ is not at equilibrium.

33. (a) $\qquad AgCl(s) + e^- \rightleftarrows Ag(s) + Cl^-$ $\qquad\qquad\qquad E_+^\circ = 0.222 \text{ V}$

$\qquad\quad - \;\; H^+ + e^- \rightleftarrows \frac{1}{2}H_2(g)$ $\qquad\qquad\qquad\qquad E_-^\circ = 0 \text{ V}$

$\qquad\quad AgCl(s) + \frac{1}{2}H_2(g) \rightleftarrows Ag(s) + H^+ + Cl^-$ $\qquad E^\circ = 0.222 \text{ V}$

$E^\circ = 0.222 - 0.059\,16 \log \dfrac{[H^+][Cl^-]}{\sqrt{P_{H_2}}}$

(b) $0.485 = 0.222 - 0.059\,16 \log \dfrac{(10^{-3.60})[Cl^-]}{\sqrt{1.00}} \Rightarrow [Cl^-] = 0.14_3 \text{ M}$

34. (a) $Hg_2Cl_2 + 2e^- \rightleftarrows 2Hg(l) + 2Cl^-$ $\qquad\qquad\qquad E_+^\circ = 0.268 \text{ V}$

$\qquad\quad -$ quinone $+ 2H^+ + 2e^- \rightleftarrows$ hydroquinone $\qquad\qquad E_-^\circ = 0.700 \text{ V}$

$\qquad\quad Hg_2Cl_2 +$ hydro $\rightleftarrows 2Hg(l) + 2Cl^- +$ quinone $+ 2H^+ \quad E^\circ = -0.432 \text{ V}$

$E = -0.432 - \dfrac{0.059\,16}{2} \log \dfrac{[\text{quinone}][H^+]^2[Cl^-]^2}{[\text{hydroquinone}]}$

(b) Setting $[Cl^-] = 0.50 \text{ M}$, we find

$\qquad E = -0.432 - 0.059\,16 \log (0.50) - 0.059\,16 \log [H^+]$

$\qquad E = -0.414 + 0.059\,16 \text{ pH} \quad (A = -0.414, B = 0.059\,16 \text{ V per pH unit })$

(c) $E = -0.414 + 0.059\,16\,(4.50) = -0.148.$

Since $E < 0$, electrons flow from right to left ($Hg \rightarrow Pt$) through the meter.

35. $H^+ (1.00 \text{ M}) + e^- \rightleftarrows \frac{1}{2} H_2 \text{ (g, 1.00 atm)}$ $E_+^\circ = 0 \text{ V}$

  $- \quad H^+ (x \text{ M}) + e^- \rightleftarrows \frac{1}{2} H_2 \text{ (g, 1.00 atm)}$ $E_-^\circ = 0 \text{ V}$

  ───

  $H^+ (1.00 \text{ M}) \rightleftarrows H^+ (x \text{ M})$ $E^\circ = 0 \text{ V}$

$$E = 0.490 = 0 - 0.059\,16 \log x \Rightarrow x = 5.2 \times 10^{-9} \text{ M}$$

$$K_b = \frac{[RNH_3^+][OH^-]}{[RNH_2]} = \frac{(0.050)(K_w/x)}{0.10} = 9.6 \times 10^{-7}$$

36. $M^{2+} + 2e^- \rightleftarrows M(s)$ $E_+^\circ = -0.266 \text{ V}$

  $- \quad 2H^+ + 2e^- \rightleftarrows H_2 \text{ (g, 0.50 atm)}$ $E_-^\circ = 0 \text{ V}$

  ───

  $H_2(g) + M^{2+} \rightleftarrows 2H^+ + M(s)$ $E^\circ = -0.266 \text{ V}$

$$E = -0.266 - \frac{0.059\,16}{2} \log \frac{[H^+]^2}{P_{H_2}[M^{2+}]}$$

The concentration of H^+ in the left half-cell is found by considering the titration of 28.0 mL of the tetraprotic pyrophosphoric acid (abbreviated H_4P) with 72.0 mL of KOH.

  28.0 mL of 0.010 0 M H_4P = 0.280 mmol

  72.0 mL of 0.010 0 M KOH = 0.720 mmol

  $H_4P \quad + \quad OH^- \quad \rightarrow \quad H_2P^{2-} \quad + \quad HP^{3-}$
  0.280 mmol   0.720 mmol   0.120 mmol   0.160 mmol

$$pH = pK_3 + \log \frac{[HP^{3-}]}{[H_2P^{2-}]} = 6.70 + \log \frac{0.160}{0.120} = 6.82 \Rightarrow [H^+] = 1.5 \times 10^{-7} \text{ M}$$

Putting the known values of $[H^+]$ and P_{H_2} into the Nernst equation gives

$$-0.246 = -0.266 - \frac{0.059\,16}{2} \log \frac{(1.5 \times 10^{-7})^2}{(0.50)[M^{2+}]} \Rightarrow [M^{2+}] = 2.1 \times 10^{-13} \text{ M}$$

In the right half-cell we have the equilibrium

	M^{2+}	$+$	F_{EDTA}	$\rightleftarrows$	MY^{2-}
initial mmol/mL	$\frac{0.280}{100}$		$\frac{0.720}{100}$		—
final mmol/mL	small		$\frac{0.440}{100}$		$\frac{0.280}{100}$

$$K_f = \frac{[MY^{2-}]}{[M^{2+}]\,\alpha_{Y^{4-}}\,F_{EDTA}} = \frac{0.280/100}{(2.1 \times 10^{-13})(0.005\,6)(0.440/100)} = 5.4 \times 10^{14}$$

37.   right half-cell : $Pb^{2+}(\text{right}) + 2e^- \rightleftarrows Pb(s)$ $E_+^\circ = -0.126 \text{ V}$

  left half-cell : $^- \; Pb^{2+}(\text{left}) + 2e^- \rightleftarrows Pb(s)$ $E_-^\circ = -0.126 \text{ V}$

  ───

  $Pb^{2+}(\text{right}) \rightleftarrows Pb^{2+}(\text{left})$ $E^\circ = 0$

Nernst equation for net cell reaction:

$$-0.001\,8 = -\frac{0.059\,16}{2}\log\frac{[Pb^{2+}(\text{left})]}{[Pb^{2+}(\text{right})]} \Rightarrow \frac{[Pb^{2+}(\text{left})]}{[Pb^{2+}(\text{right})]} = 1.15$$

For each half-cell we can write $[CO_3^{2-}] = K_{sp}$ (for $PbCO_3$) / $[Pb^{2+}]$

$$\frac{[CO_3^{2-}(\text{left})]}{[CO_3^{2-}(\text{right})]} = \frac{K_{sp}\,(\text{for }PbCO_3)/[Pb^{2+}(\text{left})]}{K_{sp}\,(\text{for }PbCO_3)/[Pb^{2+}\,(\text{right})]} = \frac{1}{1.15} = 0.87$$

In each compartment the Ca^{2+} concentration is equal to the total concentration of all carbonate species (since $PbCO_3$ is much less soluble than $CaCO_3$). Let the fraction of all carbonate species in the form CO_3^{2-} be $\alpha_{CO_3^{2-}}$

(i.e., $[CO_3^{2-}] = \alpha_{CO_3^{2-}}$ [total carbonate]). We can say that $[Ca^{2+}]$ = [total carbonate] $= [CO_3^{2-}] / \alpha_{CO_3^{2-}}$. The value of $\alpha_{CO_3^{2-}}$ is the same in both compartments, since the pH is the same. Now we can write

$$\frac{K_{sp}(\text{calcite})}{K_{sp}(\text{aragonite})} = \frac{[Ca^{2+}(\text{left})][CO_3^{2-}(\text{left})]}{[Ca^{2+}(\text{right})][CO_3^{2-}(\text{right})]} = \frac{[CO_3^{2-}(\text{left})]^2 / \alpha_{CO_3^{2-}}}{[CO_3^{2-}(\text{right})]^2 / \alpha_{CO_3^{2-}}}$$

$$= (0.87)^2 = 0.76$$

38. $Cu^{2+} + 2e^- \rightleftarrows Cu(s)$ $E_+^\circ = 0.339$ V

$\underline{\;\;Ni(s) \rightleftarrows Ni^{2+} + 2e^-}$ $\underline{E_-^\circ = -0.236\text{ V}}$

$Cu^{2+} + Ni(s) \rightleftarrows Cu(s) + Ni^{2+}$ $E^\circ = 0.575$ V

The ionic strength of each half-cell is 0.10 M and $\gamma_{Ni^{2+}} = \gamma_{Cu^{2+}} = 0.405$

$$0.489 = 0.575 - \frac{0.059\,16}{2}\log\frac{(0.025)(0.405)}{[Cu^{2+}](0.405)} \Rightarrow [Cu^{2+}] = 3.09 \times 10^{-5}\text{ M}$$

$$K_{sp} = [Cu^{2+}]\gamma_{Cu^{2+}}[IO_3^-]^2\gamma_{IO_3^-}^2 = (3.09 \times 10^{-5})(0.405)(0.10)^2(0.775)^2 = 7.5 \times 10^{-8}$$

39. $E^{\circ\prime}$ is the effective reduction potential for a half-reaction at pH 7, instead of pH 0. Since living systems tend to have a pH much closer to 7 than to 0, $E^{\circ\prime}$ provides a better indication of redox behavior in an organism.

40.   (a) $E = 0.731 - \dfrac{0.059\,16}{2}\log\dfrac{P_{C_2H_4}}{P_{C_2H_2}[H^+]^2}$

(b) $E = \underbrace{(0.731 + 0.059\,16\log[H^+]}_{\text{This is }E^{\circ\prime}\text{ when pH}=7} - \dfrac{0.059\,16}{2}\log\dfrac{P_{C_2H_4}}{P_{C_2H_2}}$

(c) $E^{\circ\prime} = 0.731 + 0.059\,16\log(10^{-7.00}) = 0.317$ V

41. $E = E° - \dfrac{0.059\ 16}{2} \log \dfrac{[HCN]^2}{P_{(CN)_2}\ [H^+]^2}$

Substituting $[HCN] = \dfrac{[H^+]\ F_{HCN}}{[H^+] + K_a}$ into the Nernst equation gives

$E = 0.373 - \dfrac{0.059\ 16}{2} \log \dfrac{[H^+]^2\ F_{HCN}^2}{([H^+] + K_a)^2\ P_{(CN)_2}\ [H^+]^2}$

$E = 0.373 + 0.059\ 16 \log\ ([H^+] + K_a) - \dfrac{0.059\ 16}{2} \log \dfrac{F_{HCN}^2}{P_{(CN)_2}}$

$\underbrace{}$

This is $E°'$ when pH = 7

Inserting $K_a = 6.2 \times 10^{-10}$ for HCN and $[H^+] = 10^{-7.00}$ gives

$E°' = 0.373 + 0.059\ 16 \log\ (10^{-7.00} + 6.2 \times 10^{-10}) = -0.041$ V

42. $H_2C_2O_4 + 2H^+ + 2e^- \rightleftarrows 2HCO_2H$

$E = 0.204 - \dfrac{0.059\ 16}{2} \log \dfrac{[HCO_2H]^2}{[H_2C_2O_4][H^+]^2}$

But $[HCO_2H] = \dfrac{[H^+]\ F_{HCO_2H}}{[H^+] + K_a}$ and $[H_2C_2O_4] = \dfrac{[H^+]^2\ F_{H_2C_2O_4}}{[H^+]^2 + K_1[H^+] + K_1K_2}$

Putting these expressions into the Nernst equation gives

$E = 0.204 - \dfrac{0.059\ 16}{2} \log \dfrac{[H^+]^2\ F_{HCO_2H}^2\ ([H^+]^2 + K_1[H^+] + K_1K_2)}{([H^+] + K_a)\ [H^+]^2\ F_{H_2C_2O_4}\ [H^+]^2}$

$E = 0.204 - \dfrac{0.059\ 16}{2} \log \dfrac{[H^+]^2 + K_1[H^+] + K_1K_2}{([H^+] + K_a)\ [H^+]^2} - \dfrac{0.059\ 16}{2} \log \dfrac{F_{HCO_2H}^2}{F_{H_2C_2O_4}}$

$\underbrace{\phantom{E = 0.204 - \dfrac{0.059\ 16}{2} \log \dfrac{[H^+]^2 + K_1[H^+] + K_1K_2}{([H^+] + K_a)\ [H^+]^2}}}$

This is $E°'$ when pH = 7

Putting in $[H^+] = 10^{-7.00}$ M, $K_a = 1.8 \times 10^{-4}$, $K_1 = 5.60 \times 10^{-2}$ and $K_2 = 5.42 \times 10^{-5}$ gives $E°' = -0.158$ V.

43. $E = E° - 0.059\ 16 \log \dfrac{[H_2Red^-]}{[HOx]}$

But $[HOx] = \dfrac{[H^+]\ F_{HOx}}{[H^+] + K_a}$ and $[H_2Red^-] = \dfrac{[H^+]^2\ F_{H_2Red^-}}{[[H^+]^2 + [H^+]K_1 + K_1K_2}$

Putting these values into the Nernst equation gives

$E = E° - 0.059\ 16 \log \dfrac{[H^+]^2\ F_{H_2Red^-}\ ([H^+] + K_a)}{[H^+]\ F_{HOx}\ ([H^+]^2 + [H^+]K_1 + K_1K_2)}$

$$E = E° - 0.059\ 16\ \log \frac{[H^+]\ ([H^+] + K_a)}{\underbrace{[H^+]^2 + [H^+]K_1 + K_1K_2}_{E°'}} - 0.059\ 16\ \log \frac{F_{H_2Red^-}}{F_{HOx}}$$

Since $E°' = 0.062$ V, we find $E° = -0.036$ V.

44. $E = E° - \dfrac{0.059\ 16}{2}\ \log \dfrac{[HNO_2]}{[NO_3^-][H^+]^3}$

But $[HNO_2] = \dfrac{[H^+]F_{HNO_2}}{[H^+] + K_a}$ and $[NO_3^-] = F_{NO_3^-}$

Putting these values into the Nernst equation gives

$$E = E° - \frac{0.059\ 16}{2}\ \log \underbrace{\frac{1}{([H^+] + K_a)[H^+]^2}}_{E°'} - \frac{0.059\ 16}{2}\ \log \frac{F_{HNO_2}}{F_{NO_3^-}}$$

$E°' = 0.433 = 0.940 - \dfrac{0.059\ 16}{2}\ \log \dfrac{1}{(10^{-7} + K_a)(10^{-7})^2} \Rightarrow K_a = 7.2 \times 10^{-4}$

45.   (a) $A = 0.500 =$

   $(1.12 \times 10^4\ M^{-1}cm^{-1})[Ox](1.00\ cm) + (3.82 \times 10^3\ M^{-1}cm^{-1})[Red](1.00\ cm)$

   But $[Ox] = 5.70 \times 10^{-5} - [Red]$. Combining these two equations gives
   $[Ox] = 3.82 \times 10^{-5}$ M and $[Red] = 1.88 \times 10^{-5}$ M.

   (b) $[S^-] = [Ox] = 3.82 \times 10^{-5}$ M and $[S] = Red = 1.88 \times 10^{-5}$ M.

   (c) $\begin{array}{ll} S + e^- \rightleftarrows S^- & E_+° = ? \\ - Ox + e^- \rightleftarrows Red & E_-° = -0.128\ V \\ \hline Red + S \rightleftarrows Ox + S^- & E°' = 0.059\ 16\ \log \dfrac{[Ox][S^-]}{[Red][S]} = 0.036\ V \end{array}$

   $E_+° = E°' + E_-° = -0.092\ V$

1. (a) $AgCl(s) + e^- \rightleftharpoons Ag(s) + Cl^-$

    $Hg_2Cl_2(s) + 2e^- \rightleftharpoons 2Hg(l) + 2Cl^-$

    (b) $E = E_+ - E_- = 0.241 - 0.197 = 0.044$ V

2. (a) 0.326 V (b) 0.086 V (c) 0.019 V (d) 0.021 V (e) - 0.021 V

3. $E = E_+ - E_-$

    $E = \left\{0.771 - 0.059\,16 \ \log \dfrac{[Fe^{2+}]}{[Fe^{3+}]}\right\} - (0.241) = 0.684$ V

4. For the saturated Ag-AgCl electrode we can write : $E = E° - 0.059\,16 \log \mathcal{A}_{Cl^-}$
    Putting in $E = 0.197$ and $E° = 0.222$ V gives $\mathcal{A}_{Cl^-} = 2.6_5$. For the S.C.E. we
    can write $E = E° - 0.059\,16 \log \mathcal{A}_{Cl^-} = 0.268$ V $- \ 0.059\,16 \log 2.6_5$
    $= 0.243$ V.

5. $E = E° - 0.059\,16 \log \mathcal{A}_{Cl^-}$

    $0.280 = 0.268 - 0.059\,16 \log \mathcal{A}_{Cl^-} \Rightarrow \mathcal{A}_{Cl^-} = 0.627$

6. (a) $Cu^{2+} + 2e^- \rightleftharpoons Cu(s)$ $\qquad\qquad$ $E° = 0.339$ V

    (b) $E_+ = 0.339 - \dfrac{0.059\,16}{2} \ \log \dfrac{1}{[Cu^{2+}]} = 0.309$ V

    (c) $E = E_+ - E_- = 0.309 - 0.241 = 0.068$ V

7. A silver electrode serves as an indicator for Ag^+ by virtue of the equilibrium
    $Ag^+ + e^- \rightleftharpoons Ag(s)$ that occurs at its surface. If the solution is saturated with silver
    halide, then $[Ag^+]$ is affected by changes in halide concentration. Therefore the
    electrode is also an indicator for halide.

8. $V_e = 20.0$ mL. $Ag^+ + e^- \rightleftharpoons Ag(s) \Rightarrow E_+ = 0.799 - 0.059\,16 \log \dfrac{1}{[Ag^+]}$

    0.1 mL: $[Ag^+] = \underbrace{\left(\dfrac{19.9}{20.0}\right)}_{\substack{\text{Fraction}\\\text{remaining}}} \underbrace{(0.050\,0 \ M)}_{\substack{\text{Original}\\\text{concentration}}} \underbrace{\left(\dfrac{10.0}{10.1}\right)}_{\substack{\text{Dilution}\\\text{factor}}} = 0.049\,3$ M

    $E = E_+ - E_- = \left\{0.799 - 0.059\,16 \log \dfrac{1}{0.049\,3}\right\} - 0.241 = 0.481$ V

    10.0 mL: $[Ag^+] = \left(\dfrac{10.0}{20.0}\right)(0.050\,0 \ M)\left(\dfrac{10.0}{20.0}\right) = 0.012\,5$ M

    $E = E_+ - E_- = \left\{0.799 - 0.059\,16 \log \dfrac{1}{0.012\,5}\right\} - 0.241 = 0.445$ V

20.0 mL: $[Ag^+] = [Br^-] \Rightarrow [Ag^+] = \sqrt{K_{sp}} = \sqrt{5.0 \times 10^{-13}} = 7.0_7 \times 10^{-7} M$

$$E = E_+ - E_- = \left\{ 0.799 - 0.059\ 16 \log \frac{1}{7.0_7 \times 10^{-7}} \right\} - 0.241 = 0.194\ V$$

30.0 mL: This is 10.0 mL past $V_e \Rightarrow [Br^-] = \left(\frac{10.0}{40.0}\right)(0.025\ 0\ M) = 0.006\ 25\ M$

$$[Ag^+] = K_{sp}/[Br^-] = (5.0 \times 10^{-13})/0.006\ 25 = 8.0 \times 10^{-11}\ M$$

$$E = E_+ - E_- = \left\{ 0.799 - 0.059\ 16 \log \frac{1}{8.0 \times 10^{-11}} \right\} - 0.241 = -0.039\ V$$

9. (a) From 0 to 50 mL, AgI is precipitating. Between 50 and 100 mL, AgCl is precipitating. At V = 25 mL,

$$[I^-] = \underbrace{\left(\frac{1}{2}\right)}_{\substack{\text{fraction} \\ \text{remaining}}} (0.100\ M) \underbrace{\left(\frac{50.0}{75.0}\right)}_{\substack{\text{dilution} \\ \text{factor}}} = 0.033\ 3\ M$$

$$[Ag^+] = K_I/[I^-] = K_I/0.033\ 3$$

(b) At V = 75.0 mL, $[Cl^-] = \left(\frac{1}{2}\right)(0.100\ M)\left(\frac{50.0}{125.0}\right) = 0.020\ 0\ M$

$$[Ag^+] = K_{Cl}/0.020\ 0$$

(c) $E = E_+ - E_-$

$$E = \left\{ 0.799 - 0.059\ 16 \log \frac{1}{[Ag^+]} \right\} - (0.241),$$

since the right half-cell reaction can be written $Ag^+ + e^- \rightleftarrows Ag(s)$

(d) At 25.0 mL : $E = 0.558 + 0.059\ 16 \log \dfrac{K_I}{0.033\ 3}$

At 75.0 mL : $E = 0.558 + 0.059\ 16 \log \dfrac{K_{Cl}}{0.020\ 0}$

Subtracting gives $\Delta E = 0.388$

$$= 0.059\ 16 \log \frac{K_{Cl}/0.020\ 0}{K_I/0.033\ 3} \Rightarrow \frac{K_{Cl}}{K_I} = 2.2 \times 10^6$$

10. The reaction in the right half-cell is $Hg^{2+} + 2e^- \rightleftarrows Hg(l)$

$E = E_+ - E_-$

$$-0.036 = 0.852 - \frac{0.059\ 16}{2} \log \frac{1}{[Hg^{2+}]} - (0.241)$$

$$\Rightarrow [Hg^{2+}] = 1.34 \times 10^{-22}\ M$$

The cell contains 5.00 mmol of EDTA and 1.00 mmol of Hg(II) in 100 mL.

$$K_f = \frac{[HgY^{2-}]}{[Hg^{2+}][Y^{4-}]} = \frac{[HgY^{2-}]}{[Hg^{2+}]\alpha_{Y4}\cdot[EDTA]}$$

$$K_f = \frac{(1.00/100)}{(1.34 \times 10^{-22})(0.36)(4.00/100)} = 5._2 \times 10^{21}$$

11.   (a) $Fe^{3+} + e^- \rightleftarrows Fe^{2+}$ $E° = 0.771$ V

(b) $E = E_+ - E_-$

$-0.126 = 0.771 - 0.059\,16 \log \dfrac{[Fe^{2+}]}{[Fe^{3+}]} - 0.241 \Rightarrow \dfrac{[Fe^{2+}]}{[Fe^{3+}]} = 1._2 \times 10^{11}$

(c) $\dfrac{K_f(FeEDTA^-)}{K_f(FeEDTA^{2-})} = \dfrac{[FeEDTA^-]}{[Fe^{3+}][EDTA^{4-}]} \div \dfrac{[FeEDTA^{2-}]}{[Fe^{2+}][EDTA^{4-}]}$

$= \dfrac{[FeEDTA^-]}{[FeEDTA^{2-}]} \cdot \dfrac{[Fe^{2+}]}{[Fe^{3+}]} = \left(\dfrac{1.00 \times 10^{-3}}{2.00 \times 10^{-3}}\right)(1._2 \times 10^{11}) = 6 \times 10^{10}$

12. $E = E_+ - E_- = -0.429 - 0.059\,16 \log \dfrac{[CN^-]^2}{[Cu(CN)_2^-]} - 0.197$

Putting in $E = -0.440$ V and $[Cu(CN)_2^-] = 1.00$ mM gives $[CN^-] = 0.847$ mM.

$pH = pK_a(HCN) + \log \dfrac{[CN^-]}{[HCN]} = 9.21 + \log \dfrac{8.47 \times 10^{-4}}{1.00 \times 10^{-3} - 8.47 \times 10^{-4}} = 9.95_4$

Now we use the pH to see how much HA reacted with KOH:

$$HA \quad + \quad OH^- \quad \rightarrow \quad A^- \quad + \quad H_2O$$

	HA	OH⁻	A⁻
initial mmol	10.0	x	—
final mmol	10.0 - x	—	x

$pH = pK_a(HA) + \log \dfrac{[A^-]}{[HA]}$

$9.95_4 = 9.50 + \log \dfrac{x}{10.0-x} \Rightarrow x = 7.4_0$ mmol of OH^-

$[KOH] = \dfrac{7.4_0 \text{ mmol}}{25.0 \text{ mL}} = 0.29_6$ M

13.   The junction potential arises because different ions diffuse at different rates across a liquid junction, leading to a separation of charge. The resulting electric field retards fast-moving ions and accelerates the slow-moving ions until a steady state junction potential is reached. This limits the accuracy of a potentiometric measurement, because we do not know what part of a measured cell voltage is due to the process of interest and what is due to the junction potential. The cell in Figure 14-3 has no junction potential because there are no liquid junctions.

14. H^+ has a greater mobility than K^+. The HCl side of the HCl | KCl junction will be negative because H^+ will diffuse into the KCl region faster than K^+ diffuses into the HCl region. K^+ has a greater mobility than Na^+, so this junction has the opposite sign. The HCl | KCl voltage is larger because the difference in mobility between H^+ and K^+ is much greater than the difference between K^+ and Na^+.

15. Relative mobilities:

$$K^+ \rightarrow 7.62 \qquad\qquad NO_3^- \rightarrow 7.40$$
$$5.19 \leftarrow Na^+ \qquad\qquad 7.91 \leftarrow Cl^-$$

Both the cation and anion diffusion cause negative charge to build up on the <u>left</u>.

16. Velocity = mobility × field = $(36.30 \times 10^{-8} \, m^2/(s \cdot V)) \times (7\,800 \, V/m) = 2.83 \times 10^{-3} \, m \, s^{-1}$ for H^+ and $(7.40 \times 10^{-8})(7\,800) = 5.77 \times 10^{-4} \, m \, s^{-1}$ for NO_3^-. To cover 0.120 m will require $(0.120 \, m)/(2.83 \times 10^{-3} \, m \, s^{-1}) = 42.4 \, s$ for H^+ and $(0.120)/(5.77 \times 10^{-4}) = 208 \, s$ for NO_3^-.

17. (a) $E° = 0.799 \, V \Rightarrow K = 10^{0.799/0.059\,16} = 3._{20} \times 10^{13}$

(b) $K' = 10^{0.801/0.059\,16} = 3._{46} \times 10^{13}$. $K'/K = 1.08$. The increase is 8%.

(c) $K = 10^{0.100/0.059\,16} = 49.0$. $K' = 10^{0.102/0.059\,16} = 53.0$
$K'/K = 1.08$. The change is still 8%.

18. Both half-cell reactions are the same $(AgCl + e^- \rightleftarrows Ag + Cl^-)$ and the concentration of Cl^- is the same on both sides. In principle, the voltage of the cell is zero if there were no junction potential. The measured voltage can be attributed to the junction potential. In practice, if both sides contained 0.1 M HCl (or 0.1 M KCl), the two electrodes would probably produce a small voltage because no two real cells are identical. This voltage can be measured and subtracted from the voltage measured with the HCl | KCl junction.

19. (a) In phase α we have 0.1 M H^+ $(u = 36.3 \times 10^{-8} \, m^2 \, s^{-1} \, V^{-1})$ and 0.1 M Cl^- $(u = 7.91 \times 10^{-8} \, m^2 \, s^{-1} \, V^{-1})$. In phase β is 0.1 M K^+ $(u = 7.62 \times 10^{-8} \, m^2 \, s^{-1} \, V^{-1})$ and 0.1 M Cl^- $(u = 7.91 \times 10^{-8} \, m^2 \, s^{-1} \, V^{-1})$.

Substituting into the Henderson equation gives

$$E_j = \frac{(36.3)[0-0.1] + (7.62)[0.1-0] - (7.91)[0.1-0.1]}{(36.3)[0-0.1] + (7.62)[0.1-0] + (7.91)[0.1-0.1]} \times$$

$$0.059\,16 \log \frac{(36.3)(0.1) + (7.91)(0.1)}{(7.62)(0.1) + (7.91)(0.1)} = 26.9 \, mV$$

(b)

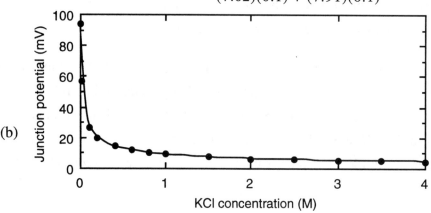

(c)	[HCl]	y M HCl ‖ 1mM KCl	y M HCl ‖ 4 M KCl
	10^{-4} M	9.1 mV	4.6 mV
	10^{-3} M	26.9 mV	3.6 mV
	10^{-2} M	57.3 mV	3.0 mV
	10^{-1} M	93.6 mV	4.7 mV

20. The electrode should be calibrated at 37° using two buffers bracketing the pH of the blood. It would be reasonable to use the MOPSO and HEPES buffers in Table 15-3 that are recommended for use with physiologic fluids. The pH of these standards at 37°C is 6.695 and 7.370. The standards should be thermostatted to 37° during calibration and the blood should also be at 37° during the measurement.

21. Uncertainty in pH of standard buffers, junction potential, alkaline or acid errors at extreme pH values, equilibration time for electrode.

22. The error shown in the graph is -0.33 pH units. The electrode will indicate 11.00 - 0.33 = 10.67.

23. Saturated potassium hydrogen tartrate and 0.05 m potassium hydrogen phthalate.

24. If the strongly alkaline solution has a high concentration of Na^+ (as in NaOH), the Na^+ cation competes with H^+ for cation exchange sites on the glass surface. The glass responds as if some H^+ were present, and the apparent pH is lower than the actual pH.

25. The junction potential changes from -6.4 mV to -0.2 mV. A change of 6.4 - 0.2 = 6.2 mV will appear to be a pH change of 6.2/59.16 = 0.10 pH units.

26. (a) (4.63)(59.16 mV) = 274 mV. The factor 59.16 mV is the value of (RT ln 10)/F at 298.15 K.

(b) At 310.15 K (37°C), (RT ln 10)/F = (8.314 4 J K^{-1})(310.15)(ln 10)/(96 485) = 61.54 mV. (4.63) (61.54 mV) = 285 mV.

27. Ion-selective electrodes have a membrane that binds the specific ion on both surfaces, and thereby builds up an electric potential difference between the two surfaces. The specific ion concentration inside the electrode is fixed, but that on the outside varies. In order to measure the potential difference, some (tiny) current must flow through the detector circuit, and therefor across the electrode membrane. There must be some mobile charge carrier capable of crossing the membrane. A compound electrode contains a second chemically active membrane outside the ion-selective membrane. The second membrane may be semipermeable and only allow the species of interest to pass through. Alternatively, the second membrane may contain a substance (such as an enzyme) that reacts with analyte to generate the species to which the ion-selective membrane responds.

28. The selectivity tells us the relative response of an ion-selective electrode to the ion of interest and an interfering ion. The smaller the value, the more selective is the electrode (smaller response to the interfering ion).

29. A mobile molecule dissolved in and confined to the membrane liquid phase binds tightly to the ion of interest and weakly to interfering ions.

30. A metal ion buffer maintains the desired (small) concentration of metal ion from a large reservoir of metal complex (ML) and free ligand (L). If you just tried to dissolve 10^{-8} M metal ion in most solutions or many containers, the metal would probably bind to the container wall or to an impurity in the solution and be lost.

31. Electrodes respond to *activity*. If the ionic strength is constant, concentration is proportional to activity, since the activity coefficient will be constant.

32. (a) $-0.230 = \text{constant} - 0.059\ 16\ \log(1.00 \times 10^{-3}) \Rightarrow \text{constant} = -0.407$ V

(b) $-0.300 = -0.407 - 0.059\ 16\ \log x \Rightarrow x = 1.5_5 \times 10^{-2}$ M

(c)
$$-0.230 = \text{constant} - 0.059\ 16\ \log(1.00 \times 10^{-3})$$
$$\underline{-0.300 = \text{constant} - 0.059\ 16\ \log\ x}$$
$$\text{subtract}:\ 0.070 = -0.059\ 16\ \log\frac{1.00 \times 10^{-3}}{x} \Rightarrow x = 1.5_2 \times 10^{-2}\ \text{M}$$

33.
$$E_1 = \text{constant} + \frac{0.059\ 16}{2}\ \log[1.00 \times 10^{-4}]$$
$$E_2 = \text{constant} + \frac{0.059\ 16}{2}\ \log[1.00 \times 10^{-3}]$$
$$\Delta E = E_2 - E_1 = \frac{0.059\ 16}{2}\ \log\frac{1.00 \times 10^{-3}}{1.00 \times 10^{-4}} = +0.029\ 6\ \text{V}$$

34. (a) K^+ has the largest selectivity coefficient and interferes the most.

(b) Group I causes more interference than Group II.

35. $\dfrac{[\text{ML}]}{[\text{M}][\text{L}]} = 4.0 \times 10^8 = \dfrac{0.030\ 0}{[\text{M}](0.020\ 0)} \Rightarrow [\text{M}] = 3.8 \times 10^{-9}$ M

36. From the graph below, $E = -22.5$ mV gives
$$\log[\text{Ca}^{2+}] = -2.62 \Rightarrow [\text{Ca}^{2+}] = 2.4 \times 10^{-3}\ \text{M}.$$
The slope is 28.14 mV $\Rightarrow 28.14 = \beta(0.059\ 16)/2 \Rightarrow \beta = 0.951\ 3.$

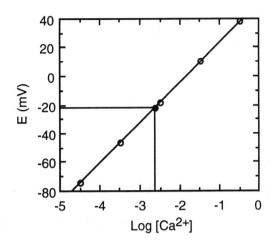

37. The least squares parameters are

$E = 51.10 (\pm 0.24) + 28.14 (\pm 0.08_5) \log [Ca^{2+}]$ $(s_y = 0.2_7)$

Putting $E = -22.5 (\pm 0.3)$ into this equation gives

$$\log [Ca^{2+}] = \frac{-22.5(\pm 0.3) - 51.10(\pm 0.24)}{28.14(\pm 0.08_5)} = \frac{-73.60(\pm 0.384)}{28.14(\pm 0.08_5)}$$

$\log [Ca^{2+}] = -2.615(\pm 0.603\%) = -2.615(\pm 0.015_8)$

From Table 3-1, we can write that if $F = 10^x$, $e_F/F = (\ln 10)e_x$.

In this problem, $F = [Ca^{2+}] = 10^{-2.615(\pm 0.015_8)}$

$\dfrac{e_F}{F} = (\ln 10)(0.015_8) = 0.036\ 4$

$e_F = (0.036\ 4)\ F = 3.64\ \%$ of $10^{-2.615} \Rightarrow F = 2.43(\pm 0.09) \times 10^{-3}$ M

38. Without calcium:

-0.333 V $=$ constant $+ 0.059\ 16 \log [3.44 \times 10^{-4}] \Rightarrow$ constant $= -0.128$ V

With calcium:

$E = -0.128 + 0.059\ 16 \log [3.44 \times 10^{-4} + (5 \times 10^{-5})(0.100)^{1/2}] = -0.332$ V

39. (a) 8.22 ppm $= 8.22 \times 10^{-6}$ g of Cl⁻ per g of solution $= 8.22 \times 10^{-3}$ g of Cl⁻ per liter of solution $= 2.32 \times 10^{-4}$ M.

(b) For a Nernstian electrode we can write

$E =$ constant $- 0.059\ 16 \log [Cl^-]$

For the first solution, $0.228\ 0 =$ constant $- 0.059\ 16 \log (2.32 \times 10^{-4})$

$\Rightarrow$ constant $= 0.013\ 0$ V.

10.0 mL of 100.0 ppm standard contains 2.82×10^{-5} moles of Cl⁻. The original 100.0 mL of solution contains 2.32×10^{-5} moles of Cl⁻. Therefore the new solution has

$$[Cl^-] = \frac{(2.32 + 2.82) \times 10^{-5}\ \text{mol}}{0.110\ \text{L}} = 4.67 \times 10^{-4}\ \text{M}$$

$E = 0.013\ 0 - 0.059\ 16 \log (4.67 \times 10^{-4}) = 210.0$ mV

(c) A 2 mV change for Cl^- corresponds to a 1 mV change for Ca^{2+}. We could use the table in the following form :

ΔE:	0	0.5	1.0	1.5
Q:	1.00	0.696	0.529	0.423

40. (a) Let C_0 = initial concentration of unknown in 100 mL and S_0 = initial concentration of standard in 10 mL.

Concentration of analyte in diluted sample $= \dfrac{100}{110} C_0 + \dfrac{10}{110} S_0 = \dfrac{10C_0+S_0}{11}$

Before standard addition : E_1 = constant + 0.059 16 log C_0

After standard addition :

$$E_2 = \text{constant} + 0.059\ 16 \log \frac{10\ C_0 + S_0}{11}$$

$$\Delta E = 0.001\ V = E_2 - E_1 = 0.059\ 16 \log \frac{10\ C_0 + S_0}{11C_0}$$

Solve for $C_0 \Rightarrow C_0 = 0.696\ S_0 \Rightarrow Q = 0.696$

(b) General solution in part a: $Q = C_0/S_0 = 1/(11a-10)$, where $a = 10^{\Delta E/0.059\ 16}$

	A	B	C
1	DeltaE (mV)	a = 10^(DeltaE/59.16)	Q = 1/(11a-10)
2	0	1.0000	1.0000
3	1	1.0397	0.6961
4	2	1.0810	0.5290
5	3	1.1239	0.4233
6	4	1.1685	0.3505
7	5	1.2148	0.2973

41. $E = \underset{A}{\text{constant}} + \underset{B}{\dfrac{\beta(0.059\ 16)}{2}} \log ([Ca^{2+}] + k_{Ca^{2+},Mg^{2+}}[Mg^{2+}])$

For the first two solutions we can write

$$-56.2\ mV = A + B\ \log (1.00 \times 10^{-6}) = A - 6\,B$$

$$+16.1\ mV = A + B\ \log (2.43 \times 10^{-4}) = A - 3.614\,B$$

Subtraction gives 68.7 mV = 2.386 B $\Rightarrow$ B = 28.80 mV

Putting this value of B back into the first equation gives A = 120.2 mV

The third set of data now gives the selectivity coefficient :

$$-38.0\ mV = 120.2 + 28.80 \log [10^{-6} + k_{Ca^{2+},Mg^{2+}} (3.68 \times 10^{-3})]$$

$\Rightarrow$ k = 6.0×10^{-4}

E = 120.2 + 28.80 log $([Ca^{2+}] + 6.0 \times 10^{-4} [Mg^{2+}])$

42. $-163.3\ mV = \text{constant} + \dfrac{56.8\ mV}{2} \log \left\{ [M^{2+}] + (0.001\ 3)(0.200)^2 \right\}$

$-158.8\ mV = \text{constant} + \dfrac{56.8\ mV}{2} \log \left\{ \dfrac{10}{11} [M^{2+}] + \dfrac{1}{11} (1.07 \times 10^{-3}) \right.$

$\left. + (0.001\ 3)(0.200)^2 \right\}$

Subtract the first equation from the second to get $4.5 \text{ mV} =$

$$\frac{56.8 \text{ mV}}{2} \log \left(\frac{\frac{10}{11} [M^{2+}] + \frac{1}{11} (1.07 \times 10^{-3}) + (0.001\ 3)(0.200)^3}{[M^{2+}] + (0.001\ 3)(0.200)^2} \right)$$

$$\Rightarrow [M^{2+}] = 1.4 \times 10^{-4} \text{ M}$$

43. (a) If the reaction of Pb^{2+} with $C_2O_4^{2-}$ goes to completion, we can say that
$[Pb(C_2O_4)_2^{2-}] = 0.100 \text{ mmol}/10.0 \text{ mL} = 0.010 \text{ M}$ and
$[C_2O_4^{2-}] = 1.80 \text{ mmol}/10.0 \text{ mL} = 0.180 \text{ M}$.

$$[Pb^{2+}] = \frac{[Pb(C_2O_4)_2^{2-}]}{\beta_2 [C_2O_4^{2-}]^2} = 8.9 \times 10^{-8} \text{ M}$$

(b) $[C_2O_4^{2-}] = \sqrt{\dfrac{[Pb(C_2O_4)_2^{2-}]}{\beta_2 [Pb^{2+}]}} = \sqrt{\dfrac{0.010\ 0}{10^{6.54}(1.0 \times 10^{-7})}} = 0.170 \text{ M}$

To produce a concentration of 0.170 M will require 1.90 mmol of $Na_2C_2O_4$, since 0.20 mmol reacts with Pb^{2+}.

44. (a) slope $= 29.58 \text{ mV} = \dfrac{E_2 - E_1}{\log \mathcal{A}_2 - \log \mathcal{A}_1} = \dfrac{(-25.90) - 2.06}{\log \mathcal{A}_2 - (-3.000)}$

$$\Rightarrow \mathcal{A}_2 = 1.13 \times 10^{-4}$$

(b) $\quad Ca^{2+} \qquad + \qquad A^{3-} \qquad \rightleftarrows \qquad CaA^{-}$

$\quad 5.00 \times 10^{-4} - x \qquad (0.998)[(5.00 \times 10^{-4}) - x] \qquad\qquad x$

But $\mathcal{A}_{Ca^{2+}} = 1.13 \times 10^{-4} = (5.00 \times 10^{-4} - x)\ (0.405)$

$\qquad\qquad\qquad\qquad\qquad\qquad\qquad\qquad \uparrow$

$\qquad\qquad\qquad\qquad\qquad\qquad \gamma$ from Table 8-1

$$\Rightarrow x = 2.2 \times 10^{-4} \text{ M}$$

$$K_f = \frac{[CaA^{-}]\gamma_{CaA^{-}}}{[Ca^{2+}]\gamma_{Ca^{2+}}[A^{3-}]\gamma_{A^{3-}}}$$

$$K_f = \frac{(2.20 \times 10^{-4})(0.79)}{(1.13 \times 10^{-4})[(0.998)(5 \times 10^{-4} - 2.20 \times 10^{-4})](0.115)}$$

$$K_f = 4.8 \times 10^4$$

45. Analyte is adsorbed on the surface of the gate and changes the electric potential of the gate. This, in turn, changes the current between the source and drain. The potential that must be applied by the external circuit to restore the current to its initial value is a measure of the change in gate potential. Following the Nernst equation, there is close to a 59 mV change in gate potential for each factor-of-10 change in activity of univalent analyte at 25°C. The key to ion-specific response is to have a chemical species on the gate that interacts selectively with one analyte.

CHAPTER 16
REDOX TITRATIONS

1. The titration reaction is the reaction between analyte and titrant. The cell reactions are the reactions between the species in the reaction solution (including titration reactants and products) and the species in the reference electrode.

2. (a) $Ce^{4+} + Fe^{2+} \rightarrow Ce^{3+} + Fe^{3+}$

   (b) $Fe^{3+} + e^- \rightleftarrows Fe^{2+}$ $\quad E° = 0.767$ V

   $Ce^{4+} + e^- \rightleftarrows Ce^{3+}$ $\quad E° = 1.70$ V

   (c) $E = \left\{ 0.767 - 0.059\ 16 \log \dfrac{[Fe^{2+}]}{[Fe^{3+}]} \right\} - \left\{ 0.241 \right\}$ $\qquad$ (A)

   $E = \left\{ 1.70 - 0.059\ 16 \log \dfrac{[Ce^{3+}]}{[Ce^{4+}]} \right\} - \left\{ 0.241 \right\}$ $\qquad$ (B)

   (d) <u>10.0 mL</u> : Use eq. (A) with $[Fe^{2+}]/[Fe^{3+}] = 40.0/10.0$, since
   $V_e = 50.0$ mL $\Rightarrow E = 0.490$ V

   <u>25.0 mL</u> : $[Fe^{2+}]/[Fe^{3+}] = 25.0/25.0 \Rightarrow E = 0.526$ V

   <u>49.0 mL</u> : $[Fe^{2+}]/[Fe^{3+}] = 1.0/49.0 \Rightarrow E = 0.626$ V

   <u>50.0 mL</u> : This is V_e, where $[Ce^{3+}] = [Fe^{3+}]$ and $[Ce^{4+}] = [Fe^{2+}]$.
   Eq. 16-11 gives $E_+ = 1.23$ V and $E = 0.99$ V.

   <u>51.0 mL</u> : Use eq. (B) with $[Ce^{3+}]/[Ce^{4+}] = 50.0/1.0 \Rightarrow E = 1.36$ V

   <u>60.0 mL</u> : $[Ce^{3+}]/[Ce^{4+}] = 50.0/10.0 \Rightarrow E = 1.42$ V

   <u>100.0 mL</u> : $[Ce^{3+}]/[Ce^{4+}] = 50.0/50.0 \Rightarrow E = 1.46$ V

3. (a) $Ce^{4+} + Cu^+ \rightarrow Ce^{3+} + Cu^{2+}$

   (b) $Ce^{4+} + e^- \rightleftarrows Ce^{3+}$ $\quad E° = 1.70$ V

   $Cu^{2+} + e^- \rightleftarrows Cu^+$ $\quad E° = 0.161$ V

   (c) $E = \left\{ 1.70 - 0.059\ 16 \log \dfrac{[Ce^{3+}]}{[Ce^{4+}]} \right\} - \left\{ 0.197 \right\}$ $\qquad$ (A)

   $E = \left\{ 0.161 - 0.059\ 16 \log \dfrac{[Cu^+]}{[Cu^{2+}]} \right\} - \left\{ (0.197) \right\}$ $\qquad$ (B)

   (d) <u>1.00 mL</u> : Use eq. (A) with $[Ce^{3+}]/[Ce^{4+}] = 1.00/24.0$, since
   $V_e = 25.0$ mL $\Rightarrow E = 1.58$ V

   <u>12.5 mL</u> : $[Ce^{3+}]/[Ce^{4+}] = 12.5/12.5 \Rightarrow E = 1.50$ V

   <u>24.5 mL</u> : $[Ce^{3+}]/[Ce^{4+}] = 24.5/0.5 \Rightarrow E = 1.40$ V

$$\underline{25.0 \text{ mL}}: \quad E_+ = 1.70 - 0.059\ 16 \log \frac{[Ce^{3+}]}{[Ce^{4+}]}$$

$$E_+ = 0.161 - 0.059\ 16 \log \frac{[Cu^+]}{[Cu^{2+}]}$$

$$2E_+ = 1.86_1 - 0.059\ 16 \log \frac{[Ce^{3+}][Cu^+]}{[Ce^{4+}][Cu^{2+}]}$$

At the equivalence point, $[Ce^{3+}] = [Cu^{2+}]$ and $[Ce^{4+}] = [Cu^+]$. Therefore the log term above is zero and $E_+ = 1.86_1/2 = 0.930$ V.

$$E = 0.930 - 0.197 = 0.733 \text{ V}$$

$\underline{25.5 \text{ mL}}:$ Use eq. (B) with $[Cu^+]/[Cu^{2+}] = 0.5/25.0 \Rightarrow E = 0.065$ V

$\underline{30.0 \text{ mL}}:$ $[Cu^+]/[Cu^{2+}] = 5.0/25.0 \Rightarrow E = 0.005$ V

$\underline{50.0 \text{ mL}}:$ $[Cu^+]/[Cu^{2+}] = 25.0/25.0 \Rightarrow E = -0.036$ V

4. (a) $Sn^{2+} + Tl^{3+} \rightarrow Sn^{4+} + Tl^+$

(b) $Sn^{4+} + 2e^- \rightleftarrows Sn^{2+} \qquad E° = 0.139$ V

$Tl^{3+} + 2e^- \rightleftarrows Tl^+ \qquad E° = 0.77$ V

(c) $\quad E = \left\{ 0.139 - \dfrac{0.059\ 16}{2} \log \dfrac{[Sn^{2+}]}{[Sn^{4+}]} \right\} - \left\{ 0.241 \right\}$ \hfill (A)

$\quad E = \left\{ 0.77 - \dfrac{0.059\ 16}{2} \log \dfrac{[Tl^+]}{[Tl^{3+}]} \right\} - \left\{ 0.241 \right\}$ \hfill (B)

(d) $\underline{1.00 \text{ mL}}:$ Use eq. (A) with $[Sn^{2+}]/[Sn^{4+}] = 4.00/1.00$, since $V_e = 5.00$ mL $\Rightarrow E = -0.120$ V

$\underline{2.50 \text{ mL}}:$ $[Sn^{2+}]/[Sn^{4+}] = 2.50/2.50 \Rightarrow E = -0.102$ V

$\underline{4.90 \text{ mL}}:$ $[Sn^{2+}]/[Sn^{4+}] = 0.10/4.90 \Rightarrow E = -0.052$ V

$\underline{5.00 \text{ mL}}: \quad E_+ = 0.139 - \dfrac{0.059\ 16}{2} \log \dfrac{[Sn^{2+}]}{[Sn^{4+}]}$

$$E_+ = 0.77 - \frac{0.059\ 16}{2} \log \frac{[Tl^+]}{[Tl^{3+}]}$$

$$2E_+ = 0.90_9 - \frac{0.059\ 16}{2} \log \frac{[Sn^{2+}][Tl^+]}{[Sn^{4+}][Tl^{3+}]}$$

At the equivalence point, $[Sn^{4+}] = [Tl^+]$ and $[Sn^{2+}] = [Tl^{3+}]$. Therefore the log term above is zero and $E_+ = 0.90_9/2 = 0.45_4$ V.

$$E = 0.45_4 - 0.241 = 0.21 \text{ V}$$

$\underline{5.10 \text{ mL}}:$ Use eq. (B) with $[Tl^+]/[Tl^{3+}] = 5.00/0.10 \Rightarrow E = 0.48$ V

$\underline{10.0 \text{ mL}}:$ Use eq. (B) with $[Tl^+]/[Tl^{3+}] = 5.00/5.00 \Rightarrow E = 0.53$ V

5. (a) $2Fe^{3+} + \text{ascorbic acid} + H_2O \rightarrow 2Fe^{2+} + \text{dehydroascorbic acid} + 2H^+$

(b) One cell reaction is based on $Fe^{3+}|Fe^{2+}$ and $Ag|AgCl$:

$$Fe^{3+} + Ag(s) + Cl^- \rightarrow Fe^{2+} + AgCl(s)$$

The other is based on ascorbic acid|dehydroascorbic acid and $Ag|AgCl$:

$$dehydro + 2H^+ + 2Ag(s) + 2Cl^- \rightarrow ascorbic\ acid + 2AgCl(s) + H_2O$$

6. (a) Titration reaction: $Cr^{2+} + Fe^{3+} \rightarrow Fe^{2+} + Cr^{3+}$

Titrant: $Fe^{3+} + e^- \rightleftarrows Fe^{2+}$ $E° = 0.767\ V$

Analyte: $Cr^{3+} + e^- \rightleftarrows Cr^{2+}$ $E° = -0.42\ V$

$\tau = 10^{(0.767 - E)/0.059\ 16}$ $\alpha = 10^{(-0.42 - E)/0.059\ 16}$

$$\phi = \frac{(1 + \tau)}{\tau\ (1 + \alpha)}$$ $V_e = 50.0\ mL$

	A	B	C	D	E	F	G
1	E°(T) =	E (vs S.H.E.)	Tau	Alpha	Phi	E (vs S.C.E.)	Volume (mL)
2	0.767	-0.538	1.15E+22	9.88E+01	0.01002	-0.779	0.501
3	E°(A) =	-0.420	1.16E+20	1.00E+00	0.50000	-0.661	25.000
4	-0.42	-0.302	1.17E+18	1.01E-02	0.98998	-0.543	49.499
5	Nernst =	-0.242	1.14E+17	9.80E-04	0.99902	-0.483	49.951
6	0.05916	0.174	1.06E+10	9.11E-11	1.00000	-0.067	50.000
7	Ve =	0.649	9.88E+01	8.52E-19	1.01013	0.408	50.506
8	50	0.708	9.94E+00	8.57E-20	1.10062	0.467	55.031
9		0.767	1.00E+00	8.63E-21	2.00000	0.526	100.000
10							
11	C2 = 10^(($A$2-B2)/$A$6)			E2 = (1+C2)/(C2*(1+D2))			
12	D2 = 10^(($A$4-B2)/$A$6)			F2 = B2-0.241		G2 = $A$8*E2	

(b) Titration reaction: $5Fe^{2+} + MnO_4^- + 8H^+ \rightarrow 5Fe^{3+} + Mn^{2+} + 4H_2O$

Titrant: $MnO_4^- + 8H^+ + 5e^- \rightarrow Mn^{2+} + 4H_2O$ $E° = 1.507\ V$

Analyte: $Fe^{3+} + e^- \rightleftarrows Fe^{2+}$ $E° = 0.68\ V$

$\tau = 10^{\{[5(1.507 - E)/0.059\ 16] - 8\ pH\}}$ $\alpha = 10^{(0.68 - E)/0.059\ 16}$

$$\phi = \frac{(1 + \tau)}{\tau\ (1 + \alpha)}$$ $V_e = 10.0\ mL$

	A	B	C	D	E	F	G
1	E°(T) =	E (vs S.H.E.)	Tau	Alpha	Phi	E (vs S.C.E.)	Volume (mL)
2	1.507	0.562	7.38E+71	9.88E+01	0.01002	0.321	0.100
3	E°(A) =	0.680	7.86E+61	1.00E+00	0.50000	0.439	5.000
4	0.68	0.798	8.36E+51	1.01E-02	0.98998	0.557	9.900
5	Nernst =	0.858	7.10E+46	9.80E-04	0.99902	0.617	9.990
6	0.05916	1.290	2.19E+10	4.89E-11	1.00000	1.049	10.000
7	Ve =	1.389	9.40E+01	1.04E-12	1.01064	1.148	10.106
8	10	1.400	1.10E+01	6.75E-13	1.09052	1.159	10.905
9	pH =	1.412	1.07E+00	4.23E-13	1.93525	1.171	19.352
10	1						
11	C2 = 10^(5*($A$2-B2)/$A$6-8*$A$10)			E2 = (1+C2)/(C2*(1+D2))			
12	D2 = 10^(($A$4-B2)/$A$6)			F2 = B2-0.241		G2 = $A$8*E2	

(c) Titration reaction: dehydro. $+ 2Fe^{2+} + 2H^+ \rightarrow$ ascorbic acid $+ 2Fe^{3+} + H_2O$

Titrant: dehydro. $+ 2H^+ + 2e^- \rightarrow$ ascorbic acid $+ H_2O$ $E° = 0.390$ V

Analyte: $Fe^{3+} + e^- \rightleftarrows Fe^{2+}$ $E° = 0.732$ V

$\tau = 10^{\{[2(0.390 - E)/0.059\,16] - 2\,pH\}}$ $\alpha = 10^{(0.732 - E)/0.059\,16}$

$\phi = \dfrac{\alpha\,(1 + \tau)}{(1 + \alpha)}$ $V_e = 30.0_6$ mL

	A	B	C	D	E	F	G
	E°(T) =	E (vs S.H.E.)	Tau	Alpha	Phi	E (vs S.C.E.)	Volume (mL)
1							
2	0.39	0.850	2.81E-18	1.01E-02	0.01002	0.609	0.301
3	E°(A) =	0.732	2.74E-14	1.00E+00	0.50000	0.491	15.030
4	0.732	0.614	2.67E-10	9.88E+01	0.98998	0.373	29.759
5	Nernst =	0.555	2.64E-08	9.81E+02	0.99898	0.314	30.029
6	0.05916	0.464	3.15E-05	3.39E+04	1.00000	0.223	30.060
7	Ve =	0.390	1.00E-02	6.04E+05	1.01000	0.149	30.361
8	30.06	0.360	1.03E-01	1.94E+06	1.10332	0.119	33.166
9	pH =	0.331	9.88E-01	6.00E+06	1.98762	0.090	59.748
10	1						
11	C2 = 10^(2*($A$2-B2)/$A$6-2*$A$10)				E2 = D2*(1+C2)/(1+D2)		
12	D2 = 10^(($A$4-B2)/$A$6)			F2 = B2-0.241		G2 = $A$8*E2	

(d) Titration reaction: $UO_2^{2+} + Sn^{2+} + 4H^+ \rightarrow U^{4+} + Sn^{4+} + 2H_2O$

Titrant: $Sn^{4+} + 2e^- \rightleftarrows Sn^{2+}$ $E° = 0.139$ V

Analyte: $UO_2^{2+} + 4H^+ + 2e^- \rightarrow U^{4+} + 2H_2O$ $E° = 0.273$ V

$\tau = 10^{[2(0.139 - E)/0.059\,16]}$ $\alpha = 10^{\{2(0.273 - E)/0.059\,16 - 4\,pH\}}$

$\phi = \dfrac{\alpha\,(1 + \tau)}{(1 + \alpha)}$ $V_e = 25.0$ mL

	A	B	C	D	E	F	G
	E°(T) =	E (vs S.H.E.)	Tau	Alpha	Phi	E (vs S.C.E.)	Volume (mL)
1							
2	0.139	0.332	2.99E-07	1.01E-02	0.01002	0.091	0.251
3	E°(A) =	0.301	3.34E-06	1.13E-01	0.10160	0.060	2.540
4	0.273	0.216	2.49E-03	8.45E+01	0.99077	-0.025	24.769
5	Nernst =	0.207	5.03E-03	1.70E+02	0.99916	-0.034	24.979
6	0.05916	0.206	5.43E-03	1.84E+02	1.00000	-0.035	25.000
7	Ve =	0.195	1.28E-02	4.33E+02	1.01046	-0.046	25.261
8	25	0.169	9.68E-02	3.28E+03	1.09645	-0.072	27.411
9	pH =	0.139	1.00E+00	3.39E+04	1.99994	-0.102	49.999
10	0						
11	C2 = 10^(2*($A$2-B2)/$A$6-4*$A$10)				E2 = D2*(1+C2)/(1+D2)		
12	D2 = 10^(2*($A$4-B2)/$A$6)			F2 = B2-0.241		G2 = $A$8*E2	

7.

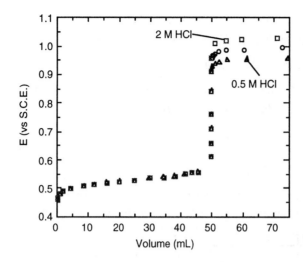

8. (a) Balanced reaction: $ClO_3^- + 6H^+ + 6e^- \rightleftharpoons Cl^- + 3H_2O$

(b) $ClO_3^- + 6H^+ + 5e^- \rightarrow \frac{1}{2}Cl_2 + 3H_2O$ $\quad\quad \Delta G_1^\circ = -5F(1.458)$

$\frac{1}{2}Cl_2 + e^- \rightarrow Cl^-$ $\quad\quad\quad\quad\quad\quad\quad\quad\quad \Delta G_2^\circ = -1F(1.360)$

$ClO_3^- + 6H^+ + 6e^- \rightleftharpoons Cl^- + 3H_2O$ $\quad \Delta G_3^\circ = \Delta G_1^\circ + \Delta G_2^\circ = -6FE_3^\circ$

$-6FE_3^\circ = -5F(1.458) - F(1.360) \Rightarrow E_3^\circ = 1.442$ V

(c) Titration reaction: $ClO_3^- + 6H^+ + 6Cr^{2+} \rightleftharpoons Cl^- + 6Cr^{3+} + 3H_2O$

Titrant: $ClO_3^- + 6H^+ + 6e^- \rightleftharpoons Cl^- + 3H_2O$ $\quad\quad E^\circ = 1.442$ V

Analyte: $Cr^{3+} + e^- \rightleftharpoons Cr^{2+}$ $\quad\quad\quad\quad\quad\quad\quad\quad E^\circ = -0.42$ V

$\tau = 10^{\{[6(1.442 - E)/0.059\,16] - 6\,pH\}}$ $\quad \alpha = 10^{(-0.42 - E)/0.059\,16}$

$\phi = \dfrac{(1 + \tau)}{\tau\,(1 + \alpha)}$

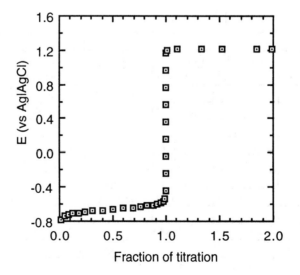

9.	A	B	C	D	E	F	G	H
1	E°(T) =	E(SHE)	Tau	Alpha1	Alpha2	Phi	E(SCE)	mL
2	1.24	0.134	6.0E+74	1.5E+00	3.2E+21	0.4039	-0.107	20.195
3	E°(A1) =	0.139	2.8E+74	1.0E+00	2.1E+21	0.5000	-0.102	25.000
4	0.139	0.454	1.4E+53	2.2E-11	4.8E+10	1.0000	0.213	50.000
5	E°(A2) =	0.752	9.9E+32	1.9E-21	4.1E+00	1.1976	0.511	59.882
6	0.77	0.770	6.0E+31	4.7E-22	1.0E+00	1.5000	0.529	75.000
7	Nernst =	1.080	6.6E+10	1.5E-32	3.3E-11	2.0000	0.839	100.000
8	0.05916	1.225	1.0E+01	1.9E-37	4.1E-16	2.1936	0.984	109.678
9	Ve =							
10	50							
11	pCl =		C2 = 10^(4*($A$2-B2)/$A$8-2*$A$12-6*$A$14)					
12	0		D2 = 10^(2*($A$4-B2)/$A$8)					
13	pH =		E2 = 10^(2*($A$6-B2)/$A$8)					
14	0		F2 = ((1+C2)/C2)*((1/(1+D2))+(($A$16/$A$18)/(1+E2)))					
15	Tl(total)=		G2 = B2-0.241					
16	0.01		H2 = $A$10*F2					
17	Sn(total)=							
18	0.01							

10. (a) The analyte with the more negative reduction potential reacts first:

First: $2T + 3A \rightarrow 2T^{3-} + 3A^{2+}$ $E° = 0.93 - (-0.13) = 1.06$ V

$K = 10^{nE°/0.059\,16} = 10^{6(1.06)/0.059\,16} = 10^{107}$

Second: $T + 3B \rightarrow T^{3-} + 3B^{+}$ $E° = 0.93 - 0.46 = 0.47$ V

$K = 10^{nE°/0.059\,16} = 10^{3(0.57)/0.059\,16} = 10^{28}$

(b) 1st: $(100.0 \text{ mL})(0.300 \text{ M}) = \frac{3}{2} V_{e1}(1.00 \text{ M}) \Rightarrow V_{e1} = 20.0 \text{ mL}$

2nd: $(100.0 \text{ mL})(0.060\,0 \text{ M}) = 3\,\Delta V(1.00 \text{ M}) \Rightarrow \Delta V = 2.00 \text{ mL}$

$\Rightarrow V_{e2} = 22.0 \text{ mL}$

(c) $\tau = \dfrac{[T^{3-}]}{[T]} \Rightarrow [T^{3-}] = \dfrac{\tau T_{total}}{1 + \tau}$

$\alpha = \dfrac{[A]}{[A^{2+}]} \Rightarrow [A^{2+}] = \dfrac{A_{total}}{1 + \alpha}; \quad \beta = \dfrac{[B]}{[B^{+}]} \Rightarrow [B^{+}] = \dfrac{B_{total}}{1 + \beta}$

From the stoichiometry of the two titration reactions, we can say

$[T^{3-}] = \dfrac{2}{3}[A^{2+}] + \dfrac{1}{3}[B^{+}]$

$\dfrac{\tau T_{total}}{1 + \tau} = \dfrac{2}{3}\dfrac{A_{total}}{1 + \alpha} + \dfrac{1}{3}\dfrac{B_{total}}{1 + \beta}$

From the stoichiometry of the first titration reaction, we define the fraction of titration as $\phi = (3/2)\, T_{total}/A_{total}$, because ϕ must be unity at V_{e1}. Rearranging the equation above gives

$\dfrac{T_{total}}{A_{total}} = \left(\dfrac{1 + \tau}{\tau}\right)\left(\dfrac{2}{3}\dfrac{1}{1 + \alpha} + \dfrac{1}{3}\dfrac{B_{total}/A_{total}}{1 + \beta}\right)$

$$\phi = \frac{3}{2}\frac{T_{total}}{A_{total}} = \left(\frac{1 + \tau}{\tau}\right)\left(\frac{1}{1 + \alpha} + \frac{1}{2}\frac{B_{total}/A_{total}}{1 + \beta}\right)$$

(d)	A	B	C	D	E	F	G
1	E°(T) =	E (S.H.E.)	Tau	Alpha	Beta	Phi	Volume
2	0.93	-0.16	1.9E+55	1.0E+01	3.0E+10	0.08824	1.765
3	E°(A) =	-0.13	5.7E+53	1.0E+00	9.4E+09	0.50000	10.000
4	-0.13	-0.10	1.7E+52	9.7E-02	2.9E+09	0.91176	18.235
5	E°(B) =	0.086	6.3E+42	5.0E-08	2.1E+06	1.00000	20.000
6	0.46	0.43	2.3E+25	1.2E-19	3.2E+00	1.02373	20.475
7	Nernst =	0.46	6.8E+23	1.1E-20	1.0E+00	1.05000	21.000
8	0.05916	0.49	2.1E+22	1.1E-21	3.1E-01	1.07627	21.525
9	Ve1 =	0.797	5.6E+06	4.6E-32	2.0E-06	1.10000	22.000
10	20	0.9	3.3E+01	1.5E-35	3.7E-08	1.13312	22.662
11	[B]/[A] =						
12	0.2		C2 = 10^(3*($A$2-B2)/$A$8)				
13			D2 = 10^(2*($A$4-B2)/$A$8)				
14			E2 = 10^(($A$6-B2)/$A$8)				
15			F2 = ((1+C2)/C2)*((1/(1+D2))+((0.5*$A$14)/(1+E2)))				
16			G2 = $A$10*F2				

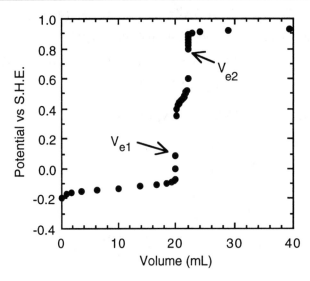

(e)   E at V_{e1} = 0.086 V; E at V_{e2} = 0.797 V

11.   Titrant: $MnO_4^- + 8H^+ + 5e^- \rightleftarrows Mn^{2+} + 4H_2O$ E° = 1.507 V

$\quad\quad \tau = 10^{\{[5(1.507 - E)/0.059\,16] - 8\,pH\}}$

First analyte: $UO_2^{2+} + 4H^+ + 2e^- \rightleftarrows U^{4+} + 2H_2O$ E° = 0.273 V

$\quad\quad \alpha_1 = 10^{\{2(0.273 - E)/0.059\,16 - 4\,pH\}}$

Second analyte: $Fe^{3+} + e^- \rightleftarrows Fe^{2+}$ E° = 0.767 V

$\quad\quad \alpha_2 = 10^{(0.767 - E)/0.059\,16}$

(a)   First reaction: $2MnO_4^- + 5U^{4+} + 2H_2O \rightarrow 2Mn^{2+} + 5UO_2^{2+} + 4H^+$

$\quad\quad$ Second reaction: $MnO_4^- + 8H^+ + 5Fe^{2+} \rightarrow Mn^{2+} + 5Fe^{3+} + 4H_2O$

(b) $[U^{4+}] = \dfrac{5}{2}\left(\dfrac{\text{mmol MnO}_4^-}{25.0 \text{ mL}}\right) = \dfrac{5}{2}\left(\dfrac{12.73 \text{ mL} \times 0.009\ 87 \text{ M}}{25.0 \text{ mL}}\right) = 0.012\ 5_6 \text{ M}$

$[Fe^{2+}] = 5\left(\dfrac{\text{mmol MnO}_4^-}{25.0 \text{ mL}}\right) = 5\left(\dfrac{(31.21\text{-}12.73) \text{ mL} \times 0.009\ 87 \text{ M}}{25.0 \text{ mL}}\right) = 0.036\ 4_8 \text{ M}$

(c) From the stoichiometry of the two reactions, we know that

$[Mn^{2+}] = \dfrac{2}{5}[UO_2^{2+}] + \dfrac{1}{5}[Fe^{3+}]$

$\dfrac{\tau Mn_{total}}{1 + \tau} = \dfrac{2}{5}\dfrac{U_{total}}{1 + \alpha_1} + \dfrac{1}{5}\dfrac{Fe_{total}}{1 + \alpha_2}$

$\phi \equiv \dfrac{5}{2}\dfrac{Mn_{total}}{U_{total}} = \left(\dfrac{1 + \tau}{\tau}\right)\left(\dfrac{1}{1 + \alpha_1} + \dfrac{1}{2}\dfrac{Fe_{total}/U_{total}}{1 + \alpha_2}\right)$

(d)	A	B	C	D	E	F	G
1	E°(T) =	E (vs S.H.E.)	Tau	Alpha1	Alpha2	Phi	Volume(mL)
2	1.507	0.273	########	1.00E+00	2.24E+08	0.50000	6.365
3	E°(A1) =	0.434	4.86E+90	3.61E-06	4.25E+05	1.00000	12.730
4	0.273	0.767	3.49E+62	1.99E-17	1.00E+00	1.72575	21.969
5	E°(A2) =	1.458	1.38E+04	8.69E-41	2.09E-12	2.45168	31.210
6	0.767						
7	Nernst =						
8	0.05916		C2 = 10^(5*($A$2-B2)/$A$8-8*$A$12)				
9	Ve1 =		D2 = 10^(2*($A$4-B2)/$A$8-4*$A$12)				
10	12.73		E2 = 10^(($A$6-B2)/$A$8)				
11	pH =		F2 = ((1+C2)/C2)*((1/(1+D2))+((0.5*$A$14)/(1+E2)))				
12	0		G2 = $A$10*F2				
13	[Fe]/[U] =						
14	2.903						

(e)

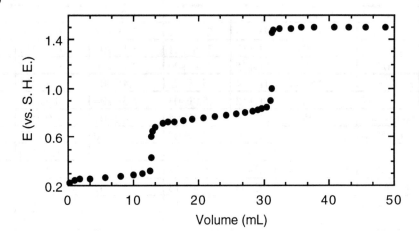

12. Diphenylamine sulfonic acid : colorless → red-violet

Diphenylbenzidine sulfonic acid : colorless → violet

tris (2,2'-bipyridine) iron : red → pale blue

Ferroin : red → pale blue

13. Standard potentials :   indigo tetrasulfonate   0.36 V

$Fe[CN]_6^{3-} \mid Fe[CN]_6^{4-}$ 0.356 V $Tl^{3+} \mid Tl^+$ 0.77 V

The end point potential will be between 0.356 and 0.77 V. Indigo tetrasulfonate changes color near 0.36 V. Therefore it will not be a useful indicator for this titration.

14. The reduction potentials are

$Sn^{4+} + 2e^- \rightleftarrows Sn^{2+}$ $E° = 0.139$ V

$Mn(EDTA)^- + e^- \rightleftarrows Mn(EDTA)^{2-}$ $E° = 0.825$ V

The end point will be between 0.139 and 0.825 V. Tris(2,2'-bipyridine) iron has too high a reduction potential (1.120 V) to be useful for this titration.

15. First reaction: $Ce^{4+} + Fe^{2+} \rightarrow Ce^{3+} + Fe^{3+}$

Second reaction $Ce^{4+} + Fe(phen)_3^{2+} \rightarrow Ce^{3+} + Fe(phen)_3^{3+}$

$V_{e1} = 200.0$ mL and $V_{e2} = 200.1$ mL

This problem is analogous to the titration of a mixture in Section 16-3, but the stoichiometry is 1:1:1 instead of 1:2:2. Equation 16-26 becomes $[Ce^{3+}] = [Fe^{3+}] + [Fe(phen)_3^{3+}]$ and Equation 16-27 becomes

$$\phi \equiv \frac{Ce_{total}}{Fe_{total}} = \left(\frac{1+\tau}{\tau}\right)\left(\frac{1}{1+\alpha_1} + \frac{Indictor_{total}/Fe_{total}}{1+\alpha_2}\right)$$

where α_1 applies to Fe^{2+} and α_2 applies to $Fe(phen)_3^{2+}$ (indicator). The spreadsheet and graph below show that ferroin is excellent for this titration.

	A	B	C	D	E	F	G
1	E°(T) =	E (S.H.E.)	Tau	Alpha1	Alpha2	Phi	Volume
2	1.7	0.631	1.2E+18	2.0E+02	5.3E+08	0.00500	1.000
3	E°(A) =	0.691	1.1E+17	1.9E+01	5.1E+07	0.04936	9.872
4	0.767	0.767	5.9E+15	1.0E+00	2.7E+06	0.50000	100.000
5	E°(B) =	0.843	3.1E+14	5.2E-02	1.4E+05	0.95064	190.128
6	1.147	0.903	3.0E+13	5.0E-03	1.3E+04	0.99500	199.000
7	Nernst =	1.055	8.0E+10	1.4E-05	3.6E+01	1.00000	200.000
8	0.05916	1.147	2.2E+09	3.8E-07	1.0E+00	1.00025	200.050
9	Ve1 =	1.326	2.1E+06	3.6E-10	9.4E-04	1.00050	200.100
10	200	1.487	4.0E+03	6.8E-13	1.8E-06	1.00075	200.150
11	[B]/[A] =	1.505	2.0E+03	3.4E-13	8.9E-07	1.00101	200.201
12	0.0005	1.561	2.2E+02	3.8E-14	1.0E-07	1.00497	200.995
13		1.623	2.0E+01	3.4E-15	9.0E-09	1.05046	210.093
14							
15	C2 = 10^(($A$2-B2)/$A$8)				E2 = 10^(($A$6-B2)/$A$8)		
16	D2 = 10^(($A$4-B2)/$A$8)				G2 = $A$10*F2		
17	F2 = ((1+C2)/C2)*(1/(1+D2)+$A$12/(1+E2))						

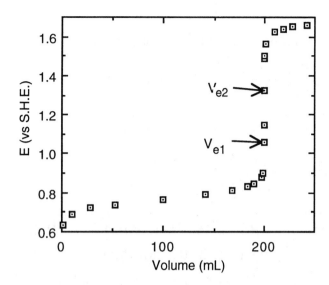

16. For the equilibrium between enzyme (E) and indicator (In) we can write

$$E(ox) + 2e^- \rightleftarrows E(red) \qquad\qquad E^{\circ\prime}(enzyme)$$

$$\underline{2[In(red) \rightleftarrows In(ox) + e^-)] \qquad\qquad\quad E^{\circ\prime} = -0.187}$$

$$E(ox) + 2In(red) \rightleftarrows E(red) + 2In(ox) \qquad E_3^{\circ\prime} = E^{\circ\prime}(enzyme) + 0.187$$

For the net reaction,

$$K = \frac{[E(red)][In(ox)]^2}{[E(ox)][In(red)]^2} = \frac{(1.80 \times 10^{-5})(3.9 \times 10^{-5})^2}{(4.2 \times 10^{-5})(5.5 \times 10^{-5})^2} = 0.21_5$$

$$E_3^{\circ\prime} = \frac{0.059\ 16}{2} \log K = -0.019\ 7\ V. \qquad E^{\circ\prime}(enzyme) = E_3^{\circ\prime} - 0.187 = -0.207\ V.$$

17. Preoxidation and prereduction refer to adjusting the oxidation state of analyte to a suitable value for a titration. The preoxidation or prereduction agent must be destroyed so it does not interfere with the titration by reacting with titrant.

18. $2S_2O_8^{2-} + 2H_2O \xrightarrow{\text{boiling}} 4SO_4^{2-} + O_2 + 4H^+$

$4Ag^{2+} + 2H_2O \xrightarrow{\text{boiling}} 4Ag^+ + O_2 + 4H^+$

$2H_2O_2 \xrightarrow{\text{boiling}} O_2 + 2H_2O$

19. A Jones reductor is a column packed with zinc granules coated with zinc amalgam. Prereduction is accomplished by passing analyte solution through the column.

20. Cr^{3+} and TiO^{2+} would interfere if they were reduced to Cr^{2+} and Ti^{3+}. In the Jones reductor, Zn is a strong enough reductant to react with Cr^{3+} and TiO^{2+}.

$$E^\circ = -0.764 \text{ for the } Zn^{2+} \mid Zn \text{ couple}$$

$E° = -0.42$ for the $Cr^{3+} | Cr^{2+}$ couple

$E° = 0.1$ for the $TiO^{2+} | Ti^{3+}$ couple

In the Walden reductor, Ag is not strong enough to reduce Cr^{3+} and TiO^{2+}:

$E° = 0.222$ for the $AgCl|Ag$ couple

21. (a) $MnO_4^- + 8H^+ + 5e^- \rightleftarrows Mn^{2+} + 4H_2O$

(b) $MnO_4^- + 4H^+ + 3e^- \rightleftarrows MnO_2(s) + 2H_2O$

(c) $MnO_4^- + e^- \rightleftarrows MnO_4^{2-}$

22. $3MnO_4^- + 5Mo^{3+} + 4H^+ \rightarrow 3Mn^{2+} + 5MoO_2^{2+} + 2H_2O$

$(16.43 - 0.04) = 16.39$ mL of 0.010 33 M $KMnO_4 = 0.169\,3$ mmol of MnO_4^-

which will react with $(5/3)(0.169\,3) = 0.282\,2$ mmol of Mo^{3+}.

$[Mo^{3+}] = 0.282\,2$ mmol/25.00 mL $= 0.011\,29$ M.

23. $2MnO_4^- + 5H_2C_2O_4 + 6H^+ \rightarrow 2Mn^{2+} + 10CO_2 + 8H_2O$

18.04 mL of 0.006 363 M $KMnO_4 = 0.114\,8$ mmol of MnO_4^- which reacts with

$(5/2)(0.114\,8) = 0.287\,0$ mmol of $H_2C_2O_4$ which came from $(2/3)(0.287\,0) =$

$0.191\,3$ mmol of La^{3+}. $[La^{3+}] = 0.191\,3$ mmol/50.00 mL $= 3.826$ mM.

24.

$$C_3H_8O_3 + 3H_2O \rightleftarrows 3HCO_2H + 8e^- + 8H^+$$

  glycerol   formic acid
  (average oxidation   (oxidation
  number of C = -2/3)   number of C = +2)

$$8Ce^{4+} + 8e^- \rightleftarrows 8Ce^{3+}$$

$$C_3H_8O_3 + 8Ce^{4+} + 3H_2O \rightleftarrows 3HCO_2H = 8Ce^{3+} + 8H^+$$

One mole of glycerol requires eight moles of Ce^{4+}.

   50.0 mL of 0.083 7 M Ce^{4+} = 4.185 mmol

   12.11 mL of 0.044 8 M Fe^{2+} = 0.543 mmol

   Ce^{4+} reacting with glycerol = 3.642 mmol

glycerol $= (1/8)\,(3.642) = 0.455_2$ mmol $= 41.9$ mg $\Rightarrow$ original solution $= 41.9$

wt% glycerol

25. 50.00 mL of 0.118 6 M Ce^{4+} = 5.930 mmol Ce^{4+}

31.13 mL of 0.042 89 M Fe^{2+} = $\underline{1.335\text{ mmol }Fe^{2+}}$

                                    4.595 mmol Ce^{4+} consumed by NO_2^-

Since two moles of Ce^{4+} react with one mole of NO_2^-, there must have been 1/2

$(4.595) = 2.298$ mmol of $NaNO_2 = 0.158\,5$ g in 25.0 mL. In 500.0 mL there

would be $\left(\dfrac{500.0}{25.0}\right)(0.158\,5) = 3.170$ g $= 78.67\%$ of the 4.030 g sample.

26.   Step 2 gives the total Cr content of the crystal, since each Cr^{x+} ion in any oxidation state is oxidized and reacts with $3Fe^{2+}$.

$$\text{Step 2: } \frac{(0.703 \text{ mL})(2.786 \text{ mM})}{0.156 \text{ 6 g of crystal}} = \frac{12.51 \text{ } \mu\text{mol Fe}^{2+}}{\text{g of crystal}}$$

$$\frac{\frac{1}{3}(12.51) \text{ } \mu\text{mol Cr}}{\text{g of crystal}} = \frac{4.169 \text{ } \mu\text{mol Cr}}{\text{g of crystal}}$$

Step 1 tells us how much Cr^{x+} is oxidized above the +3 state. Each Cr^{x+} reacts with (x-3) Fe^{2+}.

$$\text{Step 1: } \frac{(0.498 \text{ mL})(2.786 \text{ mM})}{0.437 \text{ 5 g of crystal}} = \frac{3.171 \text{ } \mu\text{mol Fe}^{2+}}{\text{g of crystal}}$$

Since one gram of crystal contains 4.169 μmol of Cr that reacts with 3.171 μm of Fe^{2+}, the average oxidation state of Cr is $3 + \frac{3.171}{4.169} = 3.761$.

Total Cr (from step 2) = 4.169 μmol Cr per gram = 217 μg per gram of crystal.

27. I^- reacts with I_2 to give I_3^-, which increases the solubility of I_2 and decreases its volatility.

28.   Reaction with standard As_4O_6. Reaction with standard $S_2O_3^{2-}$ prepared from anhydrous $Na_2S_2O_3$. Reaction of standard IO_3^- with acid plus iodide.

29.   Starch is not added until just before the end point in iodometry, so that it does not irreversibly bind to I_2 which is present during the whole titration.

30.   (a) One mole of As_4O_6 gives four moles of H_3AsO_3 which react with four moles of I_3^- (reaction 16-64). 25.00 mL of As_4O_6 reacts with

$$4\left(\frac{25.00}{100.00}\right)\left(\frac{0.366 \text{ 3 g}}{395.683 \text{ g/mol}}\right) = 9.257 \times 10^{-4} \text{ mol of I}_3^-$$

$[I_3^-]$ = 0.925 7 mmol/31.77 mL = 0.029 14 M

(b) It does not matter. Excess I_3^- is not present until after the end point. If the procedure were reversed and I_3^- were titrated with H_3AsO_3, then starch should not be added until just before the end point.

31.   (a)

$$\begin{array}{ll} I_2(aq) + 2e^- \rightleftharpoons 2I^- & E° = 0.620 \text{ V} \\ \underline{3I^- \rightleftharpoons I_3^- + 2e^-} & \underline{E° = -0.535 \text{ V}} \\ I_2(aq) + I^- \rightleftharpoons I_3^- & E° = 0.085 \text{ V} \end{array}$$

$$K = 10^{2(0.085)/0.059 \text{ 16}} = 7 \times 10^2$$

(b)

$$I_2(s) + 2e^- \rightleftarrows 2I^- \qquad\qquad\qquad E° = 0.535 \text{ V}$$
$$\underline{3I^- \rightleftarrows I_3^- + 2e^- \qquad\qquad\qquad E° = -0.535 \text{ V}}$$
$$I_2(s) + I^- \rightleftarrows I_3^- \qquad\qquad\qquad E° = 0.000 \text{ V}$$

$$K = 10^{2(-0.000)/0.059\ 16} = 1.0$$

(c)

$$I_2(s) + 2e^- \rightleftarrows 2I^- \qquad\qquad\qquad E° = 0.535 \text{ V}$$
$$\underline{2I^- \rightleftarrows I_2(aq) + 2e^- \qquad\qquad\qquad E° = -0.620 \text{ V}}$$
$$I_2(s) \rightleftarrows I_2(aq) \qquad\qquad\qquad E° = -0.085 \text{ V}$$

$$K = [I_2(aq)] = 10^{2(-0.085)/0.059\ 16} = 1.3 \times 10^{-3} \text{ M} = 0.34 \text{ g of } I_2/\text{L}$$

32. Each mole of NH_3 liberated in the Kjeldahl digestion reacts with one mole of H^+ in the standard H_2SO_4 solution. Six moles of H^+ left (3 moles of H_2SO_4) after reaction with NH_3 will react with 1 mole of iodate by Reaction 16-30 to release 3 moles of I_3^-. Two moles of thiosulfate react with one mole of I_3^- in Reaction 16-31. Therefore each mole of thiosulfate corresponds to $\frac{1}{2}$ mol of residual H_2SO_4.

$$\text{mol } NH_3 = 2 \text{ (initial mol } H_2SO_4 - \text{ final mol } H_2SO_4)$$

$$\text{mol } NH_3 = 2 \text{ (initial mol } H_2SO_4 - \frac{1}{2} \times \text{ mol thiosulfate)}$$

33. 25.00 mL of 0.020 00 M $KBrO_3$ = 0.500 0 mmol of BrO_3^- which generates 1.500 mmol of Br_2. One mole of excess Br_2 generates one mole of I_2 (from I^-) and one mole of I_2 consumes 2 moles of $S_2O_3^{2-}$. Since mmol of $S_2O_3^{2-}$ = (8.83)(0.051 13) = 0.451 5 mmol, I_2 = 0.225 7 mmol and Br_2 consumed by reaction with 8-hydroxyquinoline = 1.500 - 0.225 7 = 1.274 mmol. But one mole of 8-hydroxy-quinoline consumes 2 moles of Br_2, so 8-hydroxyquinoline = 0.637 1 mmol and Al^{3+} = 0.6371/3 = 0.212 4 mmol = 5.730 mg.

34. (a) $YBa_2Cu_3O_7$ contains 1 Cu^{3+} and 2 Cu^{2+}. $YBa_2Cu_3O_{6.5}$ contains no Cu^{3+} and 3 Cu^{2+}. The moles of Cu^{3+} in the formula $YBa_2Cu_3O_{7-z}$ are therefore 1-2z. The moles of superconductor in 1 g of superconductor are (1g)/[(666.246 - 15.999 4 z)g/mol]. The difference between experiments B and A is 5.68 - 4.55 = 1.13 mmol $S_2O_3^{2-}$/g superconductor. Since 1 mol of thiosulfate is equivalent to 1 mol of Cu^{3+}, there are 1.13 mmol Cu^{3+}/g superconductor.

$$\frac{\text{mol } Cu^{3+}}{\text{mol superconductor}} = 1 - 2z = \frac{1.13 \times 10^{-3} \text{ mol } Cu^{3+}}{\left(\dfrac{1 \text{ g superconductor}}{(666.246 - 15.999\ 4\ z) \text{ g/mol}}\right)}$$

Solving this equation gives z = 0.125. The formula is $YBa_2Cu_3O_{6.875}$.

(b) $\quad 1-2z = \dfrac{[5.68(\pm0.05) - 4.55(\pm0.10)] \times 10^{-3}}{\left(\dfrac{1}{666.246 - 15.999\ 4\ z}\right)}$

$\quad 1-2z = \dfrac{1.13\ (\pm0.112) \times 10^{-3}}{\left(\dfrac{1}{666.246 - 15.999\ 4\ z}\right)}$

$1-2z = 0.752\ 86\ (\pm0.074\ 49) - 0.018\ 079\ (\pm0.001\ 789)\ z$

$0.247\ 124\ (\pm0.074\ 488) = 1.981\ 92\ (\pm0.001\ 79)\ z$

$z = 0.125 \pm0.038.$ $\qquad$ The formula is $YBa_2Cu_3O_{6.875 \pm0.038}$

35. Denote the average oxidation number of Bi as 3+b and the average oxidation number of Cu as 2+c.

$$Bi^{3+b}_2\ Sr^{2+}_2\ Ca^{2+}\ Cu^{2+c}_2\ O_x$$

Positive charge $= 6 + 2b + 4 + 2 + 4 + 2c = 16 + 2b + 2c$

The charge must be balanced by $O^{2-} \Rightarrow x = 8 + b + c$

The formula weight of the superconductor is $760.37 + 15.999\ 4(8+b+c)$.

One gram contains $1/[760.37 + 15.999\ 4(8+b+c)]$ moles.

(a) Experiment A: Initial $Cu^+ = 0.200\ 0$ mmol; final $Cu^+ = 0.108\ 5$ mmol.

$\quad$ Therefore 102.3 mg of superconductor consumed 0.091 5 mmol Cu^+.

$\quad 2\times$mmol Bi^{5+} + mmol Cu^{3+} in 102.3 mg superconductor $= 0.091\ 5$.

$\quad$ Experiment B: Initial $Fe^{2+} = 0.100\ 0$ mmol; final $Fe^{2+} = 0.057\ 7$ mmol.

$\quad$ Therefore 94.6 mg of superconductor consumed 0.042 3 mmol Fe^{2+}.

$\quad 2\times$mmol Bi^{5+} in 94.6 mg superconductor $= 0.042\ 3$.

$\quad$ Normalizing to 1 gram of superconductor gives

$\qquad$ Expt A: $2($mmol $Bi^{5+}) +$ mmol Cu^{3+} in 1 g superconductor $= 0.894\ 43$

$\qquad$ Expt B: $2($mmol $Bi^{5+})$ in 1 g superconductor $= 0.447\ 15$

$\quad$ It is easier not to get lost in the arithmetic if we suppose that the oxidized bismuth is Bi^{4+} and equate one mole of Bi^{5+} to two moles of Bi^{4+}. Therefore we can rewrite the two equations above as

$\qquad$ mmol Bi^{4+} + mmol Cu^{3+} in 1 g superconductor $= 0.894\ 43$ $\qquad$ (1)

$\qquad$ mmol Bi^{4+} in 1 g superconductor $= 0.447\ 15$ $\qquad$ (2)

$\quad$ Subtracting (2) from (1) gives

$\qquad$ mmol Cu^{3+} in 1 g superconductor $= 0.447\ 28$ $\qquad$ (3)

$\quad$ Equations (2) and (3) tell us that the stoichiometric relationship in the formula of the superconductor is $b/c = 0.447\ 15/0.447\ 28 = 0.999\ 7$.

$\quad$ Since 1 g of superconductor contains 0.447 28 mmol Cu^{3+}, we can say

$$\frac{\text{mol } Cu^{3+}}{\text{mol solid}} = 2c$$

$$\frac{\text{mol } Cu^{3+}/\text{mol solid}}{\text{gram solid}/\text{mol solid}} = \frac{2c}{760.37+15.999\ 4(8+b+c)}$$

$$\frac{\text{mol } Cu^{3+}}{\text{gram solid}} = \frac{2c}{760.37+15.999\ 4(8+b+c)} = 4.472\ 8 \times 10^{-4} \qquad (4)$$

Substituting $b = 0.999\ 7c$ in the denominator of (4) allows us to solve for c:

$$\frac{2c}{760.37 + 15.999\ 4(8+1.999\ 7c)} = 4.472\ 8 \times 10^{-4} \Rightarrow c = 0.2001_1$$

$$\Rightarrow b = 0.999\ 7c = 0.2000_0$$

The average oxidation numbers are $Bi^{3.2000+}$ and $Cu^{2.2001+}$ and the formula of the compound is $Bi_2Sr_2CaCu_2O_{8.4001}$, since the oxygen stoichiometry derived at the beginning of the solution is $x = 8 + b + c$.

(b) Propagation of error:

Expt A: 102.3 (±0.2) mg compound consumed 0.091 5 (±0.000 7) mmol Cu^+

Expt B: 94.6 (±0.2) mg compound consumed 0.042 3 (±0.000 7) mmol Fe^{2+}

Normalizing to 1 gram of superconductor gives

Expt A: mmol Bi^{4+} + mmol Cu^{3+} in 1 g of superconductor

$$= \frac{0.091\ 5\ (\pm0.000\ 7)}{0.102\ 3\ (\pm0.000\ 2)} = 0.894\ 43\ (\pm0.007\ 06)\ \frac{\text{mmol}}{\text{gram}}$$

Expt B: mmol Bi^{4+} in 1 g of superconductor

$$= \frac{0.042\ 3\ (\pm0.000\ 7)}{0.094\ 6\ (\pm0.000\ 2)} = 0.447\ 15\ (\pm0.007\ 46)\ \frac{\text{mmol}}{\text{gram}}$$

$$\frac{\text{mmol } Cu^{3+}}{\text{g superconductor}} = 0.894\ 43\ (\pm0.007\ 06) - 0.447\ 15\ (\pm0.007\ 46)$$

$$= 0.447\ 28\ (\pm0.010\ 27)$$

$$\frac{b}{c} = \frac{0.447\ 15\ (\pm0.007\ 46)}{0.447\ 28\ (\pm0.010\ 27)} = 0.999\ 7\ (\pm0.028\ 4)$$

$$\frac{2c}{760.37 + 15.999\ 4(8+[1.999\ 7(\pm0.028\ 4)]c)} = 4.472\ 8\ (\pm0.102\ 7) \times 10^{-4}$$

$$[447\ 1.47\ (\pm102.7)]\ c = 888.365 + [31.999\ 4\ (\pm0.445)]\ c$$

$$\Rightarrow c = 0.200\ 1\ (\pm0.004\ 6)$$

The relative uncertainty in b given above as 0.007 46/0.447 15 is smaller than the relative uncertainty in c, which is 0.010 27/0.447 28.

$$\text{Uncertainty in b} = \frac{0.007\ 46/0.447\ 15}{0.010\ 27/0.447\ 28}\ (\text{uncertainty in c})$$

$$= \frac{0.007\ 46/0.447\ 15}{0.010\ 27/0.447\ 28}\ (\pm0.004\ 6) = \pm0.003\ 3$$

$$\Rightarrow b = 0.200\ 0\ (\pm0.003\ 3)$$

The average oxidation numbers are $Bi^{+3.2000(\pm0.0033)}$ and $Cu^{+3.2001(\pm0.0046)}$ and the formula of the compound is $Bi_2Sr_2CaCu_2O_{8.4001(\pm0.0057)}$.

1. A galvanic cell produces electricity by using a spontaneous chemical reaction. An electrolysis cell uses electricity to cause an otherwise nonspontaneous reaction to occur.

2. Internal cell resistance decreases the output of a galvanic cell and increases the magnitude of the voltage needed to operate an electrolysis cell.

3. When the rate of an electrochemical reaction is too fast for reactants to diffuse to the electrode and products to diffuse away from it, the concentrations at the electrode surface are not the same as those in bulk solution. This changes the concentrations (activities) in the Nernst equation and changes the cell voltage.

4. Overpotential helps to overcome the activation energy for an electrode reaction. Greater current density means a greater rate of reaction at the electrode. To make a reaction occur faster, more energy is needed to bring electrons closer to the top of the barrier. Therefore more potential (the overpotential) needs to be applied to increase the rate (the current density) of reaction.

5. $(0.100 \text{ mol})(964\,85 \text{ C/mol}) = 9.648 \times 10^3 \text{ C}$

$(9.648 \times 10^3 \text{ C})/(1.00 \text{ C/s}) = 9.648 \times 10^3 \text{ s} = 2.68 \text{ h}$

6. $E° = -\Delta G°/2F = -237.19 \times 10^3/[(2)(964\,85)] = -1.23 \text{ V}$

7. (a) $E_{eq} = E°(\text{cathode}) - E°(\text{anode}) - 0.059\,16 \log \dfrac{P_{H_2}^{1/2} [OH^-]}{[Br^-]}$

$E_{eq} = -0.828 - 1.078 - 0.059\,16 \log \dfrac{(1.0)^{1/2} (0.10)}{(0.10)} = -1.906 \text{ V}$

(b) Ohmic potential $= I{\cdot}R = (0.100 \text{ A})(2.0 \text{ }\Omega) = 0.20 \text{ V}$

(c) Applied voltage $= E_{eq} - I{\cdot}R - \text{Overpotentials}$

$= -1.906 - 0.20 - (0.20 + 0.40) = -2.71 \text{ V}$

(d) $E_{eq} = -1.906 - 0.059\,16 \log \dfrac{(1.0)^{1/2} (1.0)}{(0.010)} = -2.024 \text{ V}$

Applied voltage $= -2.024 - 0.20 - (0.20 + 0.40) = -2.82 \text{ V}$

8. (a) cathode: $Ce^{4+} + e^- \rightleftarrows Ce^{3+}$ $E° = 1.70 \text{ V}$

anode: $\underline{Fe^{3+} + e^- \rightleftarrows Fe^{2+}}$ $\underline{E° = 0.771 \text{ V}}$

net: $Ce^{4+} + Fe^{2+} \rightleftarrows Fe^{3+} + Ce^{3+}$ $E° = 1.70 - 0.771 = 0.93 \text{ V}$

$$E_{eq} = 0.93 - 0.059\ 16 \log \frac{(0.10)(0.050)}{(0.10)(0.10)} = 0.95\ V$$

$$E = E_{eq} - IR = 0.95\ V - (0.030\ 0\ A)\ (3.50\ \Omega) = 0.84\ V$$

(b) $E_{applied} = -E_{eq} - IR = -0.95\ V - (0.030\ 0\ A)\ (3.50\ \Omega) = -1.06\ V$

9. $$E_{eq} = 0.93 - 0.059\ 16 \log \frac{[Fe^{3+}]_s[Ce^{3+}]_s}{[Ce^{4+}]_s[Fe^{2+}]_s} = 0.88\ V$$

$$E = E_{eq} - IR = 0.88 - (0.100\ A)(3.50\ \Omega) = 0.53\ V$$

10.   (a)   For every mole of Hg produced, one electron flows.

1.00 mL Hg = 13.53 g Hg = 0.067 45 mol Hg = 0.067 45 mol e$^-$.

(0.067 45 mol) (96 485 C/mol) = 6 508 C.

Work = $q \cdot E$ = (6 508 C) (1.02 V) = 6.64×10^3 J.

(b)   The power is 0.209 J/min = 0.003 48 J/s.

$$P = I^2R \Rightarrow I = \sqrt{P/r} = \sqrt{(0.003\ 48\ W)/(100\ \Omega)} = 5.902\ mA.$$

In 1h the total charge flowing through the circuit is

$(5.902 \times 10^{-3}\ C/s) \cdot (3\ 600\ s) = 21.25\ C/(96\ 485\ C/mol)$

$= 2.202 \times 10^{-4}$ mol of e$^-$/h $= 1.01 \times 10^{-4}$ mol of Cd/h = 0.012 4 g Cd/h.

11.   Hydroxide generated at the cathode and Cl$^-$ in the anode compartment cannot cross the membrane. Na$^+$ from the sea water crosses from the anode to the cathode to preserve charge balance. Therefore NaOH can be formed free from Cl$^-$ contamination.

12.   In **controlled-potential** electrolysis, the potential of the working electrode is not allowed to vary. With two electrodes, the potential of the working electrode becomes more extreme as the concentration of reactant changes. Eventually the electrode potential reaches a range where other reactions can occur.

13. V_2.

14.   Cathodic depolarizer.

15.   The activity of a partial layer of Ag atoms on the Au surface is less than the activity of bulk Ag(s) (whose activity is unity). Therefore the thermodynamic driving force for depositing the first partial layer of Ag is higher than that for deposition onto bulk Ag. We can calculate the activity of Ag(s) that is deposited at +0.420 V with respect to the potential for bulk deposition:

Ag(s,partial layer) $\rightleftarrows$ Ag(s, bulk)

$$E = -0.420 \text{ V} = -0.059 \, 16 \log \frac{\mathcal{A}_{bulk}}{\mathcal{A}_{partial}}$$

$$-0.420 = -0.591 \, 6 \log \frac{1}{\mathcal{A}_{partial}} \Rightarrow \mathcal{A}_{partial} = 8 \times 10^{-8}$$

16. $Pb(lactate)_2 + 2H_2O \rightarrow PbO_2(s) + 2 \text{ lactate}^- + 4H^+ + 2e^-$

   Pb^{2+} Pb^{4+}

The lead is oxidized to PbO_2 at the anode.

The mass of lead lactate (FW 385.3) giving 0.111 1 g of PbO_2

(FW = 239.2) is (385.3/239.2)(0.111 1 g) = 0.179 0 g.

$\% \text{ Pb} = \dfrac{0.179 \, 0}{0.326 \, 8} \times 100 = 54.77\%$

17.   Cathode : $Sn^{2+} + 2e^- \rightleftarrows Sn(s) \quad E° = -0.141 \text{ V}$

$E(\text{cathode, vs S.H.E.}) = -0.141 - \dfrac{0.059 \, 16}{2} \log \dfrac{1}{1 \times 10^{-8}} = -0.378 \text{ V}$

$E(\text{cathode, vs S.C.E.}) = -0.378 - 0.241 = -0.619 \text{ V}$

The voltage will be more negative if concentration polarization occurs.

Concentration polarization means that $[Sn^{2+}]_s < 1.0 \times 10^{-8}$ M.

18.   (a)   Since Mn is oxidized, it is the anode.

   (b) $\dfrac{(2.60 \text{ C/s})(18.0 \times 60 \text{ s})}{96 \, 485 \text{ C/mol}} = 0.029 \, 10$ mol of e^- = 0.009 70 mol of M (since

one mole of M gives $3e^-$).   0.504 g/0.009 70 mol = 52.0 g/mol

   (c)   In the electrolysis 0.02910/2 = 0.014 55 mol of Mn^{2+} were produced.

   $[Mn^{2+}] = 0.025 \, 0 + 0.014 \, 55 = 0.039 \, 6$ M.

19.   anode: $H_2O \rightleftarrows 1/2 \, O_2 + 2H^+ + 2e^- \quad E°(\text{for reduction}) = -1.229 \text{ V}$

cathode: $Zn(OH^-)_4^{2-} + 2e^- \rightleftarrows Zn(s) + 4OH^- \qquad E° = -1.199 \text{ V}$

$\overline{Zn(OH^-)_4^{2-} + H_2O \rightleftarrows 1/2O_2 + 2H^+ + Zn(s) + 4OH^- \quad E°(\text{cathode}) - E°(\text{anode}) =}$

$-1.199 - 1.229 = -2.428 \text{ V}$

Note that we cannot combine the H^+ and OH^- to cancel H_2O, since that would

require a different value of E° for the reaction.

$E_{eq} = -2.428 - \dfrac{0.059 \, 16}{2} \log \dfrac{P_{O_2}{}^{1/2}[H^+]^2[OH^-]^4}{[Zn(OH)_4{}^{2-}]}$

$= -2.428 - \dfrac{0.059 \, 16}{2} \log \dfrac{(0.20)^{1/2}(10^{-13})^2(10^{-1})^4}{(0.010)} = -1.589 \text{ V}$

An alternative way to see that this voltage is correct is to write

$E_{eq} = E(\text{cathode}) - E(\text{anode}) =$

$$\left\{-1.199 - \frac{0.059\ 16}{2} \log \frac{[OH^-]^4}{[Zn(OH)_4{}^{2-}]}\right\} - \left\{1.229 - \frac{0.059\ 16}{2} \log \frac{1}{P_{O_2}{}^{1/2}[H^+]^2}\right\}$$

$$\underbrace{\qquad\qquad\qquad\qquad\qquad\qquad\qquad\qquad\qquad\qquad}$$
(anode reaction written as a reduction)

The ohmic potential is $IR = (0.20\ A)(0.35\ \Omega) = 0.070\ V$. The overpotential for O_2 evolution at a Ni surface with a current density of $100\ A\ m^{-2}$ is $0.519\ V$ in Table 17-1. $E_{applied} = E_{eq} - IR$ - overpotential $= -1.589 - 0.070 - 0.519 = -2.178\ V$

20. When 99.99% of Cd(II) is reduced, the formal concentration will be $1.0 \times 10^{-5}\ M$, and the predominant form is $Cd(NH_3)_4{}^{2+}$.

$$\beta_4 = \frac{[Cd(NH_3)_4{}^{2+}]}{[Cd^{2+}][NH_3]^4} = \frac{(1.0 \times 10^{-5})}{[Cd^{2+}](1.0)^4} \Rightarrow [Cd^{2+}] = 2.8 \times 10^{-12}\ M$$

$$Cd^{2+} + 2e^- \rightleftarrows Cd(s) \qquad\qquad E^\circ = -0.402$$

$$E(cathode) = -0.402 - \frac{0.059\ 16}{2} \log \frac{1}{[Cd^{2+}]} = -0.744\ V$$

21. Relevant information :

$$CuY^{2-} + 2e^- \rightleftarrows Cu(s) + Y^{4-} \qquad\qquad E^\circ = -0.216\ V$$

$$Co^{2+} + 2e^- \rightleftarrows Co(s) \qquad\qquad E^\circ = -0.282\ V$$

$$CoY^{2-} \quad K_f = 2.0 \times 10^{16}$$

$$\alpha_{Y^{4-}} = 3.8 \times 10^{-9} \text{ at pH } 4$$

When 99% of CuY^{2-} is reduced, $[CuY^{2-}] = 1.0 \times 10^{-8}\ M$.

$$E(cathode) = -0.216 - \frac{0.059\ 16}{2} \log \frac{[Y^{4-}]}{[CuY^{2-}]}$$

But $[Y^{4-}] = \alpha_{Y^{4-}}[EDTA] = (3.8 \times 10^{-9})(0.010\ M) = 3.8 \times 10^{-11}\ M \Rightarrow$

$$E(cathode) = -0.144\ V$$

Will this cathode potential reduce Co^{2+}?

$$\alpha_{Y^{4-}}\ K_f \text{ (for } CoY^{2-}) = \frac{[CoY^{2-}]}{[Co^{2+}][EDTA]} \Rightarrow [Co^{2+}] = 1.3 \times 10^{-8}\ M$$

$$E(cathode, Co^{2+}) = -0.282 - \frac{0.059\ 16}{2} \log \frac{1}{1.3 \times 10^{-8}} = -0.515\ V$$

The cobalt will not be reduced and the separation is feasible.

22. When excess Br_2 appears in the solution, current flows in the detector circuit by virtue of the reactions

$$\text{anode :} \qquad 2Br^- \rightarrow Br_2 + 2e^-$$

$$\text{cathode :} \qquad Br_2 + 2e^- \rightarrow 2Br^-$$

23. A mediator shuttles electrons between the analyte and electrode, without net consumption of the mediator.

24. (a) $0.005 \text{ C/s} \times 0.1 \text{ s} = 0.000\ 5 \text{ C}$

$$\frac{0.000\ 5 \text{ C}}{96\ 485 \text{ C/mol}} = 5.2 \times 10^{-9} \text{ moles of e}^-$$

(b) A 0.01 M solution of a 2 electron reductant delivers 0.02 moles of electrons/liter.

$$\frac{5.2 \times 10^{-9} \text{ moles}}{0.02 \text{ moles/liter}} = 2.6 \times 10^{-7} \text{ L} = 0.000\ 26 \text{ mL}$$

25. (a) $\text{mol e}^- = \dfrac{I \cdot t}{F} = \dfrac{(5.32 \times 10^{-3} \text{ C/s})(964 \text{ s})}{96\ 485 \text{ C/mol}} = 5.32 \times 10^{-5} \text{ mol}$

(b) One mol e$^-$ reacts with 1/2 mol Br_2, which reacts with 1/2 mol cyclohexene $\Rightarrow 2.66 \times 10^{-5}$ mol cyclohexene.

(c) $2.66 \times 10^{-5} \text{ mol}/5.00 \times 10^{-3} \text{ L} = 5.32 \times 10^{-3} \text{ M}$

26. (a) $n = \dfrac{PV}{RT} = \dfrac{(0.983 \text{ atm})(0.049\ 22 \text{ L})}{(0.082\ 06 \text{ L atm K}^{-1} \text{ mol}^{-1})(303 \text{ K})}$

$= 1.946$ mmol of H_2

(b) For every mole of H_2 produced, 2 moles of e$^-$ are consumed and one mole of Cu is oxidized. Therefore 1.946 mmol of Cu^{2+} are produced and the concentration of EDTA is 1.946 mmol/47.36 mL $= 0.041\ 09$ M.

(c) 1.946 mmol of H_2 comes from 2(1.946) $= 3.892$ mmol of e$^-$ $=$
$(3.892 \times 10^{-3})(96\ 485) = 3.755 \times 10^2$ C.
Time $= 3.755 \times 10^2$ C/(0.021 96 C/s) $= 1.710 \times 10^4$ s $= 4.75$ h.

27. Trichloroacetate is reduced at -0.90 V, consuming 224 C/(96 485 C/mol) $= 2.322$ mmol of e$^-$. This means that (1/2)(2.322) $= 1.161$ mmol of Cl_3CCO_2H (FW 163.386) $= 0.189\ 7$ g of Cl_3CCO_2H were present. The total quantity of Cl_2HCCO_2H (FW 128.943) is (1/2)[758 C/(96 485 C/mol)] $= 3.928$ mmol, of which 1.161 mmol came from reduction of Cl_3CCO_2H. Cl_2HCCO_2H in original sample $= 3.928 - 1.161 = 2.767$ mmol $= 0.356\ 8$ g.

$$\text{wt \% trichloroacetic acid} = \frac{0.189\ 7}{0.721} \times 100 = 26.3\%$$

$$\text{wt \% dichloroacetic acid} = \frac{0.356\ 8}{0.721} \times 100 = 49.5\%$$

28. $2I^- \rightarrow I_2 + 2e^- \Rightarrow$ one mole of I_2 is created when two moles of electrons flow. (812 s)(52.6 $\times 10^{-3}$ C/s)/(96 485 C/mol) $= 0.442\ 7$ mmol of e$^-$ $= 0.221\ 3$ mmol of I_2. Therefore there must have been 0.221 3 mmol of H_2S (FW 34.08) $= 7.542$ mg of H_2S/50.00 mL $= 7.542 \times 10^3$ μg of H_2S/50.00 mL $= 1.51 \times 10^2$ μg/mL

29. (a) Electron flow $= 4(25.9 \text{ nmol/s})(96\,485 \text{ C/mol}) = 1.00 \times 10^{-2}$ C/s

current density $= \dfrac{1.00 \times 10^{-2} \text{ A}}{1.00 \times 10^{-4} \text{ m}^2} = 1.00 \times 10^2$ A/m^2

$\Rightarrow$ overpotential $= 0.85$ V

(b) $E(\text{cathode}) = 0.100 - 0.059\,16 \log \dfrac{[Ti^{3+}]_s}{[TiO^{2+}]_s[H^+]^2} = -0.036$ V

(c) $O_2 + 4H^+ + 4e^- \rightleftharpoons H_2O \qquad E° = 1.229$ V

$E(\text{anode}) = 1.229 - \dfrac{0.059\,16}{4} \log \dfrac{1}{P_{O_2}[H^+]^4} = 1.160$ V

(d) $E_{\text{applied}} = E(\text{cathode}) - E(\text{anode}) - I \cdot R - \text{Overpotential}$

$= -0.036 - 1.160 - (1.00 \times 10^{-2} \text{ A})(52.4 \ \Omega) - 0.85 = -2.57$ V

30. $F = \dfrac{\text{coulombs}}{\text{mol}} = \dfrac{I \cdot t}{\text{mol}}$

$= \dfrac{[0.203\,639\,0(\pm 0.000\,000\,4) \text{ A}][18\,000.075\,(\pm 0.010)\text{ s}]}{[4.097\,900\,(\pm 0.000\,003)\text{ g}]/[107.868\,2\,(\pm 0.000\,2)\text{g/mol}]}$

$= \dfrac{[0.203\,639\,0(\pm 1.96 \times 10^{-4}\,\%)][18\,000.075\,(\pm 5.56 \times 10^{-5}\,\%)]}{[4.097\,900\,(\pm 7.32 \times 10^{-5}\,\%)]/[107.868\,2\,(\pm 1.85 \times 10^{-4}\,\%)]}$

$= 9.648\,667 \times 10^4 \ (\pm 2.85 \times 10^{-4}\,\%) = 964\,86.6_7 \pm 0.2_8$ C/mol

VOLTAMMETRY

1. Since Cu(0) dissolves in Hg, but not in Pt, the reaction product is different [Cu(Hg) instead of Cu(s)] and the reduction potential should not be the same. In Reaction 18-5 the reactant and product are the same, regardless of the electrode. The potential should be independent of the electrode.

2.

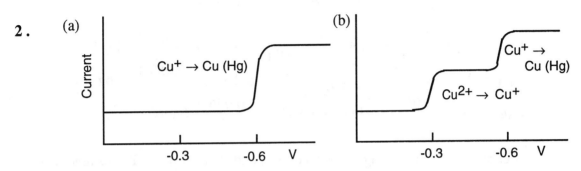

(c) The potential for the reaction Cu(I) → Cu(Hg) will change if Pt is used, since the product obviously cannot be copper amalgam.

3. To minimize convection (mechanical transport), the solution is *not* stirred and vibrations are reduced by setting the apparatus on a heavy base. Electrostatic attraction (or repulsion) of analyte ions by the electrode is reduced by a high concentration of supporting electrolyte.

4. See Figure 18-10.

5. (a) Condenser current arises from charging or discharging of the electric double layer at the electrode-solution interface. Faradaic current arises from oxidation or reduction reactions.

(b) Measurements in differential pulse polarography are made after the condenser current has decayed to near zero. The desired faradaic current is measured, nearly free of the background condenser current.

6. $\overline{I}_d = 6.07 \times 10^4 \, n \, C \, D^{1/2} \, m^{2/3} \, t^{1/6}$

$8.45 \, \mu A = (6.07 \times 10^4)(2)(0.95 \, \text{mmol/L})[D(m^2/s)]^{1/2}(2.28)$

$$\underset{\text{capillary constant}}{\uparrow}$$

$\Rightarrow D = 1.03 \times 10^{-9} \, m^2/s$

7. (a) moles of e^-/min $= \dfrac{(14 \times 10^{-6} \, \text{C/s})(60 \, \text{s/min})}{96\,485 \, \text{C/mol}} = 8.706 \times 10^{-9} \, \text{mol/min}$

moles of Cd^{2+}/min $= \frac{1}{2}$ moles of e^-/min $= 4.353 \times 10^{-9} \, \text{mol/min}$

moles of Cd^{2+} in 25 mL of 0.50 mM solution $= 1.25 \times 10^{-5}$

fraction of Cd^{2+} reduced/min $= \dfrac{4.353 \times 10^{-9}}{1.25 \times 10^{-5}} = 0.000\,34_8$ min^{-1}

(b) There appear to be 51 drops (oscillations) between -0.6 and -1.2 V in Fig. 18-2. (51 drops)(4 s/drop) $= 204$ s $= 3.4$ min.

(c) % Cd^{2+} decrease $= (0.034\,8\%/\text{min})(3.4 \text{ min}) = 0.118\%$

8. $E_{3/4} = E_{1/2} + \dfrac{0.059\,16}{n} \log \dfrac{(3/4)I_d}{I_d - (3/4)I_d} = E_{1/2} + \dfrac{0.059\,16}{n} \log 3$

$E_{1/4} = E_{1/2} + \dfrac{0.059\,16}{n} \log \dfrac{(1/4)I_d}{I_d - (1/4)I_d} = E_{1/2} + \dfrac{0.059\,1\tilde{5}}{n} \log 1/3$

$E_{3/4} - E_{1/4} = \dfrac{0.059\,16}{n} (\log 3 - \log 1/3) = \dfrac{0.056\,5}{n}$ (volts)

9. A current of 2.0 µA in Figure 18-9a corresponds to 0.17 mM. The sample was diluted by a factor of 10 for analysis, so the original concentration was 1.7 mM.

10. $\dfrac{[X]_i}{[S]_f + [X]_f} = \dfrac{I_X}{I_{S+X}}$

$\dfrac{[X]_i}{\left(\dfrac{1.00}{101.0}\right)(0.050\,0) + \left(\dfrac{100.0}{101.0}\right)[X]_i} = \dfrac{10.0}{14.0} \Rightarrow [X]_i = 1.21$ mM

11. (a) Since all nitrite is converted to NO, we can write a proportionality in terms of grams instead of moles per liter:

$\dfrac{\mu g\ NO_2^- \text{ in unknown}}{\mu g\ NO_2^- \text{ in standard} + \mu g\ NO_2^- \text{ in unknown}} = \dfrac{8.9\ \mu A}{14.6\ \mu A}$

$\dfrac{x}{5.00 + x} = \dfrac{8.9\ \mu A}{14.6\ \mu A} \Rightarrow \mu g\ NO_2^- \text{ in unknown} = 7.8_1$

This nitrite was found in a 5.00-mL aliquot, so there were $\left(\dfrac{200.0}{5.00}\right)(7.8_1) =$

312 µg in 10.0 g of bacon = 31.2 µg of nitrite per gram of bacon.

(b) In both experiments, the nitrate increased the current in step 6 by 14.3 µA over that in step 5. From the first experiment, this means that

$\dfrac{\text{moles of } NO_3^- \text{ in unknown}}{\text{moles of } NO_2^- \text{ in unknown}} = \dfrac{14.3\ \mu A}{8.9\ \mu A} = 1.60_7$

The formula weight of NO_2^- is 46.00, while that of NO_3^- is 62.00.

Since the bacon contains 31.2 µg of nitrate per gram, it must contain $\left(\dfrac{62.00}{46.00}\right)(1.60_7)(31.2) = 67.6$ µg of nitrate per gram of bacon.

12. The least squares parameters for a graph of I_d vs $[Cu^{2+}]$ are

$$\text{slope} = m = 6.616 \qquad \text{standard deviation} = 0.018$$
$$\text{intercept} = b = -0.086 \qquad \text{standard deviation} = 0.062$$
$$\sigma_y = 0.142$$

An unknown giving a current of 15.6 μA has a concentration of

$$[Cu^{2+}] = \frac{(I_d - b)}{m} = \frac{15.6 - (-0.086)}{6.616} = 2.371 \text{ mM}$$

and an uncertainty calculated with Equation 4-23:

uncertainty $= \pm 0.023 \Rightarrow [Cu^{2+}] = 2.37 \ (\pm 0.02) \text{ mM}$

13. $\dfrac{[X]_i}{[S]_f + [X]_f} = \dfrac{I_X}{I_{S+X}}$

$$\frac{x(mM)}{3.00\left(\dfrac{2.00}{52.00}\right) + x\left(\dfrac{50.0}{52.0}\right)} = \frac{0.37 \ \mu A}{0.80 \ \mu A} \Rightarrow x = 0.096 \text{ mM}$$

14. The relative heights of the signals for acetone are

$$\frac{\text{signal for unknown}}{\text{signal for unknown + standard}} = 0.259$$

$$\frac{[X]_i}{[S]_f + [X]_f} = \frac{I_X}{I_{S+X}}$$

$$\frac{x \ (wt\%)}{0.001 \ 00 + x} = 0.259 \Rightarrow x = 0.000 \ 35 \ wt\%$$

15. Mixed potential ≈ -0.01 V

16. In anodic stripping voltammetry, analyte is reduced and concentrated at the working electrode at a controlled potential for a constant time. The potential is then ramped in a positive direction to reoxidize the analyte, during which current is measured. The height of the oxidation wave is proportional to the original concentration of analyte. In potentiometric stripping, the same concentration step is carried out at a controlled potential for a constant time. Then the potentiostat is disconnected and the voltage of the working electrode is measured as a function of time while Hg^{2+} from the solution reoxidizes analyte back into solution. The time required to reach the potentiometric end point is proportional to the original concentration of analyte.

17.
$$\frac{[Cu^{2+}]_{unknown}}{[Cu^{2+}]_{unknown} + [Cu^{2+}]_{standard}} = \frac{(stripping\ time)_{unknown}}{(stripping\ time)_{unknown\ +\ standard}}$$

$$\frac{[Cu^{2+}]_{unknown}}{[Cu^{2+}]_{unknown} + 0.5\ \mu g/L} = \frac{8.6\ s}{13.4\ s} \Rightarrow [Cu^{2+}]_{unknown} = 0.90\ \mu g/L = 1.4 \times 10^{-8}\ M$$

18.

19. Peak B : $RNHOH \rightarrow RNO + 2H^+ + 2e^-$

Peak C : $RNO + 2H^+ + 2e^- \rightarrow RNHOH$

There was no RNO present before the initial scan.

20. At room temperature the interconversion between axial and equatorial conformations is much faster than the reduction, and one peak is seen (near -2.5 V). At low temperature, interconversion slows down and we observe one reduction wave for each molecule. It turns out that the -2.5 V signal is from the axial molecule and the -3.1 V signal is from the equatorial molecule. At -80° the relative heights of the waves are close to the relative equilibrium concentrations, because interconversion is much slower than reduction. At -60°, some of the axial species (-3.1 V) is converted to the equatorial species (-2.5 V) at a rate similar to the rate of reduction of the equatorial species. If the voltage scan rate is increased, less interconversion occurs and the -3.1 V signal grows relative to the -2.5 V signal.

21. Addition of 2-methylimidazole makes it easier to reduce PFe⁺ (-0.18 V shifts to -0.12 V) and harder to reduce PFe (-1.02 V shifts to -1.11 V). Therefore, Fe^{2+} is stabilized the most.

22.

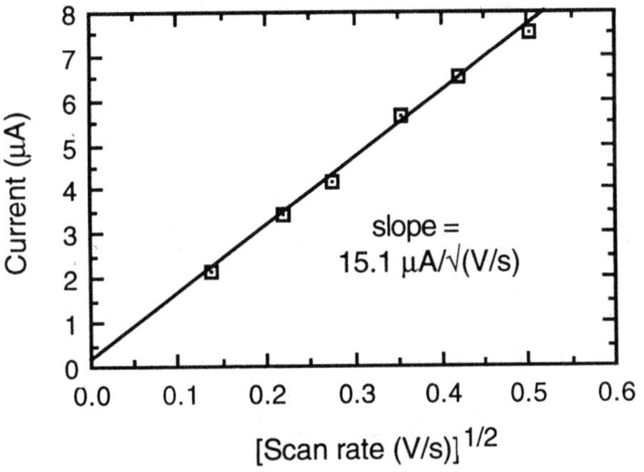

$$i_p = (2.69 \times 10^8)n^{3/2}ACD^{1/2}v^{1/2}$$

$$\text{slope} = (2.69 \times 10^8)n^{3/2}ACD^{1/2}$$

$$\Rightarrow D = \frac{\text{slope}^2}{(2.69 \times 10^8)^2n^3A^2C^2}$$

$$= \frac{(15.1 \times 10^{-6} \text{ A}/\sqrt{\text{V/s}})^2}{(2.69\times10^8)^2 1^3 (0.020\ 1\times10^{-4} \text{ m}^2)^2(1.00\times10^{-3} \text{ M})^2} = 7.8 \times 10^{-10} \text{ m}^2/\text{s}$$

23. Microelectrodes fit into small places, are useful in nonaqueous solution (because of small ohmic losses) and allow rapid voltage scans (because of small capacitance), which allows the study of short-lived species.

24. The Nafion membrane permits neutral and cationic species to pass through to the electrode, but excludes anions. It reduces the background signal from the ascorbate anion, which would otherwise swamp the signal from dopamine.

25. (a)

trace	$\left(\dfrac{A_3 - A_2}{A_2 - A_1}\right)$ measured from the figure
b	0.029
c	0.116
d	0.527
e	2.49
f	10.4

(b) A graph of $E_{applied}$ vs $\log\left(\dfrac{A_3 - A_2}{A_2 - A_1}\right)$ is linear.

(c) When $\log\left(\dfrac{A_3 - A_2}{A_2 - A_1}\right) = 0$, $E_{applied} = E' = 0.612$ V (read on the graph).

E' (vs. S.H.E.) $= 0.612 + 0.241 = 0.853$ V.

(d) Slope of graph $= 0.031\ 0$ V $= \dfrac{0.059\ 16}{n} \Rightarrow n = 1.91 \approx 2.$

(e)   A possible two-electron oxidation of tolidine is:

$$H_3\overset{+}{N}\text{—}\langle\text{ring}\rangle\text{—}\langle\text{ring}\rangle\text{—}\overset{+}{N}H_3 \rightarrow H_2\overset{+}{N}=\langle\text{ring}\rangle=\langle\text{ring}\rangle=\overset{+}{N}H_2 + 2H^+ + 2e^-$$

26.   The detection system is a pair of platinum electrodes with a voltage of 0.2 V applied between them. Prior to the end point, only Br^- is present, so there is no mechanism for substantial current to flow. After the end point, both Br_2 and Br^- are present and the two reactions below can carry substantial current.

Cathode: $Br_2 + 2e^- \rightarrow 2Br^-$

Anode: $2Br^- \rightarrow Br_2 + 2e^-$

27.   The high overpotential for reduction of H^+ at a mercury surface allows thermodynamically less favorable reductions to be occur without competitive reduction of H^+. However, Hg is too easily oxidized to be used for anodic reactions.

28.   See Box 18-3

29. $B \cdot I_2 + B \cdot SO_2 + B + H_2O \rightarrow 2BH^+I^- + {}^+B\text{-}SO_3^-$
          ${}^+B\text{-}SO_3^- + ROH \rightarrow 2BH^+ROSO_3^-$

One mole of H_2O in the first reaction allows one mole of SO_2 to be oxidized by one mole of I_2.

30.   (a)   The initial solution contains H_3AsO_3 and Br^-, neither of which can support a substantial current between two Pt electrodes. Only a small residual current is expected. As BrO_3^- is added, it is converted to Br_2 and then to Br^- by reaction with H_3AsO_3. Since the $H_3AsO_4 \mid H_3AsO_3$ does not conduct current, the current remains small. After the equivalence point, when both Br_2 and Br^- are present, substantial current flows by virtue of oxidation of Br^- at one electrode and reduction of Br_2 at the other.. The expected titration curve will have the same shape as curve (b) in Exercise 18-F.

(b)   Initially there is no easy mechanism for carrying current, so the voltage will be high. As I_2 is added, it is converted to I^- and H_3AsO_3 is oxidized to H_3AsO_4. Since the $H_3AsO_3 \mid H_3AsO_4$ couple does not carry current (see previous problem), the voltage remains high. Only after the end point, when both I_2 and I^- are present, does the voltage drop to near zero. The titration curve is expected to look like the one in Demonstration 18-1.

31.

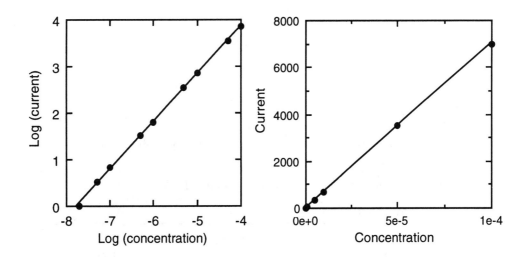

(b) $m = 7.002 \, (\pm 0.005) \times 10^7$
      $b = -1.5 \, (\pm 1.9)$
      $s_y = 5.0$

(c)   Concentration $= \dfrac{\text{current - intercept}}{\text{slope}}$

$$= \frac{300 \, (\pm 15) - [-1.5 \, (\pm 1.9)]}{7.002 \, (\pm 0.005) \times 10^7}$$

$$= \frac{301.5 \, (\pm 15.1)}{7.002 \, (\pm 0.005) \times 10^7}$$

$$= 4.3_1 \, (\pm 0.2_2) \times 10^{-6} \, M$$

CHAPTER 19
APPLICATIONS OF SPECTROPHOTOMETRY

1. Putting $b = 0.100$ cm into the determinants gives

$$[X] = \frac{\begin{vmatrix} 0.233 & 387 \\ 0.200 & 642 \end{vmatrix}}{\begin{vmatrix} 1\ 640 & 387 \\ 399 & 642 \end{vmatrix}} = 8.03 \times 10^{-5} \text{ M} \qquad [Y] = \frac{\begin{vmatrix} 1\ 640 & 0.233 \\ 399 & 0.200 \end{vmatrix}}{\begin{vmatrix} 1\ 640 & 387 \\ 399 & 642 \end{vmatrix}} = 2.62 \times 10^{-4} \text{ M}$$

2. Let $X = Cr_2O_7^{2-}$ and $Y = MnO_4^-$ in Equation 19-5.

wavelength (nm)	A_m/A_{X_s}	A_{Y_s}/A_{X_s}
266	1.868	0.102
288	2.018	0.290
320	2.671	1.063
350	2.113	0.393
360	2.022	0.309

A graph of A_m/A_{X_s} vs A_{Y_s}/A_{X_s} has a slope of 0.842 and an intercept of 1.78

Slope $= [Y]/[Y_s] \Rightarrow [Y] = [MnO_4^-] = (0.842)(1.00 \times 10^{-4} \text{ M}) = 8.42 \times 10^{-5} \text{ M}$

Intercept $= [X]/[X_s] \Rightarrow [X] = [Cr_2O_7^{2-}] = (1.78)(1.00 \times 10^{-4} \text{ M}) = 1.78 \times 10^{-4} \text{ M}$

3. If the spectra of two compounds with a constant total concentration cross at any wavelength, all mixtures with the same total concentration will go through that same point, called an isosbestic point. The appearance of isosbestic points in a chemical reaction is good evidence that we are observing one main species being converted to one other major species.

4. As VO^{2+} is added (traces 1-9) the peak at 434 decreases and a new one near 485 nm develops, with an isosbestic point at 457 nm. When VO^{2+}/xylenol orange > 1, the peak near 485 nm decreases and a new one at 566 nm grows in, with an isosbestic point at 528 nm. This sequence is logically interpreted by the sequence

$$\begin{array}{ccccc} M & + & L & \rightarrow & ML \\ & & 434 \text{ nm} & & 485 \text{ nm} \end{array}$$

$$\begin{array}{ccccc} ML & + & M & \rightarrow & M_2L \\ 485 \text{ nm} & & & & 566 \text{ nm} \end{array}$$

where M is vanadyl ion and L is xylenol orange. The structure of xylenol orange in Table 13-3 shows that it has metal-binding groups on both ends of the molecule, and could form an M_2L complex.

5.   First convert T to A (= -log T) and then convert A to ε (= A/bc = A/[(0.005)(0.01)])

	Absorbance			ε (M^{-1} cm^{-1})	
	2 022	1 993 cm^{-1}		2 022	1 993 cm^{-1}
A	0.508 6	0.098 54	A	10 170	1 971
B	0.011 44	0.699 0	B	228.8	13 980

For the mixture, $A_{2022} = -\log(0.340) = 0.468\ 5$ and $A_{1993} = -\log(0.383) = 0.416\ 8$.
Using Equations 19-7, we find $[A] = 9.11 \times 10^{-3}$ M and $[B] = 4.68 \times 10^{-3}$ M.

7.   Coefficient matrix:
$$\begin{bmatrix} 12\ 200 & 3\ 210 & 290 \\ 4\ 140 & 6\ 550 & 990 \\ 3\ 000 & 2\ 780 & 8\ 080 \end{bmatrix}$$

Constant vector:
$$\begin{bmatrix} 0.846 \\ 0.400 \\ 0.555 \end{bmatrix}$$

Solution vector:
$$\begin{bmatrix} [X] \\ [Y] \\ [Z] \end{bmatrix} = \begin{bmatrix} 6.47 \times 10^{-5} \\ 1.42 \times 10^{-5} \\ 3.98 \times 10^{-5} \end{bmatrix}$$

8.   (b) Slope = -K = -88.2 $\Rightarrow$ K = 88.2

9.   (a) We will make the substitutions [complex] = A/ε and $[I_2] = [I_2]_{tot}$ - [complex]
        in the equilibrium expression:

$$K = \frac{[\text{complex}]}{[I_2][\text{mesitylene}]} = \frac{A/\varepsilon}{([I_2]_{tot} - [\text{complex}])\,[\text{mesitylene}]}$$

$$K[I_2]_{tot} - K[\text{complex}] = \frac{A}{\varepsilon[\text{mesitylene}]}$$

Making the substitution [complex] = A/ε once more on the left hand side gives

$$K[I_2]_{tot} - \frac{KA}{\varepsilon} = \frac{A}{\varepsilon[\text{mesitylene}]}$$

Multiplying both sides by ε and dividing by $[I_2]_{tot}$ gives the desired result:

$$\varepsilon K - \frac{KA}{[I_2]_{tot}} = \frac{A}{[I_2]_{tot}\,[\text{mesitylene}]}$$

(b) The graph of $A/([\text{mesitylene}][I_2]_{tot})$ versus $A/[I_2]_{tot}$ is an excellent straight line
     with a slope of -0.464 and an intercept of 4.984×10^3. Since slope = -K, the
     equilibrium constant is 0.464. The molar absorptivity is ε = intercept/K $\Rightarrow$
     $\varepsilon = 1.074 \times 10^4$ M^{-1} cm^{-1}.

10.   (a) Maximum absorbance occurs at $X_{SCN^-} = 0.500 \Rightarrow$ stoichiometry = 1 : 1 (n = 1)

(b) The curved maximum indicates that the equilibrium constant is not very large.

(c)   The different acid concentrations give both solutions the same ionic strength
      (= 16.0 mM).

11.   (a)   Here are the results:

			[AB$_2$]		
[A]$_{total}$	[B]$_{total}$	$K = 10^6$	$K = 10^7$	$K = 10^8$	Mole fraction A
1e-5	9e-5	8.01e-8	7.27e-7	4.02e-6	0.1
2e-5	8e-5	1.26e-7	1.14e-6	6.25e-6	0.2
2.5e-5	7.5e-5	1.39e-7	1.25e-6	6.83e-6	0.25
3e-5	7e-5	1.45e-7	1.30e-6	7.12e-6	0.3
3.33e-5	6.67e-5	1.46e-7	1.31e-6	7.17e-6	0.333
4e-5	6e-5	1.42e-7	1.28e-6	6.99e-6	0.4
5e-5	5e-5	1.23e-7	1.12e-6	6.20e-6	0.5
6e-5	4e-5	9.49e-8	8.66e-7	4.97e-6	0.6
7e-5	3e-5	6.24e-8	5.78e-7	3.51e-6	0.7
8e-5	2e-5	3.18e-8	3.00e-7	2.00e-6	0.8
9e-5	1e-5	8.97e-9	8.682e-8	6.70e-7	0.9

(b)   The maximum occurs at a mole fraction of A = 1/3, since the stoichiometry is
      1:2. The greater the equilibrium constant, the greater the extent of reaction.

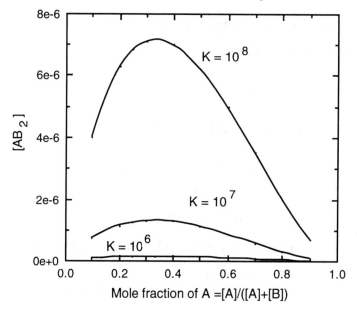

12. $n \rightarrow \pi^*(T_1)$: $E = h\nu = h\dfrac{c}{\lambda} = (6.626\,1 \times 10^{-34}\,\text{J·s})\,\dfrac{2.997\,9 \times 10^8\,\text{s}^{-1}}{397 \times 10^{-9}\,\text{m}} = 5.00 \times 10^{-19}\,\text{J}$

To convert to J/mol, multiply by Avogadro's number:

5.00×10^{-19} J/molecule $\times\ 6.022 \times 10^{23}$ molecules/mol = 301 kJ/mol.

$n \rightarrow \pi^*(S_1)$: $E = (6.626\,1 \times 10^{-34}\,\text{J·s})\dfrac{2.997\,9 \times 10^8\,\text{s}^{-1}}{355 \times 10^{-9}\,\text{m}} = 5.60 \times 10^{-19}\,\text{J} = 337$

kJ/mol.   The difference between the T_1 and S_1 states is 337 - 301 = 36 kJ/mol.

13. Fluorescence is emission of light with no change in the electronic spin state of the molecule (e. g. singlet → singlet). In phosphorescence, the electronic spin does change during emission ((e. g. triplet → singlet). Phosphorescence is less probable, so molecules spend more time in the excited state prior to phosphorescence than to fluorescence. That is, phosphorescence has a longer lifetime than fluorescence. Phosphorescence also comes at lower energy (longer wavelength) than fluorescence, because the triplet excited state is at lower energy than the singlet excited state.

14. Phosphorescence comes at longest wavelength and absorption is at shortest wavelength.

15. In an excitation spectrum the exciting wavelength (λ_{ex}) is varied while the detector wavelength (λ_{em}) is fixed. In an emission spectrum λ_{ex} is constant and λ_{em} is varied. The excitation spectrum resembles an absorption spectrum because emission intensity is proportional to absorption of the exciting radiation.

16. Quenching occurs at high concentration when a fluorescent substance absorbs too much incoming light prior to reaching the region being observed in a cell, and when the substance absorbs its own emission before the light can exit the cell. This is seen at high concentration in Figure 19-16. Absorption of incoming light is given by the $10^{-\varepsilon_{ex}b_1c}$ term in Equation 19-27 and absorption of fluorescence is given by the $10^{-\varepsilon_{em}b_3c}$ term.

17. Each molecule of analyte bound to antibody 1 also binds one molecule of antibody 2 that is linked to one enzyme molecule. Each enzyme molecule catalyzes many cycles of reaction in which a colored or fluorescent product is created. Therefore, each analyte molecule results in many product molecules.

18. In time-resolved emission measurements, the short-lived background fluorescence decays to near zero prior to recording emission from the lanthanide ion. By reducing background signal, the signal-to-noise ratio is increased.

19. The absorption spectrum has the same shape as the excitation spectrum. This is best seen for the tallest traces at an emission wavelength near 550 nm. The absorption spectrum has two main peaks near 330 and 410 nm. The emission spectrum is a plot of emission intensity versus emission wavelength, for a constant excitation wavelength of 400 nm. Eyeballing a cut through the spectra in the figure, there appears to be a single emission peak with a maximum near ~550 nm.

20.

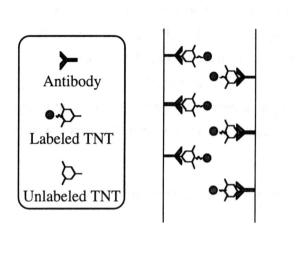

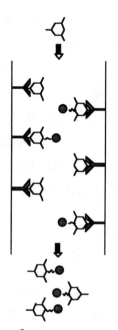

Step 1:
Column loaded with labeled TNT and excess is washed away. There is no fluorescence in liquid leaving the column.

Step 3:
Some unlabeled TNT analyte displaces labeled TNT. Fluorescence is now seen in liquid leaving the column.

21.   At the intersection of the two lines, $y = m_1 x + b_1 = m_2 x + b_2$.

$$x = \frac{b_2 - b_1}{m_1 - m_2} = \frac{82\ (\pm 2) - 6.8\ (\pm 0.7)}{1.08\ (\pm 0.02) - 0.12\ (\pm 0.02)} = \frac{75.2\ (\pm 2.82\%)}{0.96\ (\pm 2.95\%)} =$$

$78.33\ (\pm 4.08\%) = 78\ (\pm 3)\ \mu L$

22.   Move the cell diagonally toward the upper right hand side, so that only the lower left corner is illuminated. Then $b_1 = 0$ and b_3 is minimized.

23.

C(M)	I (multiples of $k'P_o$)	C	I
10^{-7}	0.000 070 4	10^{-4}	0.055 8
10^{-6}	0.000 703	10^{-3}	0.069 9
10^{-5}	0.006 88	10^{-2}	2.54×10^{-9}

24.   (a) The absorption spectrum shows that the absorbance of fluorescein decreases with increasing pH at a wavelength of 442 nm. Since the chromophore absorbs less light, the emission intensity will also decrease as pH increases. At an excitation wavelength of 488 nm, the situation in reversed: Absorbance increases with increasing pH, and so does emission intensity.

(b) For excitation at 488 nm, the ratio of emission intensities I_{540}/I_{610} is small at low pH and large at high pH. The calibration graph shows that this ratio is sensitive to pH in the range pH 6 to pH 8 and could be used to measure pH in this interval.

SPECTROPHOTOMETERS

1. Absorbance = $\log P_0/P$, where P_0 is incident radiant power and P is transmitted radiant power. Note that any effects of scatter, reflection or absorption by the cell and solvent must be compensated with a reference in order to measure absorbance of the analyte.

    Absorption coefficient is the quantity α in Equation 20-3. The internal transmittance (P_2/P_1) is exponentially related to the absorption coefficient. Absorptance is the fraction of incident light absorbed.

    Molar absorptivity is the number ε in Beer's law. Absorbance is proportional to the concentration of analyte and to the molar absorptivity.

2. Specular reflection has an angle of reflection equal to the angle of incidence. Diffuse reflection goes in all directions.

3. $T + a + R = T + 0.06 + 0.16 = 1 \Rightarrow T = 0.78$

4. $P_2/P_1 = e^{-\alpha b} = e^{-(0.100 \text{ cm}^{-1})(0.300 \text{ cm})} = 0.970$

5. $n_1 \sin \theta_1 = n_2 \sin \theta_2$, where $n_1 = 1.50$ and $n_2 = 1.33$

    (a) If $\theta_1 = 30°$, $\theta_2 = 34°$

    (b) If $\theta_1 = 0°$, $\theta_2 = 0°$ (no refraction)

6. $M = \sigma T^4 = [5.669\ 8 \times 10^{-8} \text{ W}/(\text{m}^2\text{K}^4)]T^4$

T(K)	M(W/m^2)
77	1.99
298	447

7. $R = \left(\dfrac{1 - n}{1 + n}\right)^2 = \left(\dfrac{1 - 2.4}{1 + 2.4}\right)^2 = 0.169\ 6$

    $T = \dfrac{1 - R}{1 + R} = 0.71 = 71\%$ (independent of thickness if absorption = 0)

8. Light inside the fiber strikes the wall at an angle greater than the critical angle for total reflection. Therefore all light is reflected back into the core and continues to be reflected from wall-to-wall as it moves along the fiber. If the bending angle is not too great, the angle of incidence will still exceed the critical angle and light will not leave the core.

9. (a) The critical angle, θ_c, is such that $(n_1/n_2)\sin \theta_c = 1$. For $n_1 = 1.52$ and $n_2 = 1.50$, $\theta_c = 80.7°$. That is θ must be $\geq 80.7°$ for total internal reflection.

(b) $0.010\ 0\ \dfrac{dB}{m} = \dfrac{-10 \log \left(\dfrac{power\ out}{power\ in}\right)}{length}$

For length = 20.0 m, we find $\dfrac{power\ out}{power\ in} = 0.955$

10.   (a) $\tilde{v} = 1/\lambda = 1/(24 \times 10^{-4}\ cm) = 4.2 \times 10^{2}\ cm^{-1}$

(b) $R = \left(\dfrac{1.00 - 1.47}{1.00 + 1.47}\right)^{2} = 0.036_2$

(c) $P_2/P_1 = e^{-(0.25\ cm^{-1})(0.60\ cm)} = 0.86_1$

(d) $T = \dfrac{(1 - 0.036\ 2)^2\ e^{-(0.25\ cm^{-1})(0.60\ cm)}}{1 - (0.036\ 2)^2\ e^{-2\ (0.25\ cm^{-1})(0.60\ cm)}} = 0.80$

11.   If 5.0% is scattered, 95.0% is transmitted.

$0.950 = e^{-\alpha_s b} = e^{-\alpha_s(1.00\ cm)} \Rightarrow \alpha_s = 0.051\ cm^{-1}$

12. $R = \left(\dfrac{1.50 - 1.54}{1.50 + 1.54}\right)^{2} = 2 \times 10^{-4}$

13.   (a)

λ (μm)	n	λ (μm)	n
0.2	1.550 5	2	1.438 1
0.4	1.470 1	3	1.419 2
0.6	1.458 0	4	1.389 0
0.8	1.453 3	5	1.340 4
1	1.450 4	6	1.258 0

(b) $dn/d\lambda$ is greater for blue light (~ 400 nm) than red light (~ 600 nm)

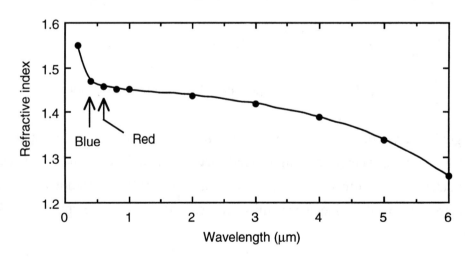

14.   (a) $n_{core} \sin \theta_i = n_{cladding} \sin \theta_r$

For total reflection, $\sin \theta_r \geq 1 \Rightarrow \sin \theta_i \geq \dfrac{n_{cladding}}{n_{core}}$

For $n_{cladding} = 1.400$ and $n_{core} = 1.600$, $\sin \theta_i \geq \dfrac{1.400}{1.600}$

$\Rightarrow \theta_i \geq 61.04°$

(b) For $n_{cladding} = 1.400$ and $n_{core} = 1.800$, $\theta_i \geq 51.06°$

15. Angle of incidence = angle of reflection = 45°. Angle of refraction $\equiv \theta$.
$n_{prism} \sin 45° = n_{air} \sin \theta$. If total reflection occurs, there is no refracted light.
This happens if $\sin \theta > 1$, or $\dfrac{n_{prism} \sin 45°}{n_{air}} > 1$. Using $n_{air} = 1$ gives
$n_{prism} > \sqrt{2}$. As long as $n_{prism} > \sqrt{2}$, no light will be transmitted through the
prism and all light will be reflected.

16. For air, $n = 1$. For AgCl, $n = 2.07$. The coating should have $n = \sqrt{1 \cdot 2.07} =$
1.44. Quartz or MgF$_2$ are close to this refractive index.

17. (a) $M_\lambda = \dfrac{2\pi h c^2}{\lambda^5} \left(\dfrac{1}{e^{hc/\lambda kT} - 1} \right)$

at $T = 1\,000$ K :

λ (μm)	M_λ (W/m^3)
2.00	8.789×10^9
10.00	1.164×10^9

(b) $M_\lambda \Delta \lambda = (8.789 \times 10^9 \text{ W/m}^3)(0.02 \times 10^{-6} \text{ m}) = 1.8 \times 10^2 \text{ W/m}^2$ at 2.00 μm

(c) $M_\lambda \Delta \lambda = (1.164 \times 10^9 \text{ W/m}^3)(0.02 \times 10^{-6}) = 2.3 \times 10^1 \text{ W/m}^2$ at 10.00 μm

(d) at $T = 100$ K :

λ (μm)	M_λ (W/m^3)
2.00	6.677×10^{-19}
10.00	2.111×10^3

$\dfrac{M_{2.00\,\mu m}}{M_{10.00\,\mu m}} = \dfrac{8.789 \times 10^9 \text{ W/m}^3}{1.164 \times 10^9 \text{ W/m}^3} = 7.551$ at 1 000 K

$\dfrac{M_{2.00\,\mu m}}{M_{10.00\,\mu m}} = \dfrac{6.677 \times 10^{-19} \text{ W/m}^3}{2.111 \times 10^3 \text{ W/m}^3} = 3.163 \times 10^{-22}$ at 100 K

At 100 K there is virtually no emission at 2.00 μm compared to 10.00 μm,
whereas at 1 000 K there is a great deal of emission at both wavelengths.

18. An excited state of the lasing material is pumped to a high population by light, an
electric discharge, or other means. Photons emitted when the excited state decays

to a less populated lower state stimulate emission from other excited molecules. The stimulated emission has the same energy and phase as the incident photon. In the laser cavity most light is retained by reflective end mirrors. Some light is allowed to escape from one end.

19. Deuterium

20. Resolution increases in proportion to the number of grooves that are illuminated and to the diffraction order. The number of grooves can be increased with a more finely ruled grating (closer grooves) and with a longer grating. The diffraction order is optimized by appropriate choice of the blaze angle of the grating.

21. To remove higher order diffraction at the same angle as the desired diffraction.

22. Advantage - increased ability to resolve closely spaced spectral peaks. Disadvantage - more noise because less light reaches detector.

23. DTGS has a permanent electric polarization. That is, one face of the crystal has a positive charge and the opposite face has a negative charge. When the temperature of the crystal changes by absorption of infrared light, the polarization (the voltage difference between the two faces) changes. The change in voltage is the detector signal.

24. The detection limit may be defined as the minimum concentration of a substance that can be reported with 99% confidence to be greater than zero. This corresponds to the value t·s, where t is Student's t for 98% confidence and s is the standard deviation for n measurements. The smaller the standard deviation, the smaller the detection limit. What this really means is that the more precisely you can measure a quantity, the more you can be certain that a nonzero reading is truly different from zero. The less precise your measurement, the less confident you can be that a positive result is not just noise superimposed on zero signal.

25. (a) $n\lambda = d(\sin \theta - \sin \phi)$

$1 \cdot 600 \times 10^{-9}$ m $= d(\sin 40° - \sin 30°) \Rightarrow d = 4.20 \times 10^{-6}$ m

Lines/cm $= 1/(4.20 \times 10^{-4}$ cm$) = 2.38 \times 10^3$ lines/cm

(b) $\lambda = 1/(1\ 000$ cm$^{-1}) = 10^{-3}$ cm $\Rightarrow d = 7.00 \times 10^{-3}$ cm $\Rightarrow 143$ lines/cm

26. 10^3 grooves/cm means $d = 10^{-5}$ m $= 10\ \mu$m

$$\text{Dispersion} = \frac{n}{d \cos \phi} = \frac{1}{(10\ \mu\text{m}) \cos 10°} = 0.102\ \frac{\text{radians}}{\mu\text{m}}$$

$$0.102 \frac{\text{radians}}{\mu m} \times \frac{180°}{\pi \text{ radians}} = 5.8°/\mu m$$

27. (a) $\text{Resolution} = \dfrac{\lambda}{\Delta\lambda} = \dfrac{512.245}{0.03} = 1.7 \times 10^4$

   (b) $\Delta\lambda = \dfrac{\lambda}{10^4} = \dfrac{512.23}{10^4} = 0.05$ nm

   (c) $\text{Resolution} = nN = (4)(8.00 \text{ cm} \times 1\,850 \text{ cm}^{-1}) = 5.9 \times 10^4$

   (d) 250 lines/mm = 4 μm/line = d

   $$\frac{\Delta\phi}{\Delta\lambda} = \frac{n}{d \cos \phi} = \frac{1}{(4 \text{ } \mu m) \cos 3°} = 0.250 \frac{\text{radians}}{\mu m} = 14.3°/\mu m$$

   For $\Delta\lambda = 0.03$ nm, $\Delta\phi = (14.3°/\mu m)(3 \times 10^{-5} \text{ } \mu m) = 4.3 \times 10^{-4}$ degrees

   For 30th order diffraction, the dispersion will be 30 times greater, or 0.013°.

28. True transmittance $= 10^{-1.500} = 0.031\,6$. With 0.50% stray light, the apparent transmittance is $0.031\,6 + 0.005\,0 = 0.036\,6$. The apparent absorbance is $-\log 0.036\,6 = 1.436$.

29. $b = \dfrac{30}{2 \cdot 1}\left(\dfrac{1}{1\,906 - 698 \text{ cm}^{-1}}\right) = 0.124\,2$ mm

   (Air between the plates has refractive index of 1.)

30. (a) $\Delta = \pm 2$ cm

   (b) Resolution refers to the ability to distinguish closely spaced peaks

   (c) $\text{Resolution} \approx 1/\Delta = 0.5 \text{ cm}^{-1}$

   (d) $\delta = 1/(2\Delta\nu) = 1/(2 \cdot 2\,000 \text{ cm}^{-1}) = 2.5 \text{ } \mu m$

31. The background transform gives the incident power P_0. The sample transform gives the transmitted power P. Transmittance is P/P_0, not P_0-P.

32. To increase the ratio from 8 to 20 (a factor of $20/8 = 2.5$) requires $2.5^2 = 6.25 \approx 7$ scans.

33. (a) $(100 \pm 1) + (100 \pm 1) = 200 \pm \sqrt{2}$, since $e = \sqrt{e_1^2 + e_2^2} = \sqrt{1^2 + 1^2} = \sqrt{2}$

   (b) $(100 \pm 1) + (100 \pm 1) + (100 \pm 1) + (100 \pm 1) = 400 \pm 2$, since

   $e = \sqrt{1^2 + 1^2 + 1^2 + 1^2} = 2$. The signal-to-noise ratio is 400:2 = 200:1.

   (c) The initial measurement has signal/noise = 100/1.

Averaging n measurements gives

$$\text{average signal} = \frac{n \cdot 100}{n} = 100$$

$$\text{average noise} = \frac{\sqrt{n}}{n} = 1/\sqrt{n}$$

$$\frac{\text{average signal}}{\text{average noise}} = \frac{100}{1/\sqrt{n}} = 100\sqrt{n} \text{, which is } \sqrt{n} \text{ times greater than}$$

the original value of signal/noise.

1. Temperature is more critical in emission spectroscopy because the small population of the excited state varies substantially as the temperature is changed. The population of the ground state does not vary much.

2. Furnaces give increased sensitivity and require smaller sample volumes, but give poorer reproducibility with manual sample introduction. Automated sample introduction gives improved precision.

3. The inductively coupled plasma operates at much higher temperature than a conventional flame. This lessens chemical interference (such as oxide formation) and allows emission instead of absorption to be used. Therefore, lamps are not required for each element. Self absorption is reduced in the plasma because the temperature is more uniform. Disadvantages of the plasma are increased cost of equipment and operation.

4. Doppler broadening occurs because an atom moving toward the radiation source sees a higher frequency than one moving away from the source. Increasing temperature gives increased speeds (more broadening) and increased mass gives decreased speeds (less broadening).

5. (a) A beam chopper alternately blocks or exposes the lamp to the flame and detector. When the lamp is blocked, signal is due to background. When the lamp is exposed, signal is due to analyte plus background. The difference is the desired analytical signal.

   (b) The flame or furnace is alternately exposed to a deuterium lamp and the hollow-cathode lamp. The absorbance from the deuterium lamp is due to background. The absorbance from the hollow-cathode lamp is due to analyte plus background. The difference is the desired signal.

   (c) In Smith-Hieftje background correction, the hollow-cathode lamp is periodically pulsed with high current to broaden the output spectrum and reduce the intensity at the analytical wavelength. The absorbance observed during the current pulse is subtracted from the absorbance observed without the pulse to obtain a corrected absorbance.

   (d) When a magnetic field parallel to the viewing direction is applied to the flame or furnace, the analytical signal is split into two components that are separated from the analytical wavelength, and one component at the analytical wavelength. The component at the analytical wavelength is not observed because of its polarization. The other two components have the wrong wavelength to be observed. The analyte is essentially "invisible" to the detector when the magnetic field is applied, and only background is seen. Corrected signal is that observed without a field minus that observed with the field.

6.   Spectral interference refers to the overlap of analyte signal with signals due to other elements or molecules in the sample or with signals due to the flame or furnace. Chemical interference occurs when a component of the sample decreases the extent of atomization of analyte through some chemical reaction. Ionization interference refers to a loss of analyte atoms through ionization.

7. $\lambda = hc/\Delta E = 5.893 \times 10^{-7}$ m $= 589.3$ nm

8.   We derive the value for 6 000 K as follows :

$$\Delta E = h\nu = \frac{hc}{\lambda} = \frac{(6.626\ 1 \times 10^{-34}\ \text{J·s})(2.997\ 9 \times 10^8\ \text{m/s})}{500 \times 10^{-9}\ \text{m}} = 3.97 \times 10^{-19}\ \text{J}$$

$$\frac{N^*}{N_0} = \frac{g^*}{g_0} e^{-\Delta E/kT} = e^{-(3.97 \times 10^{-19}\ \text{J})/(1.381 \times 10^{-23}\ \text{J/K})(6\ 000K)} = 8.3 \times 10^{-3}$$

If $g^*/g_0 = 3$, then $N^*/N_0 = 3\ (8.3 \times 10^{-3}) = 0.025$

9.   The Doppler linewidth is given by

$$\Delta\nu = \nu\ (7 \times 10^{-7})\ \sqrt{T/M} = (c/\lambda)\ (7 \times 10^{-7})\ \sqrt{T/M}$$

   $= 3._3$ GHz for Na (with $\lambda = 589 \times 10^{-9}$ m and $M = 23$)

   $= 2._6$ GHz for Hg (with $\lambda = 254 \times 10^{-9}$ m and $M = 201$)

10.   (a) $\Delta E = h\nu = \dfrac{hc}{\lambda} = \dfrac{(6.626\ 1 \times 10^{-34}\ \text{J·s})(2.997\ 9 \times 10^8\ \text{m/s})}{422.7 \times 10^{-9}\ \text{m}}$

   $= 4.699 \times 10^{-19}$ J/molecule $= 283.0$ kJ/mol

   (b) $\dfrac{N^*}{N_0} = \dfrac{g^*}{g_0} e^{-\Delta E/kT} = 3e^{-(4.699 \times 10^{-19}\ \text{J})/(1.381 \times 10^{-23}\ \text{J/K})(2\ 500K)} = 3.67 \times 10^{-6}$

   (c) At 2 515 K, $N^*/N_0 = 3.98 \times 10^{-6} \Rightarrow$ 8.4% increase from 2 500 to 2 515 K

   (d) At 6 000 K, $N^*/N_0 = 1.03 \times 10^{-2}$

11.

Element:	Na	Cu	Br
Excited state energy (eV):	2.10	3.78	8.04
Wavelength (nm):	591	328	154
Degeneracy ratio (g^*/g_0):	3	3	2/3
N^*/N_0 at 2 600 K in flame:	2.6×10^{-4}	1.4×10^{-7}	1.8×10^{-16}
N^*/N_0 at 6 000 K in plasma:	5.2×10^{-2}	2.0×10^{-3}	1.2×10^{-7}

Calculations: wavelength = $hc/\Delta E$ $N^*/N_0 = (g^*/g_0)\, e^{-\Delta E/kT}$

12.   Sensitivity = concentration of Fe giving A = 0.004 36 (= 99% T)

$$\frac{\text{[Fe] giving A} = 0.004\ 36}{\text{[Fe] giving A} = 0.055} = \frac{0.004\ 36}{0.055}$$

$$\frac{x}{1.00\ \mu g/mL} = \frac{0.004\ 36}{0.055} \Rightarrow x = 0.079\ \mu g/mL$$

13.   Since the dissociation energy of YC is greater than that of BaC, the equilibrium
BaC + Y $\rightleftarrows$ Ba + YC is driven to the right, increasing the concentration of free Ba
atoms in the gas phase.

14.   Emission is observed only if the narrow laser excitation frequency overlaps the
even narrower atomic absorption line. The laser bandwidth is narrow enough to
excite Mn without exciting Ga.

15.   The analyte and standard are lost in equal proportions, so their ratio remains
constant.

16.   (a) The concentrations of added standard are 0, 10.0, 20.0, 30.0, and 40.0
$\mu g/mL$.

(b) The graph below gives an intercept of -20.4 $\mu g/mL$, which is the
concentration of unknown after 10.00 mL has been diluted to 100.0 mL.
The original concentration of X is (20.4 $\mu g/mL$)$\left(\frac{100.0}{10.00}\right)$ = 204 $\mu g/mL$.

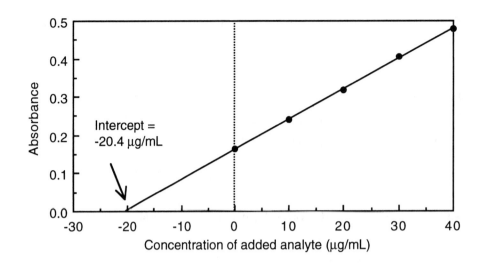

17.   (a)   [S] in unknown mixture = $(8.24\ \mu g/mL)\left(\dfrac{5.00}{50.0}\right)$ = 0.824 μg/mL

$$\frac{\text{Concentration ratio in unknown}}{\text{Concentration ratio in standard mixture}} = \frac{\text{Absorbance ratio in unknown}}{\text{Absorbance ratio in standard mixture}}$$

$$\frac{[X]/[S]}{1} = \frac{1.69}{0.93} \Rightarrow \frac{[X]}{[S]} = 1.82$$

[X] = (1.82) [S] = (1.82) (0.824 μg/mL) = 1.50 μg/mL

But X was diluted by a factor of 10.00/50.0, so the original concentration in

the unknown was $(1.50\ \mu g/mL)\left(\dfrac{50.0}{10.00}\right)$ = 7.49 μg/mL

(b) $\dfrac{[X]/[S]}{3.42} = \dfrac{1.69}{0.93} \Rightarrow \dfrac{[X]}{[S]} = 6.21 \Rightarrow [X] = (6.21)\,(0.824\ \mu g/mL) = 5.12\ \mu g/mL$

Original [X] in unknown = $(5.12\ \mu g/mL)\left(\dfrac{50.0}{10.00}\right)$ = 25.6 μg/mL

18.   A graph of intensity vs (μg K/mL) gives a straight line, from which we read

[unknown] = 17.4 μg/mL for an emission intensity of 417.

19. $\dfrac{198\ \text{units}}{807\ \text{units/ppm Ag}}$ = 0.245 ppm Ag = 0.245 μg Ag/mL = 2.27 μM Ag

For each mole of silver, there are two moles of cyanide, because the species being

analyzed is $Ag(CN)_2^-$. The concentration of CN^- in the unknown is 4.54 μM.

1. Three extractions with 100 mL are more effective than one extraction with 300 mL.

2. Adjust the pH to 3 so the acid is in its neutral form (CH_3CO_2H), rather than its anionic form ($CH_3CO_2^-$).

3. The EDTA complex is anionic (AlY^-), whereas the 8-hydroxyquinoline complex is neutral (AlL_3).

4. The complexation reaction $mHL + M^{m+} \rightleftarrows ML_m + mH^+$ is driven to the right at high pH (by consumption of H^+). This increases the fraction of metal in the form ML_m, which is extracted into organic solvent.

5. The form that is extracted into organic solvent is ML_n. The formation of ML_n is favored by increasing the formation constant (β) and by increasing the fraction of ligand in the form L^- (K_a). Increasing K_L decreases the fraction of ligand in the aqueous phase, thereby decreasing the formation of ML_n. Increasing $[H^+]$ decreases the concentration of L^- available for complexation.

6. When $pH > pK_{BH^+}$, the predominant form is B, which is extracted into the organic phase. When $pH > pK_a$ for HA, the predominant form is A^-, which is extracted into the aqueous phase.

7. (a) $S_{H_2O} \rightleftarrows S_{CHCl_3}$ $\qquad$ $K = [S]_{CHCl_3}/[S]_{H_2O} = 4.0$

$\qquad$ $[S]_{CHCl_3} = K[S]_{H_2O} = (4.0)(0.020 \text{ M}) = 0.080 \text{ M}$

(b) $\dfrac{\text{mol S in CHCl}_3}{\text{mol S in H}_2O} = \dfrac{(0.080 \text{ M})(10.0 \text{ mL})}{(0.020 \text{ M})(80.0 \text{ mL})} = 0.50$

8. Fraction remaining $= \left(\dfrac{V_1}{V_1+KV_2}\right)^n = \left(\dfrac{80.0}{80.0+(4.0)(10.0)}\right)^6 = 0.088$

9. (a) $D = \dfrac{[B]_{C_6H_6}}{[B]_{H_2O}+[BH^+]_{H_2O}}$

(b) D is the quotient of total concentrations in the phases.

$\qquad$ K is the quotient of concentrations of neutral species (B) in the phases.

(c) $D = \dfrac{K \cdot K_a}{K_a+[H^+]} = \dfrac{(50.0)(1.0 \times 10^{-9})}{(1.0 \times 10^{-9})+(1.0 \times 10^{-8})} = 4.5$

(d) D will be greater because a greater fraction of B is in the neutral form.

10. From Equation 22-12, $D \approx \dfrac{[ML_n]_{org}}{[M^{n+}]_{aq}} = K_{extraction} \dfrac{[HL]_{org}^n}{[H^+]_{aq}^n}$

Comparing this result to Equation 22-13 gives $K_{extraction} = \dfrac{K_M \beta K_a^n}{K_L^n}$

Constant	Effect on $K_{extraction}$	Reason
K_M	increase	ML_n is more soluble in organic phase.
β	increase	Ligand binds metal more tightly and ML_n is the organic-soluble form.
K_a	increase	Ligand dissociates to L^- more easily, increasing ML_n formation.
K_L	decrease	HL is more soluble in organic phase, where it is not available to react with M^{n+}(aq).

11. (a) $D = K[H^+]/([H^+] + K_a) = 3 \cdot 10^{-4.00}/(10^{-4.00} + 1.52 \times 10^{-5}) = 2.60$ at pH 4.00.
Fraction remaining in water $= q = V_1/(V_1 + DV_2) = 100/[100 + 2.60(25)] = 0.606$. Therefore the molarity in water is $0.606 (0.10 \text{ M}) = 0.060\ 6$ M. The total moles of solute in the system is $(0.100 \text{ L})(0.10 \text{ M}) = 0.010$ mol. The fraction of solute in benzene is 0.394, so the molarity in benzene is $0.394 (0.010 \text{ mol})/0.025 \text{ L} = 0.16$ M

(b) At pH 10.0: $D = 1.97 \times 10^{-5}$, $q = 0.999\ 995\ 1$, molarity in water $= 0.10$ M and molarity in benzene $= 2 \times 10^{-6}$ M.

12. $D = C/[H^+]^n$, where $C = K_M \beta K_a^n [HL]_{org}^n / K_L^n$.
$D_1 = 0.01 = C/[H^+]_1^2$ and $D_2 = 100 = C/[H^+]_2^2$.
$D_2/D_1 = 10^4 = [H^+]_1^2/[H^+]_2^2 \Rightarrow [H^+]_1/[H^+]_2 = 10^2 \Rightarrow \Delta pH = 2$ pH units

13. (a) Since there is so much more dithizone than Cu, it is safe to say that $[HL]_{org} = 0.1$ mM.

$$D = \frac{K_M \beta K_a^n}{K_L^n} \frac{[HL]_{org}^n}{[H^+]_{aq}^n} = \frac{(7 \times 10^4)(5 \times 10^{22})(3 \times 10^{-5})^2}{(1.1 \times 10^4)^2} \frac{(1 \times 10^{-4})^2}{[H^+]^2}$$

$= 2.6 \times 10^4$ at pH 1 and 2.6×10^{10} at pH 4.

(b) $q = V_1/(V_1 + DV_2) = 100/[100 + 2.6 \times 10^4 (10)] = 3.8 \times 10^{-4}$

14. (a) $D = \dfrac{[ML_2]_{org}}{[ML_2]_{aq}} = \dfrac{C_{org} V_{org}}{C_{aq} V_{aq}} \Rightarrow C_{org} = D\, C_{aq} \dfrac{V_{aq}}{V_{org}}$

$$\% \text{ extracted} = \frac{100\, C_{org}}{C_{aq} + C_{org}} = \frac{100\, D\, C_{aq} \dfrac{V_{aq}}{V_{org}}}{C_{aq} + D\, C_{aq} \dfrac{V_{aq}}{V_{org}}} = \frac{100\, D\, \dfrac{V_{aq}}{V_{org}}}{1 + D\, \dfrac{V_{aq}}{V_{org}}}$$

(b)	A	B	C	D	E
1	K(M) =	pH	H	D = Dist.coeff	% extracted
2	70000	1	1.00E-01	2.60E-02	0.05
3	Beta =	2	1.00E-02	2.60E+00	4.95
4	5E+18	2.2	6.31E-03	6.54E+00	11.57
5	Ka =	2.4	3.98E-03	1.64E+01	24.73
6	0.00003	2.6	2.51E-03	4.13E+01	45.21
7	K(L) =	2.8	1.58E-03	1.04E+02	67.46
8	11000	3	1.00E-03	2.60E+02	83.89
9	[HL]org =	3.2	6.31E-04	6.54E+02	92.90
10	0.00001	3.4	3.98E-04	1.64E+03	97.05
11	V(org) =	3.6	2.51E-04	4.13E+03	98.80
12	2	3.8	1.58E-04	1.04E+04	99.52
13	V(aq) =	4	1.00E-04	2.60E+04	99.81
14	100	5	1.00E-05	2.60E+06	100.00
15					
16	C2 = 10^-B2				
17	D2 = ($A$2*$A$4*$A$6^2*$A$10^2)/($A$8^2*C2^2)				
18	E2 = (D2*$A$12/$A$14)/(1+(D2*$A$12/$A$14))*100				

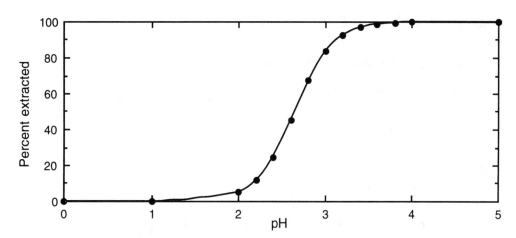

15. 1-C, 2-D, 3-A, 4-E, 5-B

16. The larger the partition coefficient, the greater the fraction of solute in the stationary phase, and the smaller the fraction that is moving through the column.

17. (a) $k' = \dfrac{\text{time solute spends in stationary phase}}{\text{time solute spends in mobile phase}} = \dfrac{t_r - t_m}{t_m} = \dfrac{t_s}{t_m}$

(b) fraction of time in mobile phase $= \dfrac{t_m}{t_m + t_s} = \dfrac{t_m}{t_m + k't_m} = \dfrac{1}{1 + k'}$

(c) $R = \dfrac{t_m}{t_r} = \dfrac{t_m}{t_m + t_s} = \dfrac{1}{1 + k'}$. Parts (b) and (c) together tell us that

$$\dfrac{\text{time for solvent to pass through column}}{\text{time for solute to pass through column}} = \dfrac{\text{time spent by solute in mobile phase}}{\text{total time on column}}$$

18. (a) Volume per cm of length $= \pi r^2 \times$ length $= \pi \left(\dfrac{0.461 \text{ cm}}{2}\right)^2 (1 \text{ cm}) = 0.167$ mL

mobile phase volume $= (0.390)(0.167 \text{ mL}) = 0.065 \ 1$ mL per cm of column

linear flow rate $= u_x = \dfrac{1.13 \text{ ml/min}}{0.065 \ 1 \text{ mL/cm}} = 17.4$ cm/min

(b) $t_m = (10.3 \text{ cm}) / (17.4 \text{ cm/min}) = 0.592$ min

(c) $k' = \dfrac{t_r - t_m}{t_m} \Rightarrow t_r = k't_m + t_m = 10(0.592) + 0.592 = 6.51$ min

19. (a) Linear flow rate $= (30.1 \text{ m}) / (2.16 \text{ min}) = 13.9$ m/min.

Inner diameter of open tube $= 530 \ \mu m - 2(3.1 \ \mu m) = 523.8 \ \mu m$

$\Rightarrow$ radius $= 261.9 \ \mu m$.

Volume $= \pi r^2 \times$ length $= \pi (261.9 \times 10^{-4} \text{ cm})^2 (30.1 \times 10^2 \text{ cm}) = 6.49$ mL

Volume flow rate $= (6.49 \text{ mL}) / (2.16 \text{ min}) = 3.00$ mL/min

(b) $k' = \dfrac{t_r - t_m}{t_m} = \dfrac{17.32 - 2.16}{2.16} = 7.02$

$k' = t_s/t_m$ (where $t_s =$ time in stationary phase)

Fraction of time in stationary phase $= \dfrac{t_s}{t_s + t_m} = \dfrac{k't_m}{k't_m + t_m} =$

$\dfrac{k'}{k' + 1} = \dfrac{7.02}{7.02 + 1} = 0.875$

(c) Volume of coating $\approx 2\pi r \times$ thickness $\times$ length

$= 2\pi[(261.9 + 1.55) \times 10^{-4} \text{ cm})](3.1 \times 10^{-4} \text{ cm})(30.1 \times 10^2 \text{ cm}) = 0.154$ mL

$k' = K\dfrac{V_s}{V_m} \Rightarrow 7.02 = K\dfrac{0.154 \text{ mL}}{6.49 \text{ mL}} \Rightarrow K = \dfrac{C_s}{C_m} = 295$

20. (a) $\dfrac{\text{Large load}}{\text{Small load}} = \left(\dfrac{\text{Large column radius}}{\text{Small column radius}}\right)^2$

$\dfrac{100 \text{ mg}}{4.0 \text{ mg}} = \left(\dfrac{\text{Large column diameter}}{0.85 \text{ cm diameter}}\right)^2 \Rightarrow$ diameter $= 4.25$ cm

Use a 40-cm-long column with a diameter near 4.25 cm.

(b) The linear flow rate should be the same. Since the cross-sectional area of the column is increased by a factor of 25, the volume flow rate should be increased by a factor of 25 $\Rightarrow u_v = 5.5$ ml/min.

(c) Volume of small column $= \pi r^2 \times$ length $= \pi(0.85/2 \text{ cm})^2(40 \text{ cm}) = 22.7$ mL

Mobile phase volume $= 35\%$ of column volume $= 7.94$ mL.

Linear flow $= \dfrac{40 \text{ cm}}{(7.94 \text{ mL})/(0.22 \text{ mL/min})} = 1.11$ cm/min for both columns

21. (a) $k' = \dfrac{9.0 - 3.0}{3.0} = 2.0$

(b) Fraction of time solute is in mobile phase $= \dfrac{t_m}{t_r} = \dfrac{3.0}{9.0} = 0.33$

(c) $K = k' \dfrac{V_m}{V_s} = (2.0) \dfrac{V_m}{0.10\, V_m} = 20.$

22. Solvent volume per cm of column length $= (0.15)\pi \left(\dfrac{0.30}{2}\right)^2 = 0.010\,6$ mL/cm.

0.20 mL corresponds to $(0.20\text{ mL})/(0.010\,6\text{ mL/cm}) = 19$ cm.

Linear flow rate $= 19$ cm/min.

23. $k' = K \dfrac{V_s}{V_m} = 3\left(\dfrac{1}{5}\right) = \dfrac{3}{5}$. For $K = 30$, $k' = 6$.

24. $K = \dfrac{V_r - V_m}{V_s} = \dfrac{76.2 - 16.6}{12.7} = 4.69$

$k' = \dfrac{V'_r}{V_m} = \dfrac{V_r - V_m}{V_m} = \dfrac{76.2 - 16.6}{16.6} = 3.59$

25. $K = k' \dfrac{V_m}{V_s}$

$k' = \dfrac{t_r - t_m}{t_m} = \dfrac{433 - 63}{63} = 5.87$

$\dfrac{V_m}{V_s} = \dfrac{\pi r^2 \times \cancel{\text{length}}}{2\pi r \times \text{thickness} \times \cancel{\text{length}}} = \dfrac{(103)^2}{2(103.25) \times 0.5} = 102.8$

(In the numerator, r refers to the radius of the open tube $= \frac{1}{2}\,(207 - 1.0)\ \mu m = 103\ \mu m$. In the denominator, r is the radius at the center of the stationary phase, which is $103 + \frac{1}{2}\,(0.5) = 103.25\ \mu m$.)

Therefore the partition coefficient is $K = k' \dfrac{V_m}{V_s} = 5.87\,(102.8) = 603$

Fraction of time in stationary phase $= \dfrac{t_s}{t_s + t_m} = \dfrac{k' t_m}{k' t_m + t_m} =$

$\dfrac{k'}{k' + 1} = \dfrac{5.87}{5.87 + 1} = 0.854$

26. The linear rate at which solution goes past the stationary phase determines how completely the equilibrium between the two phases is established. This determines the size of the mass transfer term (Cu_x) in the van Deemter equation. The extent of longitudinal diffusion depends on the time spent on the column, which is inversely proportional to linear flow rate.

27. Smaller plate height gives less band spreading: 0.1 mm

28. Diffusion coefficients of gases are 10^4 times greater than those of liquids. Therefore longitudinal diffusion occurs much faster in gas chromatography than in liquid chromatography.

29. The smaller the particle size, the more rapid is equilibration between mobile and stationary phases.

30. Minimum plate height is at 33 mL/min.

31. Silanization caps hydroxyl groups to which strong hydrogen bonding can occur.

32. Isotherms and band shapes are given in Figure 22-18. In overloading, the solute becomes more soluble in the stationary phase as solute concentration increases. This leaves little solute trailing behind the main band, and gives a non-Gaussian shape. Tailing occurs when small quantities of solute are retained more strongly than large quantities. The beginning of the band is abrupt, but the back part trails off slowly as the tightly bound solute is gradually eluted.

33. With 5.0 mg, the column may be overloaded. That is, the quantity of solute per unit length may be too great for the volume of stationary phase. This leads to the upper nonlinear isotherm in Figure 22-18, which broadens bands and decreases resolution.

34. Equation 22-26 says that the standard deviation of the band is proportional to $\sqrt{t}$. Here is what we know of the rate of diffusion:

time	standard deviation
t_1	$\sigma_1 = 1$
$t_2 = t_1 + 20$	$\sigma_2 = 2$
$t_3 = t_1 + 40$	$\sigma_3 = ?$

From the bandwidths at times t_1 and t_2 we can write

$$\frac{\sigma_2}{\sigma_1} = \sqrt{\frac{t_2}{t_1}} \Rightarrow \frac{2}{1} = \sqrt{\frac{t_1 + 20}{t_1}} \Rightarrow t_1 = \frac{20}{3} \text{ min}$$

For time t_3: $\quad \dfrac{\sigma_3}{\sigma_1} = \sqrt{\dfrac{t_3}{t_1}} \Rightarrow \dfrac{\sigma_3}{1} = \sqrt{\dfrac{\frac{20}{3} + 40}{\frac{20}{3}}} \Rightarrow \sigma_3 = 2.65 \text{ mm}$

35. (a) $N = \dfrac{5.55\, t_r^2}{w_{1/2}^2} = \dfrac{5.55\,(9.0\text{min})^2}{(2.0 \text{ min})^2} = 1.1 \times 10^2$ plates

(b) $10 \text{ cm}/1.1 \times 10^2$ plates $= 0.89$ mm

36. $N = \dfrac{41.7\,(t_r/w_{0.1})^2}{(A/B) + 1.25} = \dfrac{41.7\,(15/4.0)^2}{(3.0/1.0) + (1.25)} = 138$ plates

37. Resolution $= \dfrac{\Delta t_r}{w} = \dfrac{5\ \text{min}}{6\ \text{min}} = 0.83$. This is most like diagram b.

38. Since $w = 4V_r/\sqrt{N}$, w is proportional to V_r.

$w_2/w_1 = V_2/V_1 = 127/49 \Rightarrow w_2 = (127/49)(4.0) = 10.4$ mL.

39. $\sigma_{obs}^2 = \left(\dfrac{w_{1/2}}{2.35}\right)^2 = \left(\dfrac{39.6}{2.35}\right)^2 = 283.96\ s^2$

$\Delta t_{injection} = (0.40\ \text{mL})/(0.66\ \text{mL/min}) = 0.606\ \text{min} = 36.36\ s$

$\sigma_{injection}^2 = \dfrac{\Delta t_{injection}^2}{12} = \dfrac{36.36^2}{12} = 110.19\ s^2$

$\Delta t_{detector} = (0.25\ \text{mL})/(0.66\ \text{mL/min}) = 22.73\ s$

$\sigma_{detector}^2 = (\Delta t)_{detector}^2 /12 = 43.04\ s^2$

$\sigma_{obs}^2 = \sigma_{column}^2 + \sigma_{injection}^2 + \sigma_{detector}^2$

$283.96 = \sigma_{column}^2 + 110.19 + 43.04 \Rightarrow \sigma_{column} = 11.4\ s$

$w_{1/2} = 2.35\,\sigma_{column} = 26.9\ s$

40. $\alpha = \dfrac{t'_{r2}}{t'_{r1}} = \dfrac{k'_2}{k'_1} = \dfrac{K_2}{K_1} = \dfrac{0.18}{0.15} = 1.2_0$

$k'_2 = K_2 \dfrac{V_s}{V_m} = 0.18\left(\dfrac{1}{3.0}\right) = 0.060_0 \qquad k'_1 = 0.15\left(\dfrac{1}{3.0}\right) = 0.050_0$

$k'_{av} = \dfrac{k'_1 + k'_2}{2} = 0.055_0$

Resolution $= \dfrac{\sqrt{N}}{4}\left(\dfrac{\alpha-1}{\alpha}\right)\left(\dfrac{k'_2}{1 + k'_{av}}\right)$

$1.5 = \dfrac{\sqrt{N}}{4}\left(\dfrac{1.2_0 - 1}{1.2_0}\right)\left(\dfrac{0.060_0}{1 + 0.055_0}\right) \Rightarrow 4.0 \times 10^5$ plates

41. $\alpha = \dfrac{t'_{r2}}{t'_{r1}} = \dfrac{k'_2}{k'_1}$

(a) If $\alpha = 1.05$ and $k'_2 = 5.00$, $k'_1 = \dfrac{5.00}{1.05} = 4.76$ and $k'_{av} = \dfrac{5.00 + 4.76}{2} = 4.88$.

Resolution $= 1.00 = \dfrac{\sqrt{N}}{4}\left(\dfrac{1.05 - 1}{1.05}\right)\left(\dfrac{5.00}{1 + 4.88}\right) \Rightarrow N = 9.8 \times 10^3$ plates

(b) 2.6×10^3 plates

(c) 8.2×10^3 plates

(d) N can be increased by increasing the column length ($N \propto \sqrt{L}$). α can be increased by changing solvent and/or stationary phase to change the partition coefficients of the two components. k_2' can be increased by increasing the volume of stationary phase. In this problem α has a much larger effect than k_2'.

42. (a) C_6HF_5: $t_r' = 12.98 - 1.06 = 11.92$ min. $k' = 11.92/1.06 = 11.25$

C_6H_6: $t_r' = 13.20 - 1.06 = 12.14$ min. $k' = 12.14/1.06 = 11.45$

(b) $\alpha = 12.14/11.92 = 1.018$

(c) $w_{1/2}$ (C_6HF_5) = 0.124 min; $w_{1/2}$ (C_6H_6) = 0.121 min

$$C_6HF_5:\ N = \frac{5.55\ t_r^2}{w_{1/2}^2} = \frac{5.55\ (12.98)^2}{0.124^2} = 6.08 \times 10^4 \text{ plates}$$

$$\text{Plate height} = \frac{30.0\ m}{6.08 \times 10^4 \text{ plates}} = 0.493 \text{ mm}$$

$$C_6H_6:\ N = \frac{5.55\ (13.20)^2}{0.121^2} = 6.60 \times 10^4 \text{ plates}$$

$$\text{Plate height} = \frac{30.0\ m}{6.60 \times 10^4 \text{ plates}} = 0.455 \text{ mm}$$

(d) w (C_6HF_5) = 0.220 min; w (C_6H_6) = 0.239 min

$$C_6HF_5:\ N = \frac{16\ t_r^2}{w^2} = \frac{16\ (12.98)^2}{0.220^2} = 5.57 \times 10^4 \text{ plates}$$

$$C_6H_6:\ N = \frac{16\ (13.20)^2}{0.239^2} = 4.88 \times 10^4 \text{ plates}$$

(e) Resolution $= \dfrac{\Delta t_r}{w_{av}} = \dfrac{13.20 - 12.98}{0.229} = 0.96$

(f) $N = \sqrt{(5.57 \times 10^4)(4.88 \times 10^4)} = 5.21 \times 10^4$ plates

$$\text{Resolution} = \frac{\sqrt{N}}{4} \left(\frac{\alpha - 1}{\alpha}\right) \left(\frac{k_2'}{1 + k_{av}'}\right)$$

$$= \frac{\sqrt{5.21 \times 10^4}}{4} \left(\frac{1.018 - 1}{1.018}\right) \left(\frac{11.45}{1 + 11.35}\right) = 0.94$$

43. Initial concentratration (m) = 10 nmol/(1.96×10^{-3} m^2) = 5.09×10^{-6} mol/m^2. Diffusion will be symmetric about the origin. Only the positive axis is shown.

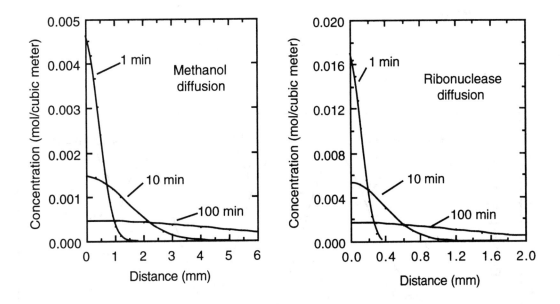

44.

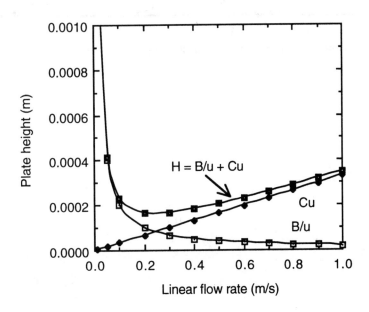

45. Inspection of Equation 4-3 shows that the general form of a Gaussian curve is $y = Ae^{-(x-x_0)^2/2\sigma^2}$, where A is a constant proportional to the area under the curve, x_0 is the abscissa of the center of the peak, and σ is the standard deviation. We can arbitrarily let $\sigma = 1$, which means that the width at the base ($w = 4\sigma$) is 4. A peak with an area of 1 centered at the origin is $y = 1*e^{-(x)^2/2}$. A curve of area 4 is $y = 4*e^{-(x-x_0)^2/2}$. The resolution is $\Delta x/w$. For a resolution of 0.5, $\Delta x = 0.5*w = 2$. That is, the second peak is centered at $x = 2$ if the resolution is 2. Its equation is $y = 4*e^{-(x-2)^2/2}$. Similarly, for a resolution of 1, $\Delta x = 1*w = 4$ and the second peak is centered at $x = 4$. For a resolution of 2, the second peak is centered at $x = 8$. The equations of the curves plotted below are:

Resolution = 0.5: $y = 1*e^{-(x)^2/2} + 4*e^{-(x-2)^2/2}$

Resolution = 1: $y = 1*e^{-(x)^2/2} + 4*e^{-(x-4)^2/2}$

Resolution = 2: $y = 1*e^{-(x)^2/2} + 4*e^{-(x-8)^2/2}$

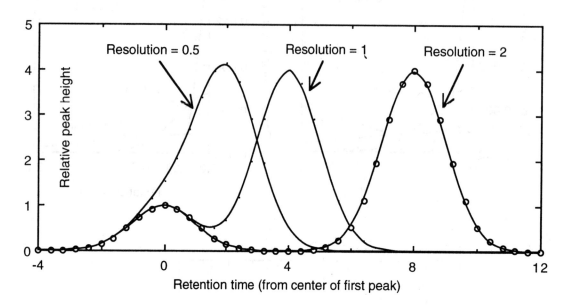

1.   (a)   Low boiling solutes are separated well at low temperature, and the retention of high boiling solutes is reduced to a reasonable time at high temperature.

(b)   Less retained solutes are eluted with a solvent of low eluent strength, and more retained solutes are eluted with increasing eluent strength.

(c)   A gradient of increasing ionic strength is used to increase the eluent strength in ion-exchange chromatography.

(d)   A gradient of increasing pressure gives increasing solvent density, which gives increasing eluent strength in supercritical fluid chromatography.

2.   (a)   Packed columns offer high sample capacity, while open tubular columns give better separation efficiency (smaller plate height), shorter analysis time and increased sensitivity to small quantities of analyte.

(b)   Wall-coated: liquid stationary phase bonded to the wall of column
Support-coated: liquid stationary phase on solid support on wall of column
Porous-layer: solid stationary phase on wall of column

(c)   The bonded stationary phase does not bleed from the column during use.

3.   (a)   Open tubular columns eliminate the multiple path term (A) from the van Deemter equation, decreasing plate height. Also, the lower resistance to gas flow allows longer columns to be used with the same elution time.

(b)   Diffusion of solute in H_2 and He is more rapid than in N_2. Therefore equilibration of solute between mobile phase and stationary phase is faster.

4.   (a)   Split injection is the ordinary mode for open tubular columns. Splitless injection is useful for trace and quantitative analysis. On-column injection is useful for thermally sensitive solutes that might decompose during a high temperature injection.

(b)   In solvent trapping, the initial column temperature is low enough to condense solvent at the beginning of the column. Solute is very soluble in the solvent and is trapped in a narrow band at the start of the column. In cold trapping, the initial column temperature is 150° lower than the boiling points of solutes, which condense in a narrow band at the start of the column. In both cases, elution occurs as the column temperature is raised.

5.   (a)   All analytes

(b)   Carbon atoms bearing hydrogen atoms

(c) Molecules with halogens, conjugated C=O, CN, NO_2

(d) P and S   (e) P and N   (f) S

6. Particles pass through 200 mesh (75 μm) sieve and are retained by 400 mesh (38 μm) sieve. 200/400 mesh particles are smaller than 100/200 mesh particles.

7. Column (a): hexane < butanol < benzene < 2-pentanone < heptane < octane
    Column (b): hexane < heptane < butanol < benzene < 2-pentanone < octane
    Column (c): hexane < heptane < octane < benzene < 2-pentanone < butanol

8. Column (a): 3,1,2,4,5,6; Column (b): 3,4,1,2,5,6; Column (c): 3,4,5,6,2,1

9. (a) $t'_r = 8.4 - 3.7 = 4.7$ min ; $k' = 4.7/3.7 = 1.3$

    (b) $k' = KV_s/V_m \Rightarrow K = (1.3)(1.4) = 1.8$

10. $I = 100 \left[8 + (9 - 8) \dfrac{\log(12.0) - \log(11.0)}{\log(14.0) - \log(11.0)} \right] = 836$

11. $\left. \begin{array}{l} \log (15.0) = \dfrac{a}{373} + b \\[2mm] \log (20.0) = \dfrac{a}{363} + b \end{array} \right\} \Rightarrow a = 1.69 \times 10^3$ K $\qquad b = -3.36$

     At 353 K : $\log t'_r = \dfrac{1.69_2 \times 10^3}{353} - 3.36 \Rightarrow t'_r = 27.1$ min

12. $\dfrac{\text{[X]/[S] in unknown}}{\text{[X]/[S] in standard}} = \dfrac{\text{area ratio in unknown}}{\text{area ratio in standard}}$

     $\dfrac{[I^-]/(160 \text{ nM})}{(63 \text{ nM}/200 \text{ nM})} = \dfrac{633/520}{395/787} \Rightarrow [I^-] = 122$ nM

     $[I^-]$ in original unknown $= \dfrac{10}{3}(122 \text{ nM}) = 0.41$ μM

13. $I = 100 \left[(7 + (10 - 7)\dfrac{\log (20.0) - \log (12.6)}{\log (22.9) - \log (12.6)} \right] = 932$

14. (a) As $k' \to 0$, $H_{min}/r = \sqrt{1/3} = 0.58$

     As $k' \to 0$, $H_{min}/r = \sqrt{\dfrac{1+6k'+11k'^2}{3(1+k')^2}} \to \sqrt{\dfrac{11k'^2}{3k'^2}} = \sqrt{\dfrac{11}{3}} = 1.9$

     (b) As $k' \to 0$, $H_{min} = 0.58\, r = 0.058$ mm
          As $k' \to \infty$, $H_{min} = 1.9\, r = 0.19$ mm

     (c) For $k' = 5$, $H_{min} = r \sqrt{\dfrac{1 + 6 \cdot 5 + 11 \cdot 25}{3(36)}} = 1.68\, r = 0.168$ mm

$$\text{Number of plates} = \frac{50 \times 10^3 \text{ mm}}{0.168 \text{ mm/plate}} = 3.0 \times 10^5$$

(d) $k' = \dfrac{2\ (0.2\ \mu m)\ (1\ 000)}{(100\ \mu m)} = 4$

15. $TZ = \dfrac{18.5 - 14.3}{0.5 + 0.5} - 1 = 3.2$

16. (a) In reverse-phase chromatography, the solutes are nonpolar and more soluble in a nonpolar mobile phase. In normal-phase chromatography, the solutes are polar and more soluble in polar mobile phase.

(b) A gradient of increasing pressure gives increasing solvent density, which gives increasing eluent strength in supercritical fluid chromatography.

17. Thin layer chromatography and column chromatography with a small-scale column.

18. Solvent is competing with solute for adsorption sites. The strength of the solvent-adsorbent interaction is independent of solute.

19. (a) Small particles give increased resistance to flow. High pressure is required to obtain a usable flow rate.

(b) A bonded stationary phase is covalently attached to the support.

(c) Efficiency increases because solute equilibrates between phases more rapidly if the thicknesses of both phases are smaller. Also, migration paths between small particles are more uniform, decreasing the multiple path term.

20. (a) Bonded reverse-phase chromatography

(b) Bonded normal-phase chromatography (Dioxane is closer to ethyl acetate than to chloroform in eluent strength in Table 23-6.)

(c) Ion-exchange or ion chromatography

(d) Molecular-exclusion chromatography

(e) Ion-exchange chromatography

(f) Molecular-exclusion chromatography

21. A Mariotte flask maintains constant eluent pressure, and hence a constant flow rate.

22. (a) Since the nonpolar compounds should become more soluble in the mobile phase, the retention time will be shorter in 90% methanol.

(b) At pH 3 the predominant forms are neutral RCO_2H and cationic RNH_3^+. The amine will be eluted first, since RNH_3^+ is insoluble in the nonpolar stationary phase.

23. Peak areas will be proportional to molar absorptivity, since the number of moles of A and B are equal.

$$\frac{\text{Area of A}}{\text{Area of B}} = \frac{2.26 \times 10^4}{1.68 \times 10^4} = \frac{hw_{1/2}}{hw_{1/2}} = \frac{(128)(10.1)}{h_B\,(7.6)} \Rightarrow h_B = 126 \text{ mm}$$

24. $$\frac{[C]/[D] \text{ in unknown}}{[C]/[D] \text{ in standard}} = \frac{\text{area ratio in unknown}}{\text{area ratio in standard}}$$

$$\frac{[C]/(12.49 \text{ mg}/25.00 \text{ ml})}{(1.03 \text{ mg/mL}/1.16 \text{ mg/mL})} = \frac{5.97/6.38}{10.86/4.37} \Rightarrow [C] = 0.167 \text{ mg/mL}$$

$$[C] \text{ in original unknown} = \frac{25}{10}(0.167 \text{ mg/mL}) = 0.418 \text{ mg/mL}$$

25. (a) Plate height for each compound depends on its rate of diffusion in the mobile and stationary phases, and the capacity factor. Since these properties differ for each solute, each has a different plate height.

(b) Tryptophan and acetone are being run above their optimum flow rate, so finite equilibration time (Cu_x) dominates the plate height expression. At sufficiently low flow rate, the curves should rise when longitudinal diffusion (B/u_x) dominates. Plate height for Blue Dextran is independent of flow rate, so the multiple path term (A) must be dominant.

26. (a) 30% tetrahydrofuran $(0.3 \times 4.4 = 1.3) > 40\%$ methanol $(0.4 \times 3.0 = 1.2)$

(b) Greater eluent strength $\Rightarrow$ lower capacity factor - 30% tetrahydrofuran

(c) $\phi\,(4.4) = (0.2)(3.1) \Rightarrow \phi = 14\%$

27. Equation 22-16: $\dfrac{t_1 - 10}{10} = 1.35 \Rightarrow t_1 = 23.5 \text{ min}$

$\alpha = 4.53 = \dfrac{t_2 - 10}{t_1 - 10} \Rightarrow t_2 = 71.2 \text{ min}$

Equation 22-23: $7.7 = \dfrac{71.2 - 23.5}{w_{av}} \Rightarrow w_{av} = 6.2 \text{ min}$

28. 10-μm-diameter spheres: volume $= \frac{4}{3}\pi r^3 = \frac{4}{3}\pi(5 \times 10^{-4} \text{ cm})^3 = 5.24 \times 10^{-10} \text{ cm}^3$

Mass of one sphere $= (5.24 \times 10^{-10} \text{ mL})(2.2 \text{ g/mL}) = 1.15 \times 10^{-9} \text{ g}$

Number of particles in 1 g $= 1 \text{ g} / (1.15 \times 10^{-9} \text{ g/particle}) = 8.68 \times 10^8$

Surface area of one particle $= 4\pi r^2 = 4\pi(5 \times 10^{-6} \text{ m})^2 = 3.14 \times 10^{-10} \text{ m}^2$

Surface area of 8.68×10^8 particles $= 0.27 \text{ m}^2$

Since the observed surface area is 300 m^2, the particles must have highly irregular shapes or be porous.

1. The separator column separates ions by ion exchange, while the suppressor exchanges the counterion to reduce the conductivity of eluent. After separating cations in the cation-exchange column, the suppressor must exchange the anion for OH^-, which makes H_2O from the HCl eluent.

2. Increased crosslinking gives decreased swelling, increased exchange capacity and selectivity, but longer equilbration time.

3. Deionized water has been passed through ion-exchange columns to convert cations to H^+ and anions to OH^-, making H_2O. Nonionic impurities (such as organic compounds) are not removed by this process.

4. One way is to wash extensively with NaOH a column containing a weighed amount of resin to load all of the sites with OH^-. After a thorough washing with water to remove excess NaOH, the column can be eluted with a large quantity of aqueous NaCl to displace the OH^-. The eluate is then titrated with standard HCl to determine the moles of displaced OH^-.

5. (a) As the pH is lowered the protein becomes protonated, so the magnitude of the negative charge decreases. The protein becomes less strongly retained.

   (b) As the ionic strength of eluent is increased, the protein will be displaced from the gel by solute ions.

6. Exchange the positions of the two buffers.

7. The pK_a values are : NH_4^+ (9.244) , $CH_3NH_3^+$ (10.64) , $(CH_3)_2NH_2^+$ (10.774) and $(CH_3)_3NH^+$ (9.800). If the four ammonium ions are adsorbed on a cation exchange resin at, say, pH 7, they might be separated by elution with a gradient of increasing pH. The anticipated order of elution is $NH_3 < (CH_3)_3N < CH_3NH_2 < (CH_3)_2NH$. We should not be surprised if the elution order were different, since steric and hydrogen bonding effects could be significant determinants of the selectivity coefficients. It is also possible that elution with a constant pH (of, say, 8) might separate all four species from each other.

8. (a) $[Cl^-]_i ([Cl^-]_i + [R^-]_i) = [Cl^-]_o^2$

   $[Cl^-]_i ([Cl^-]_i + 3.0) = (0.10)^2 \Rightarrow [Cl^-]_i = 0.003\ 33\ M$

$$\Rightarrow [Cl^-]_o/[Cl^-]_i = 0.10/0.003\ 3 = 30$$

(b) Using $[Cl^-]_o = 1.0$ in (a) gives $[Cl^-]_o/[Cl^-]_i = 1.0/0.30 = 3.3$

(c) The fraction will decrease.

9. In the 250.0 mL solution, $c = A/\varepsilon b = (0.521 - 0.049)/(900)(1.00) = 5.244 \times 10^{-4}$ M. Moles of $IO_3^- = (0.250\ L)(5.244 \times 10^{-4}\ M) = 0.131\ 1$ mmol. Therefore 0.131 1 mmol of 1,2-ethanediol must have been consumed $= 8.138$ mg of 1,2-ethanediol.

Weight percent $= 100\ \dfrac{(0.008\ 138\ g/mL)(10.0\ mL)}{0.213\ 9\ g} = 38.0\%$

10. At pH 2 (0.01 M HCl), TCA is more dissociated than DCA, which is more dissociated than MCA. The greater the average charge of the compound, the more it is excluded from the ion-exchange resin and the more rapidly it is eluted.

11. (a) Sodium octyl sulfate dissolved in the stationary phase forms an ion-pair with NE or DHBA. Other ions in the eluent compete with NE or DHBA, and slowly elute them from the column by ion exchange.

    (b) Construct a graph of (peak height ratio) vs (added concentration of NE). The x-intercept gives [NE] = 29 ng/mL.

12. This is an example of *indirect detection*. Eluent contains naphthalenetrisulfonate anion that absorbs radiation at 280 nm. Charge balance dictates that when one of the analyte anions is emerging from the column, there must be less naphthalenetri-sulfonate anion emerging. Since the analytes do not absorb as strongly at 280 nm, the absorbance is negative with respect to the steady baseline.

13. (a) There is a range in which retention volume is logarithmically related to molecular weight. The unknown is compared to a series of standards of known molecular weight.

    (b) MW 10^5 is near the middle range of the 10 μm pore size column.

14. (a) $V_t = \pi(0.80\ cm)^2\ (20.0\ cm) = 40.2$ mL

    (b) $K_{av} = \dfrac{27.4 - 18.1}{40.2 - 18.1} = 0.42$

15. The ferritin maximum is in tube 22 $(= 22 \times 0.65\ mL) = 14.3$ mL

    $V_t = \pi r^2 \times length = \pi(0.75)^2\ (37) = 65.4$ mL

    Transferrin maximum = tube 32 $= 20.8$ mL $\Rightarrow K_{av} = \dfrac{20.8 - 14.3}{65.4 - 14.3} = 0.127$

Ferric citrate maximum = tube 84 = 54.6 mL $\Rightarrow K_{av} = \dfrac{54.6 - 14.3}{65.4 - 14.3} = 0.789$

16. (a) The vertical line begins at $\log(MW) \approx 3.3 \Rightarrow MW = 2\,000$.

(b) A vertical line at 6.5 mL intersects the 10-nm calibration line at $\log(MW) \approx 2.5 \Rightarrow MW = 300$.

17. (a) The total column volume is $\pi r^2 \times length = \pi(0.39)^2 (30) = 14.3$ mL. Totally excluded molecules do not enter the pores and are eluted in the solvent volume (the void volume) outside the particles. Void volume = 40% of 14.3 mL = 5.7 mL.

(b) The smallest molecules that completely penetrate pores will be eluted in a volume that is the sum of the volumes between particles and within pores = 80% of 14.3 mL = 11.5 mL.

(c) These solutes must be adsorbed on the polystyrene resin. Otherwise they would all be eluted between 5.7 and 11.5 mL.

18. A graph of $\log(MW)$ vs V_r should be constructed.

	log(MW)	V_r(mL)
aldolase	5.199	35.6
catalase	5.322	32.3
ferritin	5.643	28.6
thyroglobulin	5.825	25.1
Blue Dextran	6.301	17.7
unknown	?	30.3

The graph of K_{av} vs $\log(MW)$ is somewhat scattered, with $\log(MW)$ for the unknown $\approx 5.5 \Rightarrow MW = 320\,000$.

19. Electroosmosis is the bulk flow of fluid in a capillary caused by migration of the dominant ion in the diffuse part of the double layer toward the anode or cathode.

20. Sample placed in port 2 is driven toward port 3 by electroosmosis when a field is imposed between these two ports. This is an example of electrokinetic injection.

21. Arginine is the only amino acid in the figure with a positively charged side chain. All of the derivatized amino acids have a net negative charge because the fluorescent group and the terminal carboxyl group are both negative. Arginine is least negative, so its electrophoretic mobility toward the anode is slowest and its net migration toward the cathode (from electroosmosis) is fastest.

22. Under ideal conditions, longitudinal diffusion is the principle source of zone broadening. Even under ideal conditions, the finite length of the injected sample and, possibly, the finite length of the detector contribute to zone broadening.

23. (a) 100×10^{-9} cm^3 (= 100 pL) = $(12 \times 10^{-4}$ cm$)(50 \times 10^{-4}$ cm$)$(length)

$\Rightarrow$ length = 0.167 mm

(b) Width of injection band = $\Delta t = \left(\dfrac{0.167 \text{ mm}}{24 \text{ mm}}\right)(8 \text{ s}) = 0.055\ 7$ s

$\sigma_{injection} = \Delta t/\sqrt{12} = 0.016$ s

(c) $\sigma_{diffusion} = \sqrt{2Dt} = \sqrt{2(1.0 \times 10^{-8} \text{ m}^2/\text{s})(8 \text{ s})} = 0.000\ 40$ s

(d) $\sigma_{total}^2 = \sigma_{diffusion}^2 + \sigma_{injection}^2 \Rightarrow \sigma_{total} = 0.016$ s

$w = 4\sigma_{total} = 0.064$ s

24. Electroosmotic flow can be reduced by (a) lowering the pH so that the charge on the capillary wall is reduced; (b) adding ions such as $^+H_3NCH_2CH_2CH_2NH_3^+$ that adhere to the capillary wall and effectively neutralize its charge; and (c) covalently attaching silanes with neutral, hydrophilic substituents to the Si—O$^-$ groups on the walls.

25. In the absence of micelles, neutral molecules are all swept through the capillary at the electroosmotic velocity. Negatively charged micelles swim upstream with some electrophoretic velocity, so they take longer than neutral molecules to reach the detector. A neutral molecule spends some time free in solution and some time dissolved in the micelles. Therefore the net velocity of the neutral molecule is reduced from the electroosmotic velocity. Because different neutral molecules have different partition coefficients between the solution and the micelles, each type of neutral molecule has its own net migration speed.

26. (a) Volume = πr^2(length) = $\pi(25 \times 10^{-6}$ m$)^2(0.006\ 0$ m$) = 1.18 \times 10^{-11}$ m^3

$\Delta P = \dfrac{128\eta L_t(\text{Volume})}{t\pi d^4} = \dfrac{128(0.001\ 0 \text{ kg/(m·s)})(0.600 \text{ m})(1.18 \times 10^{-11} \text{ m}^3)}{(4.0 \text{ s})\pi(50 \times 10^{-6} \text{ m})^4}$

$= 1.15 \times 10^4$ Pa (= 1.15×10^4 kg/(m·s^2))

(b) $\Delta P = h\rho g \Rightarrow h = \dfrac{\Delta P}{\rho g} = \dfrac{1.15 \times 10^4 \text{ kg/(m·s}^2)}{(1\ 000 \text{ kg/m}^3)(9.8 \text{ m/s}^2)} = 1.17$ m

Since the column is only 0.6 m long, we cannot raise the inlet to 1.17 m. Instead, we could use pressure at the inlet (1.15×10^4 Pa $= 0.114$ atm) or an equivalent vacuum at the outlet.

27. (a) Volume $= \pi r^2(\text{length}) = \pi(12.5 \times 10^{-6}\text{ m})^2(0.006\ 0\text{ m}) = 2.95 \times 10^{-12}\text{ m}^3$
$= 2.95\text{nL}$. Moles $= (10.0 \times 10^{-6}\text{ M})(2.95 \times 10^{-9}\text{ L}) = 29.5$ fmol.

(b) Moles injected $= \mu_{app}\left(E\ \dfrac{\kappa_b}{\kappa_s}\right)t\pi r^2 C = \mu_{app}\left(\dfrac{V}{L_t}\ \dfrac{\kappa_b}{\kappa_s}\right)t\pi r^2 C$

In order for the units to work out, we need to express the concentration, C, in mol/m^3: $(10.0 \times 10^{-6}\text{ mol/L})(1\ 000\text{ L/m}^3) = 1.00 \times 10^{-2}\text{ mol/m}^3$

$$V = \dfrac{(\text{moles})L_t(\kappa_s/\kappa_b)}{\mu_{app}\ t\pi r^2 C}$$

$$= \dfrac{(29.5 \times 10^{-15}\text{ mol})(0.600\text{ m})(1/10)}{(3.0 \times 10^{-8}\text{ m}^2/(\text{V·s}))(4.0\text{ s})\pi(12.5 \times 10^{-6}\text{ m})^2(1.00 \times 10^{-2}\text{ mol/m}^3)}$$

$$= 3.00 \times 10^3\text{ V}$$

28. Electrophoretic peak: $N = \dfrac{16\ t_r^2}{w^2} = \dfrac{16\ (6.08\text{ min})^2}{(0.080\text{ min})^2} = 9.2 \times 10^4$ plates

Chromatographic peak: $N \approx \dfrac{41.7(t_r/w_{0.1})^2}{(A/B + 1.25)}$

$$= \dfrac{41.7(6.03\text{ min}/0.37\text{ min})^2}{(1.45 + 1.25)} = 4.1 \times 10^3\text{ plates}$$

(Both plate counts from my measurements are about 1/3 lower than the values labeled in the figure from the original source.)

29. (a) Fumarate hydrodynamic radius > maleate radius $\Rightarrow$ maleate had greater mobility

(b) Since maleate moves upstream faster than fumarate, fumarate is eluted first.

(c) Since the anions move faster than the endosmotic flow, the faster anion (maleate) is eluted first.

30. (a) pH 2: $u_{neutral} = \mu_{eo}E = \left(1.3 \times 10^{-8}\ \dfrac{\text{m}^2}{\text{V·s}}\right)\left(\dfrac{27 \times 10^3\text{ V}}{0.62\text{ m}}\right) = 5.6_6 \times 10^{-4}\text{ m/s}$

Migration time $= (0.52\text{ m})/(5.6_6 \times 10^{-4}\text{ m/s}) = 9.2 \times 10^2\text{ s}$

pH 12: $u_{neutral} = \mu_{eo}E = \left(8.1 \times 10^{-8}\ \dfrac{\text{m}^2}{\text{V·s}}\right)\left(\dfrac{27 \times 10^3\text{ V}}{0.62\text{ m}}\right) = 3.5_3 \times 10^{-3}\text{ m/s}$

Migration time $= (0.52\text{ m})/(3.5_3 \times 10^{-3}\text{ m/s}) = 1.4_7 \times 10^2\text{ s}$

(b) pH 2: $\mu_{app} = \mu_{ep} + \mu_{eo} = (-1.6 + 1.3) \times 10^{-8}\ \dfrac{\text{m}^2}{\text{V·s}} = -0.3 \times 10^{-8}\ \dfrac{\text{m}^2}{\text{V·s}}$

The anion will not migrate toward the detector at pH 2.

pH 12: $\mu_{app} = \mu_{ep} + \mu_{eo} = (-1.6 + 8.1) \times 10^{-8} \dfrac{m^2}{V \cdot s} = 6.5 \times 10^{-8} \dfrac{m^2}{V \cdot s}$

$u_{anion} = \mu_{app}E = \left(6.5 \times 10^{-8} \dfrac{m^2}{V \cdot s}\right)\left(\dfrac{27 \times 10^3 \ V}{0.62 \ m}\right) = 2.8_3 \times 10^{-3} \ m/s$

Migration time $= (0.52 \ m)/(2.8_3 \times 10^{-3} \ m/s) = 1.8_4 \times 10^2 \ s$

31. At low voltage (low electric field), the number of plates increases in proportion to the voltage, as predicted by Equation 24-14. Above ~25 000 V/m, the capillary is probably overheating, which produces band broadening and decreases the number of plates.

32. $N = \dfrac{5.55 \ t_r^2}{w_{1/2}^2} = \dfrac{5.55 \ (17.12 \ min)^2}{(0.18_5 \ min)^2} = 4.7 \times 10^4 \ plates$

Plate height $= 0.60 \ m/(4.7 \times 10^4 \ plates) = 13 \ \mu m$

33. $t = \dfrac{L}{u_{net}} = \dfrac{L}{\mu_{app}E}$ (t = migration time, L = length, u = speed, E = field)

$\Rightarrow \mu_{app} = \dfrac{L}{tE} = \dfrac{L/E}{17.12}$ for Cl^- and $\mu_{app} = \dfrac{L/E}{17.78}$ for I^-

Therefore we can write

$\Delta\mu_{app}(I\text{-}Cl) = \dfrac{L/E}{17.12} - \dfrac{L/E}{17.78}$ (L/E is an unknown constant)

But we know that $\Delta\mu_{app}(I\text{-}Cl) = [\mu_{eo} + \mu_{ep}(I^-)] - [\mu_{eo} + \mu_{ep}(Cl^-)]$

$= 0.05 \times 10^{-8} \ m^2/(s \cdot V)$ in Table 15-1

For the difference between Cl^- and Br^- we can say

$\Delta\mu_{app}(Br\text{-}Cl) = \dfrac{L/E}{17.12} - \dfrac{L/E}{x}$

and we know that $\Delta\mu_{app}(Br\text{-}Cl) = 0.22 \times 10^{-8} \ m^2/(s \cdot V)$ in Table 15-1

Therefore we can set up a proportion:

$\dfrac{\Delta\mu_{app}(Br\text{-}Cl)}{\Delta\mu_{app}(I\text{-}Cl)} = \dfrac{0.22}{0.05} = \dfrac{\dfrac{L/E}{17.12} - \dfrac{L/E}{x}}{\dfrac{L/E}{17.12} - \dfrac{L/E}{17.78}} \Rightarrow x = 20.5 \ min$

The observed migration time is 19.6 min. Considering the small number of significant digits in the $\Delta\mu$ values, this is a reasonable discrepancy.

34. SO_4^{2-}: $\mu_{ep} = -8.27 \times 10^{-8} \ m^2/(s \cdot V)$ in Table 15-1

$\mu_{app} = \mu_{eo} + \mu_{ep} = 16.1 \times 10^{-8} - 8.27 \times 10^{-8} = 7.8_3 \times 10^{-8} \ m^2/(s \cdot V)$

Br^-: $\mu_{ep} = -8.13 \times 10^{-8} \ m^2/(s \cdot V)$ in Table 15-1

$$\mu_{app} = \mu_{eo} + \mu_{ep} = 16.1 \times 10^{-8} - 8.13 \times 10^{-8} = 7.9_7 \times 10^{-8} \ m^2/(s{\cdot}V)$$

$$\mu_{av} = \frac{1}{2}(7.8_3 + 7.9_7 \times 10^{-8}) = 7.9_0 \times 10^{-8} \ m^2/(s{\cdot}V)$$

$$\Delta\mu = (8.27 - 8.13) \times 10^{-8} = 0.14 \times 10^{-8} \ m^2/(s{\cdot}V)$$

$$N = \left(4 \ (Resolution) \ \frac{\mu_{av}}{\Delta\mu}\right)^2 = \left(4 \ (2.0) \ \frac{7.9_0}{0.14}\right)^2 = 2.0 \times 10^5 \ plates$$

35. In the absence of micelles, the expected order of elution is cations before neutrals before anions: thiamine < (niacinamide + riboflavin) < niacin. Since thiamin is eluted last, it must be most soluble in the micelles.

36. For the acid H_2A, the average charge is $\alpha_{HA^-} + 2\alpha_{A^{2-}}$, where α is the fraction in each form. From our study of acids and bases, we know that

$$\alpha_{HA^-} = \frac{K_1[H^+]}{[H^+]^2 + K_1[H^+] + K_1K_2} \quad \text{and} \quad \alpha_{A^{2-}} = \frac{K_1K_2}{[H^+]^2 + K_1[H^+] + K_1K_2}$$

where K_1 and K_2 are the acid dissociation constants of H_2A. The spreadsheet below computes the average charge of malonic acid (designated H_2M) and phthalic acid (designated H_2P) and finds that the maximum difference between them occurs at pH 5.55.

	A	B	C	D	E	F	G	H	I	J
				Alpha	Alpha	Alpha	Alpha	Average charges		Charge
1	Malonic:			HM-	M2-	HP-	P2-	Malonate	Phthalate	Difference
2	K1 =	pH	[H+]							
3	1.42E-03	5.52	3.0E-06	0.600	0.399	0.436	0.563	-1.398	-1.562	-0.16392
4	K2 =	5.53	3.0E-06	0.594	0.405	0.430	0.569	-1.403	-1.567	-0.16405
5	2.01E-06	5.54	2.9E-06	0.589	0.410	0.425	0.574	-1.409	-1.573	-0.16413
6	Phthalic:	5.55	2.8E-06	0.583	0.416	0.419	0.580	-1.415	-1.579	-0.16418
7	K1 =	5.56	2.8E-06	0.577	0.421	0.413	0.585	-1.420	-1.584	-0.16417
8	1.12E-03	5.57	2.7E-06	0.572	0.427	0.408	0.591	-1.426	-1.590	-0.16413
9	K2 =	5.58	2.6E-06	0.566	0.433	0.402	0.597	-1.432	-1.596	-0.16404
10	3.90E-06	5.59	2.6E-06	0.561	0.438	0.397	0.602	-1.437	-1.601	-0.16391
11										
12	D3 = $A$3*C3/(C3^2+$A$3*C3+$A$3*$A$5)						C3 = 10^-B3			
13	E3 = $A$3*$A$5/(C3^2+$A$3*C3+$A$3*$A$5)						H3 = -D3-2*E3			
14	F3 = $A$8*C3/(C3^2+$A$8*C3+$A$8*$A$10)						I3 = -F3-2*G3			
15	G3 = $A$8*$A$10/(C3^2+$A$8*C3+$A$8*$A$10)						J3 =I3-H3			

GRAVIMETRIC AND COMBUSTION ANALYSIS

1. See glossary.

2. An ideal gravimetric precipitate should be insoluble, easily filterable, pure and possess a known, constant composition.

3. High relative supersaturation often leads to formation of colloidal product with a large amount of impurities.

4. Relative supersaturation can be decreased by increasing temperature (for most solutions), mixing well during addition of precipitant and using dilute reagents. Homogeneous precipitation is also an excellent way to control relative supersaturation.

5. Washing with electrolyte preserves the electric double layer and prevents peptization.

6. The volatile HNO_3 bakes off during drying. $NaNO_3$ is nonvolatile and will lead to a high mass for the precipitate.

7. During the first precipitation, the concentration of unwanted species in the solution is high, giving a relatively high concentration of impurities in the precipitate. In the reprecipitation, the level of solution impurities is reduced, giving a purer precipitate.

8. In thermogravimetric analysis the mass of a sample is measured as the sample is heated. The mass lost during decomposition provides some information about the composition of the sample.

9. $$\frac{0.214\ 6\ g\ AgBr}{187.772\ g\ AgBr/mol} = 1.142\ 9 \times 10^{-3}\ mol\ AgBr$$

   $$[NaBr] = \frac{1.142\ 9 \times 10^{-3}\ mol}{50.00 \times 10^{-3}\ L} = 0.022\ 86\ M$$

10. $$\frac{0.104\ g\ CeO_2}{172.114\ g\ CeO_2/mol} = 6.043 \times 10^{-4}\ mol\ CeO_2 = 6.043 \times 10^{-4}\ mol\ Ce$$

    $$= 0.084\ 66\ g\ Ce$$

    weight % Ce $= \dfrac{0.084\ 66\ g}{4.37\ g} \times 100 = 1.94\ wt\ \%$

11. One mole of product (206.243 g) comes from one mole of piperazine (86.137 g). Grams of piperazine in sample =

    (0.712 9 g of piperazine / g of sample) × (0.050 02 g of sample) = 0.035 66.

Mass of product $= \left(\dfrac{206.243}{86.137}\right)(0.035\ 66) = 0.085\ 38$ g.

12. 2.500 g bis(dimethylglyoximate) nickel (II) $= 8.653\ 2 \times 10^{-3}$ mol Ni $= 0.507\ 85$ g Ni $= 50.79\%$ Ni.

13. Formula weights: $CaC_{14}H_{10}O_6 \cdot H_2O$ (332.32), $CaCO_3$ (100.09), CaO (56.08).

At 550°, $CaC_{14}H_{10}O_6 \cdot H_2O$ is converted to $CaCO_3$.

332.32 g of starting material will produce 100.09 g of CaO.

Mass at 550° $= (100.09/332.32)(0.635\ 6$ g$) = 0.191\ 4$ g. At 1 000° C, the

product is CaO and the mass is $(56.08/332.32)(0.635\ 6$ g$) = 0.107\ 3$ g.

14. 2.378 mg CO_2 / (44.010 g/mol) $= 5.403\ 3 \times 10^{-5}$ mol $CO_2 = 5.403\ 3 \times 10^{-5}$ mol C
$= 6.490\ 0 \times 10^{-4}$ g C.

ppm C $= 10^6 (6.490\ 0 \times 10^{-4} / 6.234) = 104.1$ ppm

15. 2.07% of 0.998 4 g $= 0.020\ 67$ g of Ni $= 3.520 \times 10^{-4}$ mol of Ni.

This requires $(2)(3.520 \times 10^{-4})$ mol of DMG $= 0.081\ 75$ g.

A 50.0% excess is $(1.5)(0.081\ 75$ g$) = 0.122\ 6$ g. The mass of solution containing

0.122 6 g is 0.122 6 g DMG / (0.021 5 g DMG/g solution) $= 5.704$ g of solution.

The volume of solution is 5.704 g/(0.790 g/mL) $= 7.22$ mL.

16. Moles of Fe in product (Fe_2O_3) = moles of Fe in sample.

Because 1 mole of (Fe_2O_3) contains 2 moles of Fe, we can write the equation

$$\frac{2\ (0.264\ \text{g})}{159.69\ \text{g/mol}} = 3.306 \times 10^{-3}\ \text{mol of Fe.}$$

This many moles of Fe equals 0.919 g of $FeSO_4 \cdot 7\ H_2O$.

17. Let x = mass of NH_4Cl and y = mass of K_2CO_3.

For the first part, 1/4 of the sample (25 mL) gave 0.617 g of precipitate containing
both products:

$$\frac{1}{4}\left[\underbrace{\overbrace{\left(\frac{x}{53.492}\right)}^{\text{mol } NH_4Cl}(337.27)}_{\text{g }\phi_4BNH_4} + \underbrace{\overbrace{\left(\frac{2y}{138.21}\right)}^{\text{mol } K_2CO_3 \times 2}(358.33)}_{\text{g }\phi_4BK \quad (\phi = \text{phenyl} = C_6H_5)}\right] = 0.617$$

We multiplied the moles of K_2CO_3 by 2 because one mole of K_2CO_3 gives 2 moles
of ϕ_4BK. In the second part, 1/2 of the sample (50 mL) gave 0.554 g of ϕ_4BK:

$$\underbrace{\frac{1}{2}\underbrace{\left(\frac{\overbrace{2y}^{\text{mol K}_2\text{CO}_3 \times 2}}{138.21}\right)}_{g\,\phi_4\text{BK}}(358.33)=0.554}_{} \quad \Rightarrow \quad y = 0.2137\ g = 14.49\ \text{wt}\%\ K_2CO_3$$

Putting this value of y into the first equation gives $x = 0.216\ g = 14.6\ \text{wt}\%\ NH_4Cl$

18.
$$\underbrace{Fe_2O_3 + Al_2O_3}_{2.019} \xrightarrow[H_2]{\text{heat}} \underbrace{Fe + Al_2O_3}_{1.774\ g}$$

The mass of oxygen lost is 2.019 - 1.774 = 0.245 g, which equals 0.015 31 moles of oxygen atoms. For every 3 moles of oxygen there is 1 mole of Fe_2O_3, so moles of $Fe_2O_3 = \frac{1}{3}(0.015\ 31) = 0.005\ 105$ mol of Fe_2O_3. This much Fe_2O_3 equals 0.815 g, which is 40.4 wt % of the original sample.

19. Let $x = g$ of $FeSO_4 \cdot (NH_4)_2\ SO_4 \cdot 6H_2O$ and $y = g$ of $FeCl_2 \cdot 6H_2O$.
We can say that $x + y = 0.548\ 5$ g. The moles of Fe in the final product (Fe_2O_3) must equal the moles of Fe in the sample.
The moles of Fe in $Fe_2O_3 = 2$ (moles of Fe_2O_3) $= 2\left(\frac{0.167\ 8}{159.69}\right) = 0.002\ 101\ 6$ mol.

Mol Fe in $FeSO_4 \cdot (NH_4)_2\ SO_4 \cdot 6\ H_2O = x\ /\ 392.13$ and
mol Fe in $FeCl_2 \cdot 6H_2O = y\ /\ 234.84$.

$$0.002\ 101\ 6 = \frac{x}{392.13} + \frac{y}{234.84} \qquad (1)$$

Substituting $x = 0.548\ 5 - y$ into eq. (1) gives $y = 0.411\ 46$ g of $FeCl_2 \cdot 6H_2O$.
Mass of $Cl = 2\left(\frac{35.453}{234.84}\right)(0.411\ 46) = 0.124\ 23\ g = 22.65\ \text{wt}\%$

20. (a) Let x = mass of $AgNO_3$ and $(0.432\ 1 - x)$ = mass of $Hg_2(NO_3)_2$.

$$\underbrace{\frac{1}{3}\left(\overbrace{\frac{x}{169.873}}^{\text{mol Ag}_3\text{Co(CN)}_6}\right)(538.643)}_{\text{mass of Ag}_3\text{Co(CN)}_6} + \underbrace{\frac{1}{3}\left(\overbrace{\frac{0.432\ 1-x}{525.19}}^{\text{mol (Hg}_2)_3[\text{Co(CN)}_6]_2}\right)(1\ 633.62)}_{\text{mass of (Hg}_2)_3[\text{Co(CN)}_6]_2} = 0.451\ 5$$

$\Rightarrow\ x = 0.173\ 1\ g = 40.05\ \text{wt}\%$

(b) 0.30% error in 0.451 5 g $= \pm 0.001\ 35$ g. This changes the equation of (a) to:

$$\frac{1}{3}\left(\frac{x}{169.873}\right)(538.643) + \frac{1}{3}\left(\frac{0.432\ 1 - x}{525.19}\right)(1\ 633.62) = 0.451\ 5\ (\pm 0.001\ 35)$$

Solving for x gives: $x = \dfrac{0.003\ 479\ 79(\pm 0.001\ 35)}{0.020\ 108\ 542}$

$$= \frac{0.003\ 479\ 79(\pm 38.9\%)}{0.020\ 108\ 542} = 0.173\ 1\ g \pm 38.9\%$$

Relative error = 39%.

21. (a) Balanced equation for overall (31.8%) mass loss:

$$Y_2(OH)_5Cl \cdot xH_2O \xrightarrow{\text{31.8\% mass loss}} Y_2O_3 + \underbrace{xH_2O + 2H_2O} + HCl$$

FW 298.30 + x(18.015)   FW 225.81   FW (2+x)(18.015))   FW 36.461

$$\underbrace{(2+x)(18.015) + 36.461}_{\text{mass lost}} = \underbrace{(0.318)[298.30 + x(18.015)]}_{\text{31.8\% of original mass}} \Rightarrow x = 1.82$$

(b) Logical molecular units that could be lost are H_2O and HCl. At ~8.1% mass loss, the product is $Y_2(OH)_5Cl$. Loss of 2 more H_2O would give a total mass loss of

$$\frac{1.82H_2O + 2H_2O}{Y_2(OH)_5Cl \cdot 1.82H_2O} = \frac{68.82}{331.09} = 20.8\%$$

Loss of HCl from $Y_2(OH)_5Cl$ would give a total mass loss of

$$\frac{1.82H_2O + HCl}{Y_2(OH)_5Cl \cdot 1.82H_2O} = \frac{69.25}{331.09} = 20.9\%$$

The composition at the ~19.2% plateau could be either $Y_2O_2(OH)Cl$ (from loss of $2H_2O$) or $Y_2O(OH)_4$ (from loss of HCl).

22. (a) Formula weight of $YBa_2Cu_3O_{7-x}$ = 666.194 - (15.999 4) x

$$\text{mmol of } YBa_2Cu_3O_{7-x} \text{ in experiment} = \frac{34.397\ mg}{[666.194 - (15.999\ 4)x]mg/mmol}$$

$$\text{mmol of oxygen atoms lost in experiment} = \frac{(34.397 - 31.661)\ mg}{15.999\ 4\ mg/mmol}$$

$$= 0.171\ 006\ 4\ mmol$$

From the stoichiometry of the reaction, we can write

$$\frac{\text{mmol oxygen atoms lost}}{\text{mmol } YBa_2Cu_3O_{7-x}} = \frac{3.5-x}{1}$$

$$\frac{0.171\ 006\ 4}{34.397\ /\ [666.194 - (15.999\ 4)x]} = 3.5 - x \Rightarrow x = 0.204\ 2$$

(without regard to significant figures)

(b) Now let the uncertainty in each mass be 0.002 mg and let all atomic and molecular weights have negligible uncertainty.

The mmol of oxygen atoms lost are:

$$\frac{[34.397(\pm0.002) - 31.661(\pm0.002)]\text{mg}}{15.999\ 4\ \text{mg / mmol}} = \frac{2.736(\pm0.002\ 8)}{15.999\ 4}$$

$$= 0.171\ 006\ 4\ (\pm0.103\%)$$

The relative error in the mass of starting material is $\frac{0.002}{34.397} = 0.005\ 8\%$

The master equation becomes

$$\frac{0.171\ 006\ 4\ (\pm0.103\%)}{34.397\ (\pm0.005\ 8\%)/[666.194 - (15.999\ 4)\ x]} = 3.5 - x$$

$$0.171\ 006\ 4\ (\pm0.103\%)[666.194 - (15.999\ 4)\ x] = (3.5 - x)[34.397\ (\pm0.005\ 8\%)]$$

$$113.923\ 4\ (\pm0.117) - [2.735\ 999\ (\pm0.002\ 82)]\ x$$

$$= 120.389\ 5\ (\pm0.006\ 98) - [34.397\ (\pm0.002)]\ x$$

$$[31.66\ (\pm0.003\ 46)]\ x = 6.466\ 1\ (\pm0.117)$$

$$= 0.204\ 2\ (\pm1.813\%) = 0.204 \pm 0.004$$

23. (a) $70\ \text{kg} \left(\frac{6.3\ \text{g P}}{\text{kg}}\right) = 441\ \text{g P in } 8.00 \times 10^3\ \text{L}$. This corresponds to

$$\frac{441\ \text{g P}}{8.00 \times 10^3 \text{L}} = 0.055\ 1\ \text{g/L or } 5.5_1\ \text{mg/100 mL.}$$

(b) Fraction of P in one formula mass is $\frac{2(30.974)}{3\ 596.46} = 1.722\%$.

P in 0.338 7 g of $P_2O_5 \cdot 24\ MoO_3 = (0.017\ 22)(0.338\ 7) = 5.834\ \text{mg}$

This is near the amount expected from a dissolved man.

24. In *combustion*, a substance is heated in the presence of excess O_2 to convert carbon to CO_2 and hydrogen to H_2O. In *pyrolysis*, the substance is decomposed by heating in the absence of added O_2. All oxygen in the sample is converted to CO by passage through a suitable catalyst.

25. WO_3 catalyzes the complete combustion of C to CO_2 in the presence of excess O_2.

26. The tin capsule melts and is oxidized to SnO_2 to liberate heat and crack the sample. Tin uses the available oxygen immediately, ensures that sample oxidation occurs in the gas phase, and acts as an oxidation catalyst.

27. By dropping the sample in before very much O_2 is present, pyrolysis of the sample to give gaseous products occurs prior to oxidation. This minimizes the formation of nitrogen oxides.

28. $\quad C_6H_5CO_2H + \frac{15}{2}O_2 \quad \rightarrow \quad 7\ CO_2 + 3\ H_2O$

FW $\quad$ 122.123 $\qquad\qquad\qquad\qquad$ 44.010 $\quad$ 18.015

One mole of $C_6H_5CO_2H$ gives 7 moles of CO_2 and 3 moles of H_2O.
4.635 mg of benzoic acid = 0.037 95 mmol which gives 0.265 7 mmol CO_2
(= 11.69 mg CO_2) and 0.113 9 mmol H_2O (= 2.051 mg H_2O).

29. $C_8H_7NO_2SBrCl + 9\frac{1}{4}O_2 \rightarrow 8CO_2 + \frac{5}{2}H_2O + \frac{1}{2}N_2 + SO_2 + HBr + HCl$

30. 100 g of compound contains 46.21 g C, 9.02 g H, 13.74 g N and 31.04 g O. The atomic ratios are

$$C:H:N:O = \frac{46.21\text{ g}}{12.011\text{ g/mol}} : \frac{9.02\text{ g}}{1.008\text{ g/mol}} : \frac{13.74\text{ g}}{14.007\text{ g/mol}} : \frac{31.04\text{ g}}{15.999\text{ g/mol}}$$

$$= 3.847 : 8.94_9 : 0.981\ 0 : 1.940$$

Dividing by the smallest factor (0.981 0) gives the ratios $C:H:N:O = 3.922 : 9.12 : 1 : 1.978$. The empirical formula is probably $C_4H_9NO_2$.

31. $\qquad C_6H_{12} \quad + \quad C_2H_4O \quad \rightarrow \quad CO_2 \quad + \quad H_2O$

FW   84.161   44.053   44.010

Let x = mg of C_6H_{12} and y = mg of C_2H_4O

$\qquad$ x + y = 7.290.

We also know that moles of CO_2 = 6 (moles of C_6H_{12}) + 2 (moles of C_2H_4O), by conservation of carbon atoms.

$$6\left(\frac{x}{84.161}\right) + 2\left(\frac{y}{44.053}\right) = \frac{21.999}{44.010}$$

Making the substitution x = 7.290 - y allows us to solve for y.

$\qquad$ y = 0.767 mg = 10.5 wt %.

32. The atomic ratio H:C is

$$\frac{\left(\dfrac{6.76 \pm 0.12\text{ g}}{1.008\text{ g/mol}}\right)}{\left(\dfrac{71.17 \pm 0.41\text{ g}}{12.011\text{ g/mol}}\right)} = \frac{6.706 \pm 0.119}{5.925 \pm 0.034\ 1} = \frac{6.706 \pm 1.78\%}{5.925 \pm 0.576\%} = 1.132 \pm 0.021$$

If we define the stoichiometry coefficient for C to be 8, then the stoichiometry coefficient for H is $8(1.132 \pm 0.021) = 9.06 \pm 0.17$.

The atomic ratio N:C is

$$\frac{\left(\dfrac{10.34 \pm 0.08\text{ g}}{14.007\text{ g/mol}}\right)}{\left(\dfrac{71.17 \pm 0.41\text{ g}}{12.011\text{ g/mol}}\right)} = \frac{0.738\ 2 \pm 0.005\ 7}{5.925 \pm 0.034\ 1} = \frac{0.738\ 2 \pm 0.774\%}{5.925 \pm 0.576\%}$$

$$= 0.124\ 6 \pm 0.001\ 2$$

If we define the stoichiometry coefficient for C to be 8, then the stoichiometry coefficient for N is $8(0.124\,6 \pm 0.001\,2) = 0.996\,8 \pm 0.009\,6$.

The empirical formula is reasonably expressed as $C_8H_{9.06\pm0.17}N_{0.997\pm0.010}$

33.   The reaction between H_2SO_4 and NaOH can be written

$$H_2SO_4 + 2NaOH \rightarrow 2H_2O + Na_2SO_4$$

One mole of H_2SO_4 requires two moles of NaOH. In 3.01 mL of 0.015 76 M NaOH there are $(0.003\,01\,L)(0.015\,76\,mol/L) = 4.74_4 \times 10^{-5}$ mol of NaOH. The moles of H_2SO_4 must have been $\left(\frac{1}{2}\right)(4.74_4 \times 10^{-5}) = 2.37_2 \times 10^{-5}$ mol.

Because one mole of H_2SO_4 contains one mole of S, there must have been $2.37_2 \times 10^{-5}$ mol of S $(= 0.760_4$ mg). The percentage of S in the sample is

$$\frac{0.760_4\text{ mg S}}{6.123\text{ mg sample}} \times 100 = 12.4\text{ wt \%}$$

1. (a) A homogeneous sample, such as filtered water, has the same composition everywhere. The composition of a heterogeneous sample, such as soil, varies from place to place.

(b) The composition of a random heterogeneous sample varies randomly from place to place. In a segregated material there are obviously different sections, such as layers or pockets, with distinctive compositions.

2. There is no point analyzing a sample if you do not know that it was selected in a sensible way and stored so that its composition did not change after it was taken.

3. (a) $s_o^2 = s_a^2 + s_s^2 = 3^2 + 4^2 \Rightarrow s_o = 5\%$.

(b) $s_s^2 = s_o^2 - s_a^2 = 4^2 - 3^2 \Rightarrow s_s = 2.6\%$.

4. Locate the different regions and make an estimate of their relative sizes. Then divide each region into a series of real or imaginary segments. Using a table of random numbers, select a number of segments from each region such that the number of segments is proportional to the size of the region. The resulting composite sample should be representative of the original lot.

5. $mR^2 = K_s$. $m(6^2) = 36 \Rightarrow m = 1.0$ g

6. Pass the powder through a 120 mesh sieve and then through a 170 mesh sieve. Sample retained by 170 mesh sieve has a size between 90 and 125 μm. It would be called 120/170 mesh.

7. 11.0×10^2 g will contain 10^6 total particles, since 11.0 g contains 10^4 particles.

$n_{KCl} = np = (10^6)(0.01) = 10^4$.

Relative standard deviation $= \sqrt{npq} / n_{KCl} = \sqrt{(10^6)(0.01)(0.99)} / 10^4 = 0.99\%$.

8. (a) $\sqrt{(10^3)(0.5)(0.5)} = 15.8$.

(b) We are looking for the value of z whose area is 0.45 (since the area from -z to+z is 0.90). The value lies between z = 1.6 and 1.7, whose areas are 0.445 2 and 0.455 4, respectively. Linear interpolation:

$$\frac{z - 1.6}{1.7 - 1.6} = \frac{0.45 - 0.445\ 2}{0.455\ 4 - 0.445\ 2} \Rightarrow z = 1.647.$$

(c) Since $z = (x-\bar{x})/s$, $x = \bar{x} \pm zs = 500 \pm (1.647)(15.8) = 500 \pm 26$.
The range 474-526 will be observed 90% of the time.

9. Use Equation 26-7 with $s_s = 0.05$ and $e = 0.04$. The initial value of t for 95% confidence in Table 4-2 is 1.960. $n = t^2 s_s^2 / e^2 = 6.0$ For n = 6, there are 5 degrees of freedom, so t = 2.571, which gives n = 10.3. For 9 degress of freedom, t = 2.262, which gives n = 8.0. Continuing, we find t = 2.365 $\Rightarrow$ n = 8.74. This gives t = 2.306 $\Rightarrow$ n = 8.30. Use <u>8 samples</u>. For 90% confidence, the initial t is 1.645 in Table 4-2 and the same series of calculations gives n = <u>6 samples</u>.

10. (a) $mR^2 = K_S$. For R = 2 and $K_S = 20$ g, we find m = 5.0 g.

(b) Use Equation 26-7 with $s_s = 0.02$ and $e = 0.015$. The initial value of t for 90% confidence in Table 4-2 is 1.645. $n = t^2 s_s^2 / e^2 = 4.8$ For n = 5, there are 4 degrees of freedom, so t = 2.132, which gives n = 8.1. For 7 degress of freedom, t = 1.895, which gives n = 6.4. Continuing, we find t = 2.015 $\Rightarrow$ n = 7.2. This gives t = 1.943 $\Rightarrow$ n = 6.7. Use <u>7 samples</u>.

11. (a) Volume = $(4/3)\pi r^3$, where r = 0.075 mm = 7.5×10^{-3} cm.
Volume = 1.767×10^{-6} mL.
Na_2CO_3 mass = $(1.767 \times 10^{-6}$ mL$)(2.532$ g/mL$) = 4.474 \times 10^{-6}$ g.
K_2CO_3 mass = $(1.767 \times 10^{-6}$ mL$)(2.428$ g/mL$) = 4.291 \times 10^{-6}$ g.
Number of particles of Na_2CO_3 = (4.00 g)/(4.474×10^{-6} g/particle) = 8.941×10^5.
Number of particles of K_2CO_3 = (96.00 g)/(4.291×10^{-6} g/particle) = 2.237×10^7.
The fraction of each type (which we will need for part c) is
$p_{Na_2CO_3}$ = $(8.941 \times 10^5)/(8.941 \times 10^5 + 2.237 \times 10^7) = 0.038\ 4$
$q_{K_2CO_3}$ = $(2.237 \times 10^7)/(8.941 \times 10^5 + 2.237 \times 10^7) = 0.962$.

(b) Total number of particles in 0.100 g is n = 2.326×10^4.

(c) Expected number of Na_2CO_3 particles in 0.100 g is 1/1000 of number in 100 grams = 8.94×10^2.
Expected number of K_2CO_3 particles in 0.100 g is 1/1000 of number in 100 grams = 2.24×10^4.
Sampling standard deviation = $\sqrt{npq} = \sqrt{(2.326 \times 10^4)(0.038\ 4)(0.962)}$ = 29.3.
Relative sampling standard deviation for $Na_2CO_3 = \dfrac{29.3}{8.94 \times 10^2} = 3.28\ \%$
Relative sampling standard deviation for $K_2CO_3 = \dfrac{29.3}{2.24 \times 10^4} = 0.131\ \%$.

12. Metals with reduction potentials below zero [for the reaction $M^{n+} + ne^- \rightarrow M(s)$]

are expected to dissolve in acid. These are Zn, Fe, Co and Al.

13. HNO_3 was used first to oxidize any material that could be easily oxidized. This helps prevent the possibility that an explosion will occur when $HClO_4$ is added.

14. Barbital has a higher affinity for the octadecyl phase than for water, so it is retained by the column. The drug dissolves readily in acetone/chloroform, which elutes it from the column.

15. An elemental standard is used when you need a known quantity of an element. A matrix matching standard is used to provide a matrix free of the intended analyte.

16. Pickling (washing with dilute acid) removes surface oxides and debris from the cutting tool.

17. The product gas stream is passed through an anion-exchange column, on which SO_2 is absorbed by the following reactions:

$$SO_2 + H_2O \rightarrow H_2SO_3$$
$$2Resin^+OH^- + H_2SO_3 \rightarrow (Resin^+)_2SO_3^{2-} + H_2O$$

The sulfite is eluted with Na_2CO_3/H_2O_2, which oxidizes it to sulfate that can be measured by ion chromatography.

18. (a) Highest concentration of Ni $\approx$ 80 ng/mL. A 10 mL sample contains 800 ng Ni $= 1.36 \times 10^{-8}$ mol Ni. To this is added 50 μg Ga $= 7.17 \times 10^{-7}$ mol Ga. Atomic ratio Ga/Ni $= (7.17 \times 10^{-7})/(1.36 \times 10^{-8}) = 53$.

   (b) Apparently all of the Ni and Mn is in solution because filtration does not decrease their total concentration. Since filtration removes most of the Fe, it must be present as a suspension of solid particles.

19. One fourth of the sample (25 mL out of 100 mL) required $(0.011\ 44\ M)(0.032\ 49\ L) = 3.717 \times 10^{-4}$ mol EDTA $\Rightarrow (3.717 \times 10^{-4})(4) = 1.487 \times 10^{-3}$ mol Ba^{2+} in sample $= 0.204\ 2$ g Ba $= 64.90$ wt %.

20. Balanced combustion equation: $CH_2O + O_2 \rightarrow CO_2 + H_2O$
           FW 30.026

0.100 g $CH_2O = 3.330$ mmol, requiring 3.330 mmol O_2. This volume of O_2 is contained in

$$V = \frac{nRT}{P} = \frac{(3.330 \times 10^{-3}\ \text{mol})(0.082\ 06\ \text{L·atm/mol·K})(298\ \text{K})}{1.00\ \text{atm}} = 81.4\ \text{mL}$$

S1. Write the following quantities in exponential notation, with one digit to the left of the decimal point (e.g., 17 fC $= 1.7 \times 10^{-14}$ C):

(a) 2 TJ   (c) 37 Mm   (e) 842 pF

(b) 37 mm   (d) 4 dK   (f) 18.4 kPa

S2. Express the following quantities with abbreviations for units and prefixes from Tables 1-1 through 1-3:

(a) 8×10^{-5} moles   (d) 3×10^{-2} meters

(b) 1×10^{10} watts   (e) 1.8×10^{14} hertz

(c) 4×10^{-7} liters   (f) 537×10^{10} ohms

S3. The lowest temperature attained in the laboratory in 1990 was 800 pK (for the nuclei of silver atoms). Solid ^{3}He has been cooled to 43 μK. Express the quotient 800 pK/43 μK in exponential notation (ie $a \times 10^b$).

S4. Use Table 1-4 to confirm that there are 760 torr in 1 atm.

S5. If 0.250 L of aqueous solution with a density of 1.00 g/mL contins 13.7 μg of pesticide, express the concentration of pesticide in (a) ppm and (b) ppb.

S6. Find the molarity of pyridine (C_5H_5N) if 5.00 g is dissolved in butanol to give a total volume of 457 mL.

S7. A 95.0 wt % solution of ethanol (CH_3CH_2OH, MW 46.07) in water has a density of 0.804 g/mL.

(a) Find the mass of 1.00 L of this solution and the number of grams of ethanol per liter.

(b) What is the molar concentration of ethanol in this solution?

(c) Find the molality of ethanol in this solution, considering H_2O to be the solvent (even though H_2O is really the solute in this case).

S8. (a) How many grams of the element nickel are contained in 10.0 g of a 10.2 wt % solution of nickel sulfate hexahydrate, $NiSO_4 \cdot 6H_2O$ (FW 262.85)?

(b) The concentration of this solution is 0.412 M. Find the density.

S9. Describe how to prepare exactly 100 mL of 1.00 M HCl from 12.1 M HCl reagent.

S10. Describe how to prepare approximately 100 mL of 0.082 m $NaClO_4$ (FW 122.44).

S11. A 40.0 wt % solution of CsCl (FW 168.37) has a density of 1.43 g/mL, while a 20.0 wt % solution has a density of 1.18 g/mL.

    (a) Find the molarity of CsCl in each solution.

    (b) Find the molality of CsCl in each solution.

    (c) How many mL of each solution should be diluted to 500 mL to make 0.100 M reagent? Why doesn't it take twice as much of the 20.0 wt % solution as the 40.0 wt % solution?

CHAPTER 2: SUPPLEMENTARY PROBLEMS
TOOLS OF THE TRADE

S1. Find the true mass of benzene (C_6H_6, density = 0.88 g/mL) if the apparent mass in air is 9.947 g. Assume that the air density is 0.001 2 g/mL and the balance weight density is 8.0 g/mL.

S2. An aqueous solution prepared when the lab temperature was 19°C had a concentration 0.027 64 M. What is the concentration of the same solution when used outdoors in the summer at 35°C?

S3. Water from a 5-mL pipet was drained into a weighing bottle whose empty mass was 9.974 g to give a new mass of 14.974 g at 26°C. Find the volume of the pipet at 26°C and at 20°C.

S1. Indicate how many significant figures there are in:

    (a) 0.305 0   (b) 0.003 050   (c) 1.003×10^4

S2. Round each number as indicated:

    (a) 5.124 8 to 4 significant figures   (d) 0.135 237 1 to 4 significant figures

    (b) 5.124 4 to 4 significant figures   (e) 1.525 to 3 significant figures

    (c) 5.124 5 to 4 significant figures   (f) 1.525 007 to 3 significant figures

S3. Write each answer with the correct number of digits:

    (a) 3.021 + 8.99 = 12.011   (e) $\log (2.2 \times 10^{-18})$ = ?

    (b) 12.7 - 1.83 = 10.87   (f) antilog (-2.224) = ?

    (c) 6.345×2.2 = 13.959 0   (g) $10^{-4.555}$ = ?

    (d) $0.030\ 2 \div (2.114\ 3 \times 10^{-3})$ = 14.283 69

S4. Using the correct number of significant figures, find the formula weight of $C_6H_{13}B$.

S5. Find the absolute and percent relative uncertainty and express each answer with a reasonable number of significant figures.

    (a) 3.4 (±0.2) + 2.6 (±0.1) = ?   (c) $[3.4\ (\pm0.2) \times 10^{-8}] \div [2.6\ (\pm0.1) \times 10^3]$ = ?

    (b) 3.4 (±0.2) ÷ 2.6 (±0.1) = ?   (d) [3.4 (±0.2) - 2.6 (±0.1)] × 3.4 (±0.2) = ?

S6. Express the molecular weight (± uncertainty) of benzene, C_6H_6, with the correct number of significant figures.

S7. (a) A solution is prepared by dissolving 0.222 2 (±0.000 2) g of KIO_3 [FW 214.001 0 (±0.000 9)] in 50.00 (±0.05) mL. Find the molarity and its uncertainty with an appropriate number of significant figures.

    (b) Would the answer be affected significantly if the reagent were only 99.9% pure?

S8. Find the absolute and percent relative uncertainty and express each answer with a reasonable number of significant figures.

    (a) $\sqrt{3.4\ (\pm0.2)}$ = ?   (c) $10^{3.4\ (\pm0.2)}$ = ?   (e) log [3.4 (±0.2)] = ?

    (b) $[3.4\ (\pm0.2)]^2$ = ?   (d) $e^{3.4\ (\pm0.2)}$ = ?   (f) ln [3.4 (±0.2)] = ?

S9. The value of Boltzmann's constant (k) listed on the inside front cover of the book is calculated from the quotient R/N, where R is the gas constant and N is Avogadro's number. If the uncertainty in R is 0.000 070 J/(mol·K) and the uncertainty in N is $0.000\ 003\ 6 \times 10^{23}$/mol, find the uncertainty in k.

CHAPTER 4: SUPPLEMENTARY PROBLEMS
STATISTICS

S1. Consider Rayleigh's data for the mass of nitrogen from air in Table 4-3. Find the
(a) mean   (b) standard deviation   (c) variance

S2. Suppose that a Gaussian population of measurements has a mean of 1 000 and a standard deviation of 50. What fraction of the population lies in the following intervals: (a) >1 000   (b) 950-1 050   (c) 850-1 150
                   (d) <900   (e) 930-1 030   (e) 930-1 030

S3. Find the 95 and 99% confidence intervals for the mean mass of nitrogen from chemical sources in Table 4-3.

S4. Two methods were used to measure the specific activity (units of enzyme activity per milligram of protein) of an enzyme. One unit of enzyme activity is defined as the amount of enzyme that catalyzes the formation of one micromole of product per minute under specified conditions.

Enzyme activity (five replications)

Method 1:	139	147	160	158	135
Method 2:	148	159	156	164	159

Is the mean value of method 1 significantly different from the mean value of method 2 at the 95% confidence level?

S5. It is known from many careful measurements that the concentration of magnesium in a material is 0.137 wt %. Your new analytical procedure gives values of 0.129, 0.133, 0.136, 0.130, 0.128 and 0.131 wt %. Do your results differ from the expected result at the 95% confidence level?

S6. Using the Q test, decide whether the value 0.195 should be rejected from the set of results 0.217, 0.224, 0.195, 0.221, 0.221, 0.223.

S7. (a) Use the method of least squares to calculate the equation of the best straight line going through the points (1,3), (3,2), and (5,0). Express your answer in the form $y = [m(\pm\sigma_m)]x + [b(\pm\sigma_b)]$, with a reasonable number of significant figures.
    (b) Use the method of Equation 4-23 to find the value of x (and its uncertainty) corresponding to y = 1.00.
    (c) Use Equation 4-24 to find the uncertainty in x for y = 1.00. This is the correct method.

CHAPTER 5: SUPPLEMENTARY PROBLEMS
CHEMICAL EQUILIBRIUM

S1. Write the expression for the equilibrium constant for each of the following reactions. Write the pressure of a gaseous molecule, X, as P_X.

(a) $Cl_2(g) + 2OH^-(aq) \rightleftarrows Cl^-(aq) + OCl^-(aq) + H_2O(l)$

(b) $Hg(l) + I_2(g) \rightleftarrows HgI_2(s)$

S2. Suppose that the following reaction has come to equilibrium:

$$Br_2(l) + I_2(s) + 4Cl^-(aq) \rightleftarrows 2Br^-(aq) + 2ICl_2^-(aq)$$

If more $I_2(s)$ is added, will the concentration of ICl_2^- in the aqueous phase increase, decrease, or remain unchanged?

S3. From the equations

$$CuN_3(s) \rightleftarrows Cu^+ + N_3^- \qquad K = 4.9 \times 10^{-9}$$

$$HN_3 \rightleftarrows H^+ + N_3^- \qquad K = 2.2 \times 10^{-5}$$

find the value of K for the reaction $Cu^+ + HN_3 \rightleftarrows CuN_3(s) + H^+$. All species are aqueous unless otherwise indicated.

S4. For the reaction $H_2O(l) \rightleftarrows H^+(aq) + OH^-(aq)$, $K = 1.0 \times 10^{-14}$ at 25°C. The concentrations in a system out of equilibrium are $[H^+] = 3.0 \times 10^{-5}$ M and $[OH^-] = 2.0 \times 10^{-7}$ M. Will the reaction proceed to the left or to the right to reach equilibrium?

S5. For the sum of two reactions, we know that $K_3 = K_1K_2$.

$$A + B \rightleftarrows C + D \qquad K_1$$

$$\underline{D + E \rightleftarrows B + F} \qquad K_2$$

$$A + E \rightleftarrows C + F \qquad K_3$$

Show that this implies that $\Delta G_3^\circ = \Delta G_1^\circ + \Delta G_2^\circ$

S6. Find ΔG° for the reactions

(a) $Ca(OH)_2(s) \rightleftarrows Ca^{2+} + 2OH^-$ $\qquad K = 6.5 \times 10^{-5}$

(b) $Mg(OH)_2(s) \rightleftarrows Mg^{2+} + 2OH^-$ $\qquad K = 7.1 \times 10^{-12}$

S7. For the reaction $Mg^{2+} + Cu(s) \rightleftarrows Mg(s) + Cu^{2+}$, $K = 10^{-92}$ and $\Delta S^\circ = +18$ J/(K·mol).

(a) Under standard conditions, is ΔG° positive or negative? The term "standard" conditions" means that reactants and products are in their standard states.

(b) Under standard conditions, is the reaction endothermic or exothermic?

S8. Use the solubility product to calculate the solubility of Ag_2CrO_4 (FW 331.73)

$(\rightarrow 2Ag^+ + CrO_4^{2-})$ in water expressed as (a) moles per liter, (b) g/100 mL and (c) ppm $Ag^+(\approx \mu g\ Ag^+/mL)$.

S9. The solubility product for CuCl is 1.9×10^{-7}. The equilibrium constant for the reaction $Cu(s) + Cu^{2+} \rightleftarrows 2Cu^+$ is 9.6×10^{-7}. Calculate the equilibrium constant for the reaction $Cu(s) + Cu^{2+} + 2Cl^- \rightleftarrows 2CuCl(s)$

S10. How many grams of PbI_2 (FW 461.0) will dissolve in 0.500 L of (a) water and (b) 0.063 4 M NaI?

S11. What concentration of Ca^{2+} must be added to 0.010 M oxalate $(C_2O_4^{2-})$ to precipitate 99.0% of the oxalate?

S12. Is it possible to separate 99.9% of 0.020 M Mg^{2+} from 0.10 M Ca^{2+} without precipitation of $Ca(OH)_2$ by addition of NaOH?

S13. Using Equations 5-11 to 5-15, calculate the concentrations of Pb^{2+}, PbI^+, $PbI_2(aq)$, PbI_3^- and PbI_4^- in a solution whose total I^- concentration is somehow fixed at 0.050 M. Compare your answers to Figure 5-1.

S14. Consider the following equilibria:

$$AgCl(s) \rightleftarrows Ag^+ + Cl^- \qquad K_{sp} = 1.8 \times 10^{-10}$$
$$AgCl(s) + Cl^- \rightleftarrows AgCl_2^- \qquad K_2 = 1.5 \times 10^{-2}$$
$$AgCl_2^- + Cl^- \rightleftarrows AgCl_3^{2-} \qquad K_3 = 0.49$$

Find the total concentration of silver-containing species in a silver-saturated, aqueous solution containing the following concentrations of Cl^-:
(a) 0.010 M   (b) 0.20 M   (c) 2.0 M

S15. Identify the Brønsted-Lowry acids on both sides of the reaction
$NaHSO_3 + NaOH \rightleftarrows Na_2SO_3 + H_2O$

S16. Identify the conjugate acid-base pairs in the reaction

$$H_2NCH_2CH_2NH_2 + H_2O \rightleftarrows H_3\overset{+}{N}CH_2CH_2NH_2 + OH^-$$
Ethylenediamine

S17. Calculate the concentration of H^+ and the pH of:
(a) 0.001 0 M $HClO_4$ (d) 3.0 M NaOH
(b) 0.050 M HBr   (e) 0.005 0 M $[(CH_3CH_2)_4N^+]OH^-$
(c) 0.050 M LiOH   tetraethylammonium hydroxide

S18. Write the K_a reaction for formic acid, HCO_2H and for the methylammonium ion, $CH_3NH_3^+$.

S19. Write the K_b reactions for piperidine and benzoate.

Piperidine   Benzoate

S20. Write the K_a and K_b reactions of K_2HPO_4.

S21. Write the stepwise acid-base reactions for the following species in water. Write the correct symbol (e.g., K_{b1}) for the equilibrium constant for each reaction.

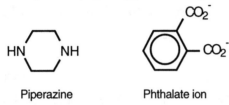

Piperazine   Phthalate ion

S22. Use Appendix G to decide which is the stronger acid: 3-nitrophenol or 4-nitrophenol. Write the acid dissociation reaction of each.

S23. Which is the stronger base: cyclohexylamine or imidazole? Write the base hydrolysis reaction of each. In the case of imidazole, the nitrogen atom without a hydrogen is the one that accepts H^+.

Cyclohexylamine   Imidazole
$pK_B = 3.36$ $pK_B = 7.01$

S24. Write the K_b reaction of hypochlorite, OCl^-. Given that the K_a value for $HOCl$ is 3.0×10^{-8}, calculate K_b for OCl^-.

S25. Write the K_{a2} reaction of H_2SO_4 and the K_{b2} reaction the trisodium salt below.

S26. From the K_a values for citric acid in Appendix G, find K_{b1}, K_{b2} and K_{b3} for trisodium citrate.

CHAPTER 6: SUPPLEMENTARY PROBLEMS
A FIRST LOOK AT SPECTROPHOTOMETRY

S1. Calculate the frequency (in hertz), wavenumber (in cm^{-1}) and energy (in joules per photon and kJ per mole of photons) of (a) ultraviolet light with a wavelength of 250 nm and (b) infrared light with a wavelength of 10 μm.

S2. A 15.0-mg sample of a compound with a molecular weight of 384.63 was dissolved in a 5-mL volumetric flask. A 1.00-mL aliquot was withdrawn, placed in a 10-mL volumetric flask, and diluted to the mark.
  (a) Find the concentration of sample in the 5-mL flask.
  (b) Find the concentration in the 10-mL flask.
  (c) The 10-mL sample was placed in a 0.500-cm cuvet and gave an absorbance of 0.634 at 495 nm. Find the molar absorptivity (ε_{495}, with units of M^{-1}·cm^{-1}) at this wavelength.

S3. A 0.267 g quantity of a compound with a molecular weight of 337.69 was dissolved in 100.0 mL of ethanol. Then 2.000 mL was withdrawn and diluted to 100.0 mL. The spectrum of this solution exhibited a maximum absorbance of 0.728 at 438 nm in a 2.000 cm cell. Find the molar absorptivity of the compound.

S4. During an assay of the thiamine (vitamin B$_1$) content of a pharmaceutical preparation, the percent transmittance scale was accidently read, instead of the absorbance scale of the spectrophotometer. One sample gave a reading of 82.2% T, and a second sample gave a reading of 50.7% T at a wavelength of maximum absorbance. What is the ratio of concentrations of thiamine in the two samples?

S5. (a) A 3.73×10^{-5} M solution of Compound A from a spectrophotometric analysis has a maximum absorbance of 0.494 at 401 nm in a 1.000-cm cell, while a reagent blank from the same analysis has an absorbance of 0.053 at 401 nm. Find the molar absorptivity of Compound A.

  (b) A 5.00 mL aliquot of unknown solution containing Compound A was mixed with color forming reagents and diluted to a final volume of 250.0 mL to give an absorbance of 0.777 at 401 nm in a 1.000-cm cell. The reagent blank had an absorbance of 0.053. Find the concentration of Compound A in the unknown solution.

S6. Spectrophotometric analysis of phosphate can be performed by the following procedure:

Standard solutions

    A. KH_2PO_4 (FW 136.09): 81.37 mg dissolved in 500.0 mL of water

    B. $Na_2MoO_4 \cdot 2H_2O$ (sodium molybdate): 1.25 g in 50 mL of 5 M H_2SO_4.

    C. $H_3NNH_3^{2+}SO_4^{2-}$ (hydrazine sulfate): 0.15 g in 100 mL of H_2O

Procedure

Place the sample (either an unknown or the standard phosphate solution, A) in a 5-mL volumetric flask, and add 0.500 mL of B plus 0.200 mL of C. Dilute to almost 5 mL with water, and heat at 100°C for 10 minutes to form a blue product ($H_3PO_4(MoO_3)_{12}$, 12-molybdophosphoric acid). Cool the flask to room temperature, dilute to the mark with water, mix well, and measure the absorbance at 830 nm in a 1.00-cm cell.

(a) When 0.140 mL of solution A was analyzed, an absorbance of 0.829 was recorded. A blank carried through the same procedure gave an absorbance of 0.017. Find the molar absorptivity of blue product.

(b) A solution of the phosphate-containing iron-storage protein ferritin was analyzed by this procedure. The unknown contained 1.35 mg of ferritin, which was digested in a total volume of 1.00 mL to release phosphate from the protein. Then 0.300 mL of this solution was analyzed by the procedure above and found to give an absorbance of 0.836. A blank carried through this procedure gave an absorbance of 0.038. Find the weight percent of phosphorus in the ferritin.

S7.   The signal (peak area) measured for different standard concentrations of silver in atomic absorption is given below:

Ag (µg/mL):	0	8.0	16.0	32.0
Peak area:	0.002	0.076	0.147	0.282

(a) Construct a calibration curve and find the slope and intercept .

(b) Find the concentration of an unknown that gives a peak area of 0.160 when the blank peak area is 0.004. That is, the corrected absorbance is 0.156.

S8.   *Propagation of uncertainty with a calibration curve.* Refer to the data in the previous problem. Find the concentration (and its uncertainty) of an unknown with a corrected peak area of 0.156 ($\pm s_y$).

S9. *Standard addition.* An unknown sample of dopamine gave a current of 34.6 nA in an electrochemical analysis. Then 2.00 mL of solution containing 0.015 6 M dopamine was mixed with 90.0 mL of unknown, and the mixture was diluted to 100.0 mL in a volumetric flask. The signal from the new solution was 58.4 nA.

(a) Denoting the initial, unknown concentration as [dopamine]$_i$, write an expression for the final concentration, [dopamine]$_f$, after dilution.

(b) In a similar manner, write the final concentration of added standard dopamine, designated as [S]$_f$.

(c) Use Equation 6-7 to find [dopamine]$_i$ in the unknown.

S10. *Internal standard.* A mixture containing 80.0 nM iodoacetone (designated A) and 64.0 nM p-dichlorobenzene (designated B) gave the relative detector response (peak area of A)/(peak area of B) = 0.71 in a chromatography experiment. A solution containing an unknown quantity of A plus 930 nM B gave a relative detector response (peak area of A)/(peak area of B) = 1.21. Find the concentration of A in the unknown.

S11. *Standard addition.* The quantity of acetylene (HC≡CH) in a gas mixture was measured by mass spectrometry, in which the signal for mass number 26 is proportional to the volume percent of acetylene.

Gas	Signal (mV)
Blank containing no acetylene	0.2
Unknown	10.8
Unknown + 0.072 vol % C_2H_2	17.1
Unknown + 0.121 vol % C_2H_2	20.2
Unknown + 0.200 vol % C_2H_2	30.0
Unknown + 0.364 vol % C_2H_2	44.6

(a) Subtract the blank value (0.2 mV) from each measured signal. Prepare a graph similar to Figure 6-10 to find the volume percent of acetylene in the unknown.

(b) Use the method of least squares to find the uncertainties in slope and intercept of your graph and to estimate the uncertainty in the answer to part a.

S12. *Derivation of standard addition graphical treatment.* Figure 6-10 is a graph of I_{S+X} vs. [S]$_f$. Rearrange Equation 6-7 to the form $I_{S+X} = m[S]_f + b$, where m is the slope and b is the intercept. The straight line in Figure 6-10 crosses the x axis when $I_{S+X} = 0$. Use your equation to show that when $I_{S+X} = 0$, $[S] = -[X]_f$.

S1. How many milligrams of oxalic acid dihydrate, $H_2C_2O_4 \cdot 2H_2O$ (FW 126.07), will react with 1.00 mL of 0.027 3 M ceric sulfate ($Ce(SO_4)_2$) if the reaction is
$$H_2C_2O_4 + 2Ce^{4+} \rightarrow 2CO_2 + 2Ce^{3+} + 2H^+?$$

S2. A mixture weighing 27.73 mg containing only $FeCl_2$ (FW 126.75) and KCl (FW 74.55) required 18.49 mL of 0.022 37 M $AgNO_3$ for complete titration of the chloride. Find the mass of $FeCl_2$ and the weight percent of Fe in the mixture.

S3. The chloride content of blood serum, cerebrospinal fluid or urine can be measured by titration of the chloride with mercuric ion: $Hg^{2+} + 2Cl^- \rightarrow HgCl_2(aq)$. When the reaction is complete, excess Hg^{2+} reacts with the indicator, diphenylcarbazone, which forms a violet-blue color.

(a) Mercuric nitrate was standardized by titrating a solution containing 147.6 mg of NaCl, which required 28.06 mL of $Hg(NO_3)_2$ solution. Find the molarity of the $Hg(NO_3)_2$.

(b) When this same $Hg(NO_3)_2$ solution was used to titrate 2.000 mL of urine, 22.83 mL was required. Find the concentration of Cl^- (mg/mL) in the urine.

S4. Consider the titration of 20.00 mL of 0.053 20 M KBr with 0.051 10 M $AgNO_3$. Calculate pAg^+ at the following volumes of added $AgNO_3$:

(a) 20.00 mL   (b) V_e (c) 22.60 mL

S5. Consider the titration of 50.00 mL of 0.024 6 M $Hg(NO_3)_2$ with 0.104 M KSCN. Calculate the value of pHg^{2+} at each of the following points and sketch the titration curve: $0.25V_e$, $0.5V_e$, $0.75V_e$, V_e, $1.05V_e$, $1.25V_e$.

S6. Calculate the value of pSO_4^{2-} at each of the following points in the titration of 100.0 mL of 0.050 0 M Sr^{2+} plus 0.050 0 M Ra^{2+} with 0.250 M SO_4^{2-}. Sketch the titration curve.

(a) 10.00 mL   (d) 30.00 mL
(b) 19.00 mL   (e) 40.00 mL
(c) 21.00 mL   (f) 50.00 mL

S7. Use Equation 7-20 to calculate the titration curve for 100.0 mL of 0.100 0 M $C_2O_4^{2-}$ titrated with 0.100 0 M La^{3+} ($2La^{3+} + C_2O_4^{2-} \rightarrow La_2(C_2O_4)_3(s)$).

CHAPTER 8: SUPPLEMENTARY PROBLEMS
ACTIVITY

S1. Calculate the ionic strength of

(a) 0.008 7 M $AgNO_3$

(b) 4 mM $Mg(ClO_4)_2$

(c) 1 mM $Na_4[Fe(CN)_6]$

(d) 13 mM NH_4Cl + 2 mM $ZnSO_4$

S2. Find the activity coefficient of each ion at the indicated ionic strength.:

(a) S^{2-} ($\mu = 0.001$ M)

(b) PO_4^{3-} ($\mu = 0.001$ M)

(c) Sn^{4+} ($\mu = 0.05$ M)

(d) $H_2NCH_2CO_2^-$ ($\mu = 0.01$ M)

S3. Use interpolation in Table 8-1 to find the activity coefficient of OH^- when $\mu = 0.030$ M.

S4. Calculate the activity coefficient of formate, HCO_2^-, when $\mu = 0.038$ M by

(a) using Equation 8-6

(b) using linear interpolation with Table 8-1

S5. Calculate the solubility of Ag_2CrO_4 (expressed as moles of Ag^+ per liter) in

(a) 0.05 M $KClO_4$

(b) 0.001 67 M K_2CrO_4

S6. Calculate the concentration of Tl^+ in a saturated solution of TlBr in water.

S7. Find the pH of (a) 0.050 M $HClO_4$ and (b) 0.050 M $HClO_4$ plus 0.050 M HBr.

CHAPTER 9: SUPPLEMENTARY PROBLEMS
SYSTEMATIC TREATMENT OF EQUILIBRIUM

S1. Write a charge balance for an aqueous solution of glycine, which reacts as follows:

$$^{+}H_3NCH_2CO_2^- \rightleftharpoons H_2NCH_2CO_2^- + H^+$$

glycine

$$^{+}H_3NCH_2CO_2^- + H_2O \rightleftharpoons {}^{+}H_3NCH_2CO_2H + OH^-$$

S2. Write a charge balance for a solution of $Al(OH)_3$ dissolved in 1 M KOH. Possible species are Al^{3+}, $AlOH^{2+}$, $Al(OH)_2^+$, $Al(OH)_3$ and $Al(OH)_4^-$.

S3. Write a mass balance for a 0.05 M solution of glycine (Problem S1) in water.

S4. Suppose that 0.30 g of AlOOH (FW 59.99) plus 150 mL of 3.0 M KOH are diluted to 1.00 L to give the same species produced by $Al(OH)_3$ in Problem S2. Write mass balance equations for aluminum and potassium.

S5. Use the systematic treatment of equilibrium to calculate the concentration of Hg_2^{2+} in a saturated aqueous solution of $(Hg_2)_3[Co(CN)_6]_2$ which dissociates into mercurous ion and $Co(CN)_6^{3-}$ (cobalticyanide).

S6. A solution is prepared by mixing M_t moles of the salt MCl_2 (which dissociates completely to $M^{2+} + 2Cl^-$) and L_t moles of the ligand HL in 1 L. The following reactions may occur:

$$M^{2+} + L^- \rightleftharpoons ML^+ \qquad K = 1.0 \times 10^8$$

$$HL(aq) \rightleftharpoons L^- + H^+ \qquad K_a = 1.0 \times 10^{-5}$$

(a) Write a mass balance for the metal species.

(b) Write a mass balance for the ligand species.

(c) Write a charge balance.

(d) Suppose that $M_t = L_t = 0.1$ M (exactly) and the pH is somehow fixed at 5.00. (This means that the charge balance no longer applies.) Use the 2 equilibrium expressions and the two mass balances to find the concentrations of ML^+, M^{2+}, L^-, and HL.

S7. Use the procedure in Section 9-4 to find the concentrations of Mg^{2+}, F^-, and HF in a saturated aqueous solution of MgF_2 held at pH 3.00.

S8. The acid HA has a solubility of 0.008 5 M in water at 25°C.

$$HA(s) \overset{K_s}{\rightleftharpoons} HA(aq) \qquad K_s = [HA(aq)] = 0.008\,5 \qquad (a)$$

If NaOH is added to a suspension of solid HA in water, more acid dissolves because of the reaction

$$K = 6.3 \times 10^5$$
$$HA(aq) + OH^- \quad \rightleftarrows \quad A^-(aq) + H_2O \tag{b}$$

Consider a saturated solution, whose pH is somehow fixed at 10.00, in contact with excess solid HA. Calculate the total concentration of HA + A$^-$.

S9. Consider a saturated solution of SrSO$_4$ in which the following reactions can occur:

$$SrSO_4(s) \rightleftarrows Sr^{2+} + SO_4^{2-} \qquad K_{sp} = 3.2 \times 10^{-7}$$
$$SO_4^{2-} + H_2O \rightleftarrows HSO_4^- + OH^- \qquad K_b = 9.8 \times 10^{-13}$$

(a) Write mass and charge balances for this solution.

(b) Find the concentration of Sr^{2+} in the solution if the pH is fixed at 2.50.

S10. Consider a saturated solution of calcium oxalate, CaC$_2$O$_4$, in which the following reactions can occur:

$$CaC_2O_4(s) \rightleftarrows Ca^{2+} + C_2O_4^{2-} \qquad K_{sp} = 1.3 \times 10^{-8}$$
$$C_2O_4^{2-} + H_2O \rightleftarrows HC_2O_4^- + OH^- \qquad K_{b1} = 1.8 \times 10^{-10}$$
$$HC_2O_4^- + H_2O \rightleftarrows H_2C_2O_4 + OH^- \qquad K_{b2} = 1.8 \times 10^{-13}$$

(a) Write mass and charge balances for this solution.

(b) Find the concentration of Ca^{2+} in the solution if the pH is fixed at 2.30.

S11. Consider a saturated solution of zinc arsenate, Zn$_3$(AsO$_4$)$_2$, in which the following reactions can occur:

$$Zn_3(AsO_4)_2(s) \rightleftarrows 3Zn^{2+} + 2AsO_4^{3-} \qquad K_{sp} = 1.0 \times 10^{-27}$$
$$AsO_4^{3-} + H_2O \rightleftarrows HAsO_4^{2-} + OH^- \qquad K_{b1} = 3.1 \times 10^{-3}$$
$$HAsO_4^{2-} + H_2O \rightleftarrows H_2AsO_4^- + OH^- \qquad K_{b2} = 9.1 \times 10^{-8}$$
$$H_2AsO_4^- + H_2O \rightleftarrows H_3AsO_4 + OH^- \qquad K_{b3} = 1.7 \times 10^{-12}$$

(a) Write mass and charge balances for this solution.

(b) Find the concentration of Zn^{2+} in the solution if the pH is fixed at 6.00.

S12. Use the data in Problem S6 to construct a graph showing the concentrations of M^{2+}, ML$^+$, L$^-$ and HL as a function of pH from 0 to 14 in 0.5 increments.

MONOPROTIC ACID-BASE EQUILIBRIA

S1. Calculate the pH of (a) 5.0×10^{-4} M HNO_3 and (b) 5.0×10^{-4} M $(CH_3)_4N^+OH^-$.

S2. Calculate the pH of 2.0×10^{-7} M $(CH_3)_4N^+OH^-$. What fraction of the total OH^- in this solution is derived from dissociation of water?

S3. Using activity coefficients correctly, calculate the pH of
(a) 0.050 M HBr (b) 0.050 M NaOH

S4. Using Appendix G, write structures of pyridine and pyridinium nitrate (pyridine·HNO_3). Write the K_b reaction for pyridine and find the values of K_b and pK_b.

S5. Find the pH and fraction of dissociation (α) of a 0.010 0 M solution of the weak acid HA with $K_a = 1.00 \times 10^{-4}$.

S6. Calculate the pH of 0.085 0 M pyridinium bromide, $C_5H_5NH^+Br^-$.

S7. Find the pH and concentrations of cyclohexylamine ($C_6H_{11}NH_2$) and cyclohexylammonium ion ($C_6H_{11}NH_3^+$) in a 0.020 M solution of cyclohexylammonium iodide.

S8. A 0.100 M solution of the weak acid HA has a pH of 2.36. Calculate pK_a for HA.

S9. (a) Calculate the pH and fraction of dissociation of $10^{-2.00}$ M hexane-2,4-dione.
(b) Calculate the pH and fraction of dissociation of $10^{-9.00}$ M hexane-2,4-dione.

S10. Compound A reacts with H_2O as follows:

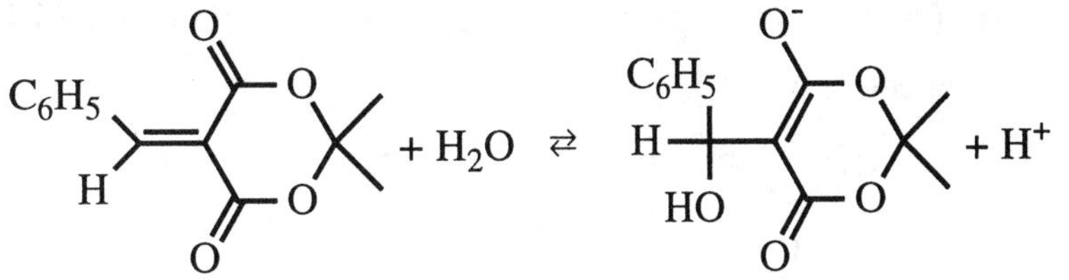

The equilibrium constant (in aqueous methanol solution) is $10^{-5.4}$. Suppose that this same equilibrium constant applies in pure water. Find the pH of a 0.020 M solution of compound A.

S11. Find the pH and fraction of association (α) of a 0.050 M solution of the weak base B with $K_b = 1.00 \times 10^{-4}$.

S12. Find the pH and concentrations of $(CH_3CH_2)_2NH$ and $(CH_3CH_2)_2NH_2^+$ in a 0.030 M solution of diethylamine.

S13. Find the pH and fraction of association (α) of 0.026 M NaOCl.

S14. Calculate the fraction of association (α) for 1.00×10^{-1}, 1.00×10^{-2}, and 1.00×10^{-12} M sodium formate.

S15. If a 0.030 M solution of a base has pH = 10.50, find K_b for the base.

S16. If a 0.030 M solution of a base is 0.27% hydrolyzed ($\alpha = 0.002\ 7$), find K_b for the base.

S17. Calculate the pH and fraction of association of $10^{-2.00}$ M sodium hexane-2,4-dionate. (This salt is derived from the acid hexane-2,4-dione in Appendix G.)

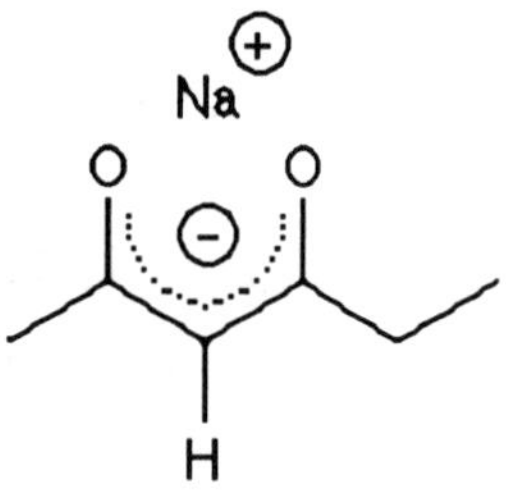

sodium hexane-2,4-dionate

S18. Which buffer system will have the greatest buffer capacity at pH 8.5?

    (a) dimethylamine / dimethylammonium ion
    (b) ammonia / ammonium ion
    (c) hydroxylamine / hydroxylammonium ion
    (d) 3-nitrophenol / 3-nitrophenolate ion

S19. Find the pH of a solution prepared from 2.53 g of oxoacetic acid, 5.13 g of potassium oxoacetate and 103 g of water.

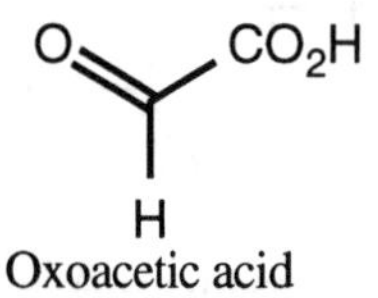

Oxoacetic acid
MW 74.036

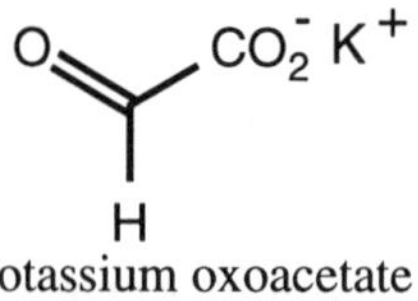

Potassium oxoacetate
MW 112.126

S20. Write the Henderson-Hasselbalch equation for a solution of methylamine. Calculate the quotient $[CH_3NH_2]/[CH_3NH_3^+]$ at (a) pH 4.00, (b) pH 10.64 and (c) pH 12.00.

S21. Given that pK_b for iodate ion IO_3^- is 13.83, find the quotient $[HIO_3]/[IO_3^-]$ in a solution of sodium iodate at (a) pH 7.00 and (b) pH 1.00.

S22. (a) Calculate the pH of a solution prepared by dissolving 10.0 g of tris(hydroxymethyl)aminomethane ("tris") plus 10.0 g of tris hydrochloride in 0.250 L of water.

(b) What will be the pH if 10.5 mL of 0.500 M NaOH is added?

S23. How many milliliters of 0.113 M HBr should be added to 52.2 mL of 0.013 4 M morpholine to give a pH of 8.00?

S24. (a) Calculate how many milliliters of 0.100 M HCl should be added to how many grams of sodium acetate dihydrate ($NaOAc \cdot 2H_2O$, FW 118.06) at 5°C to prepare 250.0 mL of 0.100 M buffer, pH 5.00. At 5°C, $pK_w = 14.734$ and pK_a for acetic acid is 4.770.

(b) If you mixed what you calculated in part a, the pH would not be 5.00. Describe how you would actually prepare this buffer in the lab.

S25. Use Equations 10-20 and 10-21 to find the concentrations of B and BH^+ in a solution prepared by mixing 0.000 100 mol of propylamine plus 0.000 100 mol of propylammonium chloride in 1.00 L of water.

CHAPTER 11: SUPPLEMENTARY PROBLEMS
POLYPROTIC ACID-BASE EQULIBRIA

S1. Write the chemical reactions whose equilibrium constants are K_{b1} and K_{b2} for the amino acid serine. Find the values of K_{b1} and K_{b2}.

S2. Consider the diprotic acid H_2A with $K_1 = 1.00 \times 10^{-5}$ and $K_2 = 1.00 \times 10^{-9}$. Find the pH and concentrations of H_2A, HA^-, and A^{2-} in case of the following solutions:
    (a) 0.100 M H_2A
    (b) 0.100 M NaHA
    (c) 0.100 M Na_2A

S3. Find the pH of 0.150 M piperazine monohydrochloride, piperazine · HCl. Calculate the concentration of each form of piperazine in this solution.

S4. Write down, but do not attempt to solve, the exact equations needed to calculate the composition of one liter of solution containing F_1 mol of HCl, F_2 mol of disodium ascorbate (Na_2A, the salt of a weak acid whose two K_a values may be called K_1 and K_2), and F_3 mol of trimethylamine (a weak base, B, whose equilibrium constant should be called K_b). Include activity coefficients wherever appropriate.

S5. How many mL of 0.423 M KOH should be added to 5.00 g of tartaric acid (2,3-dihydroxybutanedioic acid, FW 150.08) before diluting to 50 mL to give a buffer of (a) pH 3.00 and (b) pH 4.00?

S6. How many mL of 0.421 M HCl should be added to 50.0 mL of 0.055 5 M disodium malonate ($NaO_2CCH_2CO_2Na$, FW 148.03, the salt of malonic acid) to adjust the pH to (a) 6.00 and (b) 3.20?

S7. How many grams of oxalic acid (FW 90.04) should be mixed with 5.00 g of $K_2C_2O_4$ (FW 166.22) to give a pH of 3.20 when diluted to 250 mL?

S8. Starting with the fully protonated species, write the stepwise acid dissociation reactions of the amino acids aspartic acid and arginine. Be sure to remove the protons in the correct order. Which species are the neutral molecules that we call aspartic acid and arginine?

S9. (a) Find the quotient $[H_2His^+]/[HHis]$ in a 0.050 0 M histidine solution.
    (b) Find the same quotient for 0.050 0 M histidine monohydrochloride (His · HCl).

S10. Find the pH and concentration of each species of arginine in 0.012 0 M arginine·HCl solution.

S11. Consider the neutral base pyrrolidine, C_4H_9N, designated B.

    (a) Which is the predominant species, B or BH^+, at pH 11? at pH 12?

    (b) At what pH is $[BH^+] = [B]$?

    (c) What is the quotient $[B]/[BH^+]$ at pH 12.00? at pH 2.00?

S12. Which is the predominant form of sulfurous acid at pH (a) 2, (b) 4, (c) 6 and (d) 8?

S13. What is the charge of the predominant form of citric acid at pH 5.00?

S14. The acid HA has $pK_a = 7.00$. Use Equations 11-17 and 11-18 to find the fraction in the form HA and the fraction in the form A^- at pH = 8.00. Does your answer agree with what you expect for the quotient $[A^-]/[HA]$ at pH 8.00?

S15. A dibasic compound, B, has $pK_{b1} = 2.00$ and $pK_{b2} = 8.00$. Find the fraction in the form BH_2^{2+} at pH 9.00 using Equation 11-19. Note that K_1 and K_2 in Equation 11-19 are acid dissociation constants for BH_2^{2+} ($K_1 = K_w/K_{b2}$ and $K_2 = K_w/K_{b1}$).

S16. What fraction of 1,6-hexanedoic acid (adipic acid) is in each form (H_2A, HA^-, A^{2-}) at pH 5.00? at pH 6.00?

S17. Calculate α_{H_2A}, α_{HA^-}, and $\alpha_{A^{2-}}$ for butane-2,3-dione dioxime (dimethylglyoxime) at pH 10.00, 10.66, 11.00, 12.00, and 12.50.

S18. Calculate the isoelectric and isoionic pH of 0.010 M 8-hydroxyquinoline.

S19. Derive Equations 11-24 and 11-25. Also, show that $k_ak_{ab} = k_bk_{ba}$. That is, there are only three independent microequilibrium constants. If you know three of them, you can calculate the fourth one.

S20. Consider the diprotic acid H_2A, with two chemically distinguishable sites of protonation. We define the fraction of protonation at sites a and b as follows:

$$\text{Fraction of protonation at site a} = f_a = \frac{[H_2A] + [H_aA^-]}{[H_2A] + [H_aA^-] + [H_bA^-] + [A^{2-}]}$$

$$\text{Fraction of protonation at site a} = f_b = \frac{[H_2A] + [H_bA^-]}{[H_2A] + [H_aA^-] + [H_bA^-] + [A^{2-}]}$$

Show that

$$f_a = \frac{[H^+]^2 + k_b[H^+]}{[H^+]^2 + k_b[H^+] + k_a[H^+] + k_ba k_b}$$

$$f_b = \frac{[H^+]^2 + k_a[H^+]}{[H^+]^2 + k_b[H^+] + k_a[H^+] + k_ba k_b}$$

where k_a, k_b, k_{ab} and k_{ba} are microequilibrium constants. In the denominator of the f_b expression, we made use of the relation $k_a k_{ab} = k_b k_{ba}$ from the previous problem.

S21. (a) Using the microequilibrium constants for 2,3-diphosphoglycerate in Section 11-7, calculate K_1 and K_2.

(b) Using the expressions for f_a and f_b from the previous problem, calculate the fraction of protonation at each site at pH 7.00.

(c) Show that $[A^{2-}]/[H_2A] = k_{ba} k_b/[H^+]^2$ and find this quoteint at pH 7.00. Based on this composition, estimate how many equivalents of OH⁻ in Figure 11-4 give pH 7.00.

S22. *Separation by capillary zone electrophoresis.* In *electrophoresis*, charged molecules are separated by their differential abilities to migrate in an electric field. Benzoic acid containing ordinary ^{16}O can be separated from benzoic acid containing heavy ^{18}O by electrophoresis at a suitable pH because they have slightly different acid dissociation constants. The mixture of isotopic acids is applied to one end of a 50-μm-diameter × 75-cm-long fused silica tube filled with a buffer solution. Application of a 40-kV electric field causes negatively charged solute to migrate from the application end of the tube to the positive pole of the electric field at the other end of the tube. For isotopic acids, the difference in mobility is caused by the different fraction of each acid in the anionic form, A⁻. Calling this fraction α, we can write

$$H^{16}A \underset{}{\overset{^{16}K}{\rightleftharpoons}} H^+ + {}^{16}A^- \qquad\qquad H^{18}A \underset{}{\overset{^{18}K}{\rightleftharpoons}} H^+ + {}^{18}A^-$$

$$^{16}\alpha = \frac{^{16}K}{^{16}K + [H^+]} \qquad\qquad\qquad ^{18}\alpha = \frac{^{18}K}{^{18}K + [H^+]}$$

where K is the equilibrium constant. The greater the fraction of acid in the form A⁻, the faster it will migrate in the electric field. It can be shown that for electrophoresis the maximum separation will occur when $\Delta\alpha / \sqrt{\bar{\alpha}}$ is a maximum. In this expression, $\Delta\alpha = {}^{16}\alpha - {}^{18}\alpha$, and $\bar{\alpha}$ is the average fraction of dissociation $[= \frac{1}{2}({}^{16}\alpha + {}^{18}\alpha)]$.

(a) Let us denote the ratio of acid dissociation constants as $R = {}^{16}K/{}^{18}K$. In general, R will be close to unity. For benzoic acid R = 1.020. Abbreviate ^{16}K as K and write $^{18}K = K/R$. Derive an expression for $\Delta\alpha / \sqrt{\bar{\alpha}}$ in terms of K, [H$^+$], and R. Because both equilibrium constants are nearly equal (R is close to unity), set $\bar{\alpha}$ equal to $^{16}\alpha$ in your expression.

(b)   Find the maximum value of $\Delta\alpha / \sqrt{\bar{\alpha}}$ by taking the derivative with respect to [H$^+$] and setting it equal to zero. Show that the maximum difference in mobility of isotopic benzoic acids occurs when

$$[H^+] = \frac{K + K\sqrt{1 + 8R}}{2R}$$

(c)   Show that for R ≈ 1, this expression simplifies to [H$^+$] = 2K, or pH = pK - 0.30. That is, the maximum electrophoretic separation should occur when the column buffer has pH = pK - 0.30, regardless of the exact value of R.

S1. Consider the titration of 25.0 mL of 0.050 0 M $HClO_4$ with 0.100 M KOH. Find the pH at the following volumes of base added and make a graph of pH versus V_b: V_b = 0, 1, 5, 10, 12.4, 12.5, 12.6 and 13 mL.

S2. A volume of 50.0 mL of 0.050 0 M weak acid HA (pK_a = 4.00) was titrated with 0.500 M $[CH_3]_4N^+OH^-$ (tetramethylammonium hydroxide, a strong base). Find the pH at the following volumes of base added and make a graph of pH versus V_b: V_b = 0, 1, 2.5, 4, 4.9, 5, 5.1, and 6 mL.

S3. Using the same instructions as Problem 5 in Chapter 12, sketch the titration curve for the reaction of 50.0 mL of 0.050 0 M 4-nitrophenol with 0.100 M NaOH.

S4. When 16.24 mL of 0.064 3 M KOH was added to 25.00 mL of 0.093 8 M weak acid, HA, the observed pH was 3.62. Find pK_a for the acid.

S5. A 50.0-mL aliquot of 0.050 M weak base B (pK_b = 4.00) was titrated with 0.500 M HNO_3. Find the pH at the following volumes of acid added and make a graph of pH versus V_a: V_a = 0, 1, 2.5, 4, 4.9, 5, 5.1, and 6 mL.

S6. A solution of 100.00 mL of 0.040 0 M sodium propanoate (the sodium salt of propanoic acid) was titrated with 0.083 7 M HCl. Calculate the pH at the points V_a = 0, $\frac{1}{4}V_e$, $\frac{1}{2}V_e$, $\frac{3}{4}V_e$, V_e, and 1.1 V_e.

S7. A volume of 50.0 mL of the dibasic compound B (0.050 0 M, pK_{b1} = 5.00, pK_{b2} = 9.00) was titrated with 0.500 M HCl. Find the pH at the following volumes of acid added and make a graph of pH versus V_a: V_a = 0, 1, 2.5, 4, 4.8, 5, 5.2, 6, 7.5, 9, 9.8,10, 10.2, 11 and 12 mL.

S8. A 50.0-mL aliquot of 0.050 0 M diprotic acid H_2A (pK_1 = 5.00, pK_2 = 9.00) was titrated with 0.500 M NaOH. Find the pH at the following volumes of base added and make a graph of pH versus V_b: V_b = 0, 1, 2.5, 4, 4.8, 5, 5.2, 6, 7.5, 9, 9.8,10, 10.2, 11 and 12 mL.

S9. Calculate the pH at 2-mL intervals (from 0 to 12 mL) in the titration of 25.0 mL of 0.100 M cyclohexylamine with 0.250 M HI. Make a graph of pH versus V_a.

S10. A solution containing 0.010 0 M tyrosine was titrated to the first equivalence point with 0.004 00 M KOH.

  (a) Draw the structures of reactants and products.

  (b) Calculate the pH at the first equivalence point.

S11. How many milliliters of 0.043 1 M NaOH should be added to 59.6 mL of 0.122 M leucine to obtain a pH of 8.00?

S12. Consider the neutral form of the amino acid histidine, which we will abbreviate HA.

  (a) Write the sequence of reactions that occurs when HA is titrated with $HClO_4$. Draw structures of reactants and products.

  (b) How many mL of 0.050 0 M $HClO_4$ should be added to 25.0 mL of 0.040 0 M HA to give a pH of 3.00?

S13. What color change would you expect to see during the upper titration in Figure 12-4 with bromothymol blue indicator?

S14. Cresol purple has *two* transition ranges listed in Table 12-4. What color would you expect it to be at the following pH values?

  (a) 1   (b) 3   (c) 7   (d) 10

S15. Why would an indicator end point not be very useful in the titration curve for $pK_a = 10.00$ in Figure 12-3?

S16. Consider the titration of 0.10 M pyridinium bromide (the salt of pyridine plus HBr) by 0.10 M NaOH. Sketch the titration curve using calculated pH values for the volumes 0.99 V_e, V_e, and 1.01 V_e. Select an indicator from Table 12-4 that would be suitable for this titration and state what color change will be used.

S17. A very weak basic aromatic amine ($pK_b = 14.79$) was used to measure the pH of a concentrated acid. A solution was prepared by dissolving 6.390 mg of the amine (MW 278.16) in 100.0 mL of the acid. The absorbance measured at 385 nm in a 1.000-cm cell was 0.350. Find the pH of the solution

$$B \quad + \quad H^+ \quad \rightleftarrows \quad BH^+$$
$$\varepsilon_{385} = 2\ 860\ M^{-1}\ cm^{-1} \qquad\qquad \varepsilon_{385} = 937\ M^{-1}\ cm^{-1}$$

S18. An aqueous solution containing ~1 g of oxobutanedioic acid (MW 132.073) per 100 mL was titrated with 0.094 32 M NaOH to measure the acid molarity.

(a) What will be the pH at each equivalence point?

(b) Which equivalence point would be best to use in this titration?

(c) You have the indicators erythrosine, ethyl orange, bromocresol green, bromothymol blue, thymolphthalein, and alizarin yellow. Which indicator will you use and what color change will you look for?

S19. Borax (Table 12-5) was used to standardize a solution of HNO_3. Titration of 0.261 9 g of borax required 21.61 mL. What is the molarity of the HNO_3?

S20. Derive the equation in Table 12-6 for titrating dibasic B with strong acid.

S21. *Activity coefficients in titration equations.* In the titration of a weak acid by a strong base, activity coefficients do not enter the charge or mass balances and therefore do not appear in Equation 12-9. In a rigorous treatment, the only way activity coefficients enter any of the equations in Table 12-6 is in the fractional composition equations. Using activity coefficients correctly, derive the following expressions for α_{HA} and α_{A^-}. These are the expressions that should be used with Equation 12-9 in a rigorous treatment of a titration.

$$\alpha_{HA} = \frac{[H^+]\gamma_{H^+}\gamma_{A^-}}{[H^+]\gamma_{H^+}\gamma_{A^-} + K_a\gamma_{HA}} \qquad \alpha_{A^-} = \frac{K_a\gamma_{HA}}{[H^+]\gamma_{H^+}\gamma_{A^-} + K_a\gamma_{HA}}$$

S1. Calculate $\alpha_{Y^{4-}}$ for EDTA at pH 6.62.

S2. (a) Find the conditional formation constant for $Ca(EDTA)^{2-}$ at pH 10.00.

(b) Find the concentration of free Ca^{2+} in 0.050 M $Na_2[Ca(EDTA)]$ at pH 10.00.

S3. *A brainbuster!* What is the quotient $[MgY^{2-}]/[NaY^{3-}]$ in a solution prepared by mixing 0.100 M Na_2EDTA with an equal volume of 0.100 M $Mg(NO_3)_2$? Assume that the pH is high enough that there is a negligible amount of unbound EDTA. You can approach this problem by realizing that nearly all Mg^{2+} will be bound to EDTA and nearly all Na^+ will be free.

S4. Consider the titration of 100.0 mL of 0.050 0 M EDTA at pH 10.00 with 0.100 M metal ion, M^{n+}.

(a) What is the equivalence volume, V_e, in milliliters?

(b) Calculate the concentration of total free EDTA at $V = \frac{1}{2}V_e$.

(c) What fraction $(\alpha_{Y^{4-}})$ of free EDTA is in the form Y^{4-} at pH 10.00?

(d) The formation constant (K_f) is $10^{8.00}$. Calculate the value of the conditional formation constant $K_f' (= \alpha_{Y^{4-}} K_f)$.

(e) Calculate the concentration of free metal ion at $V = V_e$.

(f) What is the concentration of total free EDTA ion at $V = 1.100\ V_e$?

S5. Calculate pFe^{2+} at each of the following points in the titration of 25.00 mL of 0.020 26 M EDTA by 0.038 55 M Fe^{2+} at pH 6.00:

(a) 12.00 mL   (b) V_e (c) 14.00 mL

S6. Consider the titration of 50.0 mL of 0.011 1 M Y^{3+} (Y = yttrium) with 0.022 2 M EDTA at pH 5.00. Calculate pY^{3+} at the following volumes of added EDTA and sketch the titration curve:

(a) 0 mL	(d) 24.0 mL	(g) 25.1 mL
(b) 10.0 mL	(e) 24.9 mL	(h) 26.0 mL
(c) 20.0 mL	(f) 25.0mL	(i) 30.0 mL

S7. Calculate pCd^{2+} at each of the following points in the titration of 10.00 mL of 0.001 00 M Cd^{2+} with 0.002 00 M EDTA at pH 12.00 in a solution whose NH_3 concentration is somehow *fixed* at 0.200 M:

(a) 0 mL	(c) 4.90 mL	(e) 5.10 mL
(b) 1.00 mL	(d) 5.00 mL	(f) 6.00 mL

S8. *Allosteric interactions.* The molecule drawn below contains two large rings with oxygen atoms capable of binding metal atoms, one on each ring.

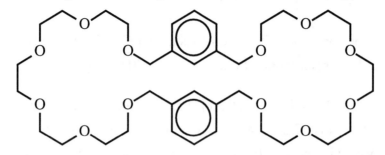

Calling the molecule L, we can represent the metal-binding reactions as

$$L + M \rightleftharpoons LM \qquad\qquad K_1 = \frac{[LM]}{[L][M]} \qquad\qquad \text{(A)}$$

$$LM + M \rightleftharpoons LM_2 \qquad\qquad K_2 \; \frac{[LM_2]}{[LM][M]} \qquad\qquad \text{(B)}$$

If binding at one site influences binding at the other site, there is said to be an *allosteric interaction* between the sites. If biding at one site makes binding at the other site more favorable than it was in the absence of the first binding, there is said to be *positive cooperativity* between the sites. If binding at one site makes biding at the second site less favorable, there is *negative cooperativity* between the sites. If there is no interaction between sites, binding is said to be *noncooperative*. This means that a metal at one site has no effect on metal binding at the other site.

The binding of $Hg(CF_3)_2$ to the molecule above in benzene solution was found to have $K_1 = 4.0 \; (\pm 0.1) \; K_2$ [J. Rebek, Jr., T. Costello, L. Marshall, R. Wattley, R. C. Gadwood, and K. Onan, *J. Amer. Chem. Soc.* **1985**, *107*, 7481]. Show that $K_1 = 4K_2$ corresponds to noncooperative binding.

Hint: If the two binding sites are represented as ⬜⬜ we can represent the equilibria as follows:

$$\text{⬜⬜} \; \overset{K_1}{\rightleftharpoons} \; \begin{array}{c} \text{⬛⬜} \\ + \\ \text{⬜⬛} \end{array} \; \overset{K_2}{\rightleftharpoons} \; \text{⬛⬛}$$

where ⬛ represents metal bound at one site. In noncooperative binding, the four populations ⬜⬜, ⬛⬜, ⬜⬛, and ⬛⬛ must be equal when the ligand is 50% saturated with metal.

S9. EDTA at pH 5 was titrated with standard Pb^{2+} using xylenol orange as indicator (Table 13-3). Which is the principal species of the indicator at pH 5? What color was observed before the equivalence point? after the equivalence point? What would be the color change if the titration were conducted at pH 8 instead of pH 5?

S10. A 25.00-mL sample containing Fe^{3+} was treated with 10.00 mL of 0.036 7 M EDTA to complex all the Fe^{3+} and leave excess EDTA in solution. The excess EDTA was then back-titrated, requiring 2.37 mL of 0.046 1 M Mg^{2+}. What was the concentration of Fe^{3+} in the original solution?

S11. Express the hardness of water containing 3.2 mM Ca^{2+} + 1.1 mM Mg^{2+} in terms of mg $CaCO_3$/L. (FW $CaCO_3$ 100.09)

S12. Cyanide can be determined indirectly by EDTA titration. A known excess of Ni^{2+} is added to the cyanide to form tetracyanonickelate:

$$4CN^- + Ni^{2+} \rightarrow Ni(CN)_4^{2-}$$

When the excess Ni^{2+} is titrated with standard EDTA, $Ni(CN)_4^{2-}$ does not react. In a cyanide analysis 12.7 mL of cyanide solution was treated with 25.0 mL of standard solution containing excess Ni^{2+} to form tetracyanonickelate. The excess Ni^{2+} required 10.1 mL of 0.013 0 M EDTA for complete reaction. In a separate experiment, 39.3 mL of 0.013 0 M EDTA was required to react with 30.0 mL of the standard Ni^{2+} solution. Calculate the molarity of CN^- in the 12.7-mL sample of unknown.

S13. A mixture of Mn^{2+}, Mg^{2+}, and Zn^{2+} was analyzed as follows: The 25.00-mL sample was treated with 0.25 g of $NH_3OH^+Cl^-$ (hydroxylammonium chloride, a reducing agent that maintains manganese in the +2 state), 10 mL of ammonia buffer (pH 10), and a few drops of eriochrome black T indicator and then diluted to 100 mL. It was warmed to 40°C and titrated with 39.98 mL of 0.045 00 M EDTA to the blue end point. Then 2.5 g of NaF was added to displace Mg^{2+} from its EDTA complex. The liberated EDTA required 10.26 mL of standard 0.020 65 M Mn^{2+} for complete titration. After this second end point was reached, 5 mL of 15% (wt/wt) aqueous KCN was added to displace Zn^{2+} from its EDTA complex. This time the liberated EDTA required 15.47 mL of standard 0.020 65 M Mn^{2+}. Calculate the number of milligrams of each metal (Mn^{2+}, Zn^{2+}, and Mg^{2+}) in the 25.00-mL sample of unknown.

S14. Here is a procedure for the consecutive determination of Bi^{3+}, Ti^{4+} and Al^{3+} in a mixture that might arise in the analysis of aluminum ore, clays or cements [M. A. El-Hamied Hafez, *Talanta* **1992**, *39*, 1189].

(1) The solution is acidified to pH 1-2 with HNO_3, at which only Bi^{3+} has a large

enough conditional formation constant to be titrated. The indicator semixylenol orange is added and the solution is titrated with standard EDTA until just reaching the end point color change from red (MIn) to yellow (In). This gives the Bi^{3+} content of the solution and leaves Ti^{4+} and Al^{3+} uncomplexed.

(2) Then excess EDTA is added and the pH is raised to 5 with hexamine, which serves as a buffer and an auxiliary complexing ligand.

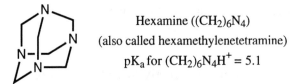

Hexamine $((CH_2)_6N_4)$
(also called hexamethylenetetramine)
pK_a for $(CH_2)_6N_4H^+ = 5.1$

The solution is boiled to complete the reaction with EDTA, cooled and titrated with Zn^{2+} until just reaching the end point color change from yellow (In) to red (MIn). Now the solution contains $Bi(EDTA)^-$, $Ti(EDTA)$ and $Al(EDTA)^-$, with no excess EDTA.

(3) Excess $H_2PO_4^-$ is added and the solution is boiled to displace Ti^{4+} from EDTA. Titration with standard Zn^{2+} until the end point color change is just reached gives the Ti^{4+} content.

(4) Finally, excess F^- is added and the solution is boiled to displace Al^{3+} from EDTA. Titration of the hot solution with standard Zn^{2+} until the end point color change is just reached gives the Al^{3+} content.

In step 1, 25.00 mL of unknown required 16.43 mL of 0.010 44 M EDTA. Step 3 required 4.22 mL of 0.012 76 M Zn^{2+} and step 4 consumed 25.92 mL of 0.012 76 M Zn^{2+}. Find the molarity of each cation in the original unknown. What color change is observed in steps (3) and (4)?

S15. **A Spectrophotometric Metal-Ligand Binding Problem Using Nonlinear Least-Squares Curve Fitting**

The spectrum on the next page shows changes that occur as metal M is titrated with ligand L.

$$M + L \overset{K}{\rightleftharpoons} ML \qquad\qquad K = \frac{[ML]}{[M][L]}$$

Curve 0 is the spectrum of M and curves 1 through 5 result from additions of L. The species M and ML absorb in the region shown, but L has no absorbance. Both the initial solution and the titrant solution contain 2.12×10^{-4} M metal ion, so the total concentration of M plus ML remains constant throughout the experiment:

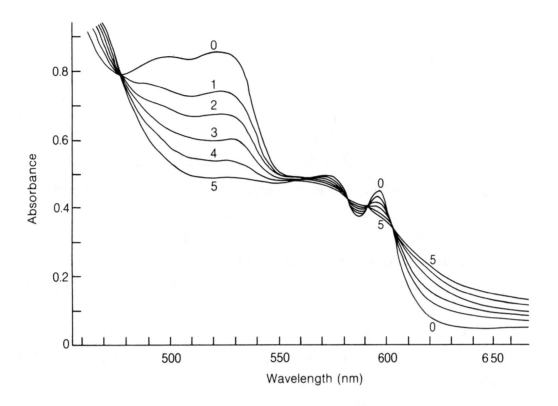

Spectrophotometric titration of species P (curve 0) with ligand X (curves 1-5).
[From J. R. Long and R.S. Drago, *J. Chem. Ed.* **1982**, *59*, 1037.]

Mass balance for M: $C_m = [M] + [ML] = 2.12 \times 10^{-4}$ M

Mass balance for L: $C_l = [L] + [ML]$

where C_l is the total ligand concentration after each addition. Substituting $[M] = C_m - [ML]$ and $[L] = C_l - [ML]$ into the equilibrium constant above, you can solve for $[ML]$:

$$[ML] = \frac{1}{2}\left[\left(C_m + C_l + \frac{1}{K}\right) - \sqrt{\left(C_m + C_l + \frac{1}{K}\right)^2 - 4C_m C_l}\right]$$

Now we want to express the absorbance changes as a function of K and various known quantities. For a 1-cm-pathlength cell, you can show from Beer's law that the change of absorbance (ΔA) at a given wavelength can be written

$$\Delta A = \Delta\varepsilon[ML]$$

where $\Delta\varepsilon$ is the change in molar absorptivity between product and reactant ($\Delta\varepsilon = \varepsilon_{ML} - \varepsilon_M$). Substituting in the expression for $[ML]$ above gives

$$\Delta A = \frac{\Delta\varepsilon}{2}\left[\left(C_m + C_l + \frac{1}{K}\right) - \sqrt{\left(C_m + C_l + \frac{1}{K}\right)^2 - 4C_m C_l}\right]$$

The least-squares problem: The table below gives ΔA as a function of C_l.

Curve	C_l (M)	Absorbance at 525 nm	ΔA
0	0	0.859	---
1	0.515×10^{-3}	0.745	-0.114
2	1.03×10^{-3}	0.676	-0.183
3	2.06×10^{-3}	0.609	-0.250
4	4.11×10^{-3}	0.544	-0.315
5	9.23×10^{-3}	0.494	-0.365

Your job is to use the computer program for least squares curve fitting at the beginning of the Solutions Manual *to find the best values of $\Delta\varepsilon$ and K in the equation* $\Delta A = f(C_l, \Delta\varepsilon, K)$ where C_l is the independent variable (x), ΔA is the dependent variable (y), and the adjustable parameters are $\Delta\varepsilon$ and K. In the notation of the least squares program, C_l is the variable x[i], $\Delta\varepsilon$ is the constant C[1] and K is the constant C[2].

To begin the least-squares procedure, we need estimates of $\Delta\varepsilon$ and K. For this purpose, let's pretend that curve 5 corresponds to complete conversion of M to ML. Therefore

$$A_{ML} = \varepsilon_{ML}[ML]$$

$$0.494 \approx \varepsilon_{ML}[2.12 \times 10^{-4} \text{ M}] \Rightarrow \varepsilon_{ML} \approx 2.33 \times 10^3 \text{ M}^{-1}\cdot\text{cm}^{-1}$$

From curve 0 we know $\varepsilon_M = A/[M] = 0.859/2.12 \times 10^{-4} = 4.05 \times 10^3$ M$^{-1}\cdot$cm^{-1}. Therefore, $\Delta\varepsilon = \varepsilon_{ML} - \varepsilon_M \approx -1.72 \times 10^3$ M$^{-1}\cdot$cm^{-1}.

To estimate K, suppose that the conversion of M to ML for curve 1 is in proportion to the absorbance change:

$$[ML] \approx \left(\frac{A_1 - A_0}{A_5 - A_0}\right)C_m \approx \left(\frac{0.745 - 0.859}{0.494 - 0.859}\right)(2.12 \times 10^{-4}) = 6.61 \times 10^{-4} \text{ M}$$

From the mass balance for metal, we estimate $[M] = C_m - [ML] \approx 1.46 \times 10^{-4}$ M. The mass balance for ligand gives $[L] = C_l - [ML] \approx 0.513 \times 10^{-3} - 6.61 \times 10^{-5} = 4.49 \times 10^{-4}$ M. Our estimate of the equilibrium constant is therefore

$$K = \frac{[ML]}{[M][L]} \approx \frac{6.61 \times 10^{-4}}{(1.46 \times 10^{-4})(4.49 \times 10^{-4})} = 1.0 \times 10^3$$

Happy hunting! The values produced by your least-squares program should be $\Delta\varepsilon = -1\,944\ (\pm 25)$ M$^{-1}\cdot$cm^{-1} and K = 829 ($\pm$35).

If you crave more, you can find spectrophotometric data for the sequence $M + L \rightarrow ML \rightarrow ML_2$ given by N. K. Kildahl, *J. Chem. Ed.* **1992**, *69*, 591. Try deriving equations to fit these data with your least squares program and find the formation constants K_1 and K_2.

S1. (a) Identify the oxidizing and reducing agents among the reactants below and write a balanced half-reaction for each.

$$2S_2O_4^{2-} + TeO_3^{2-} + 2OH^- \rightleftarrows 4SO_3^{2-} + Te(s) + H_2O$$

     Dithionite   Tellurite   Sulfite

    (b) How many coulombs of charge are passed from reductant to oxidant when 1.00 g of Te is deposited?

    (c) If Te is created at a rate of 1.00 g/h, how much current is flowing?

S2. Draw a picture of each of the following cells, showing the location of each chemical species. For each cell, write an oxidation for the left half-cell and a reduction for the right half-cell.

    (a) $Au(s) \mid Fe(CN)_6^{4-}$ (aq), $Fe(CN)_6^{3-}$ (aq) $\parallel$ $Ag(CN)_2^-$ (aq), $KCN(aq) \mid Ag(s)$

    (b) $Pt(s) \mid Hg(l) \mid Hg_2Cl_2(s) \mid KCl(aq) \parallel ZnCl_2(aq) \mid Zn(s)$

S3. Find the theoretical electrical storage capacity of cells that make use of the following chemical reactions. Express your answer in ampere · hours per kilogram of reactants, where 1 A·h provides 1 A for 1 h. Thus if consumption of 0.5 kg of reactants produces 3 A·h, the storage capacity would be 3 A·h/0.5 kg = 6 A·h/kg. Which cell produces the most electricity per kilogram, and which the least?

    (a) Lead-acid battery: $Pb + PbO_2 + 2H_2SO_4 \rightarrow 2PbSO_4 + 2H_2O$
        (MW of reactants = $207.2 + 239.2 + 2 \times 98.079 = 642.6$.)

    (b) Carbon-zinc dry cell: $Zn + 2NH_4Cl + 2MnO_2 \rightarrow ZnCl_2(NH_3)_2 + 2MnO(OH)$
        (MW of reactants = 346.25.)

    (c) Nickel-cadmium cell: $2Ni(OH)_2 + Cd(OH)_2 \rightarrow NiO(OH) + Cd + 2H_2O$
        (MW of reactants = 331.84.)

    (d) Sulfur-aluminum battery: $2Al + 3S + 3KOH + 3H_2O \rightarrow 2Al(OH)_3 + 3KHS$
        (MW of reactants = 372.525.)

    (e) Hydrogen-oxygen fuel cell: $2H_2 + O_2 \rightarrow 2H_2O$ (MW of reactants = 36.031)

S4. Which will be the strongest reducing agent under standard conditions (all activities = 1): Se, Sn^{4+}, Cr^{2+}, Mg^{2+}, or $Fe(CN)_6^{4-}$?

S5. Use Le Châtelier's principle and half-reactions from Appendix H to find which of the following become stronger reducing agents when the solution becomes more basic. Which are unchanged, and which become weaker?

Cl_2	Al	H_2S	MnO_4^{2-}	$S_2O_3^{2-}$
Chlorine	Aluminum	Hydrogen sulfide	Manganate	Thiosulate

S6. Consider the cell

Pt(s) | H$_2$(g, 0.100 atm) | H$^+$(aq,pH = 2.54) ‖ Cl$^-$(aq, 0.200 M) | Hg$_2$Cl$_2$(s) | Hg(l) | Pt(s)

    (a) Write a reduction reaction and Nernst equation for each half-cell. For the Hg$_2$Cl$_2$ half-reaction, E° = 0.268 V.

    (b) Find E for the net cell reaction and state whether reduction will occur at the left- or right-hand electrodes.

S7. Consider a circuit in which the left half-cell was prepared by dipping a Pt wire in a beaker containing an equimolar mixture of Cr^{2+} and Cr^{3+}. The right half-cell contained a Tl rod immersed in 1.00 M TlClO$_4$.

    (a) Use line notation to describe this cell.

    (b) Calculate the cell voltage.

    (c) Write the spontaneous net cell reaction.

    (d) When the two electrodes are connected by a salt bridge and a wire, which terminal (Pt or Tl) will be the anode?

S8. Write a balanced chemical equation (in acid solution) for the reaction represented by the question mark below. Calculate E° for the reaction.

$$NO_3^- \xrightarrow{0.773} NO_2(g) \xrightarrow{1.108} HNO_2 \xrightarrow{?} NO$$
$$NO_3^- \xrightarrow{\qquad\qquad 0.955 \qquad\qquad} NO$$

S9. Calculate E°, ΔG°, and K for each of the following reactions.

    (a) Cu(s) + Cu^{2+} ⇄ 2Cu$^+$

    (b) 2F$_2$(g) + H$_2$O ⇄ F$_2$O(g) + 2H$^+$ + 2F$^-$

S10. A solution contains 0.010 0 M IO$_3^-$, 0.010 0 M I$^-$, 1.00 × 10^{-4} M I$_3^-$, and pH 6.00 buffer. Consider the reactions

$$2IO_3^- + I^- + 12H^+ + 10e^- \rightleftarrows I_3^- + 6H_2O \qquad\qquad E° = 1.210\ V$$
$$I_3^- + 2e^- \rightleftarrows 3I^- \qquad\qquad E° = 0.535\ V$$

    (a) Write a balanced net reaction that can occur in this solution.

    (b) Calculate ΔG° and K for the reaction.

    (c) Calculate E for the conditions given above.

    (d) Calculate ΔG for the conditions given above.

    (e) At what pH would the concentrations of IO$_3^-$, I$^-$, and I$_3^-$ listed above be in equilibrium at 298 K?

S11. From the half-reactions below, calculate the solubility product of Mg(OH)$_2$.

$$Mg^{2+} + 2e^- \rightleftarrows Mg(s) \qquad\qquad E° = -2.360 \text{ V}$$
$$Mg(OH)_2(s) + 2e^- \rightleftarrows Mg(s) + 2OH^- \qquad E° = -2.690 \text{ V}$$

S12. The standard free energy of vaporization of $Cl_2(aq)$ is $\Delta G° = -6.9$ kJ/mol at 298 K. Given that E° for the reaction $Cl_2(g) + 2e^- \rightleftarrows 2Cl^-(aq)$ is 1.360 V, find E° for the reaction $Cl_2(aq) + 2e^- \rightleftarrows 2Cl^-(aq)$.

S13. Consider the cell S.H.E. ‖ $Ag(S_2O_3)_2^{3-}$(aq,0.010 M), $S_2O_3^{2-}$(aq,0.050 M) ∣ Ag(s)

    (a) Using the half-reaction $Ag(S_2O_3)_2^{3-} + e^- \rightleftarrows Ag(s) + 2S_2O_3^{2-}$, calculate the cell voltage (E, not E°).

    (b) Alternatively, the cell could have been described with the half-reaction $Ag^+ + e^- \rightleftarrows Ag(s)$. Using the cell voltage from part a, calculate $[Ag^+]$ in the right half-cell.

    (c) Use the answer to part b to find the formation constant for the reaction

$$Ag^+ + 2S_2O_3^{2-} \overset{K_f}{\rightleftarrows} Ag(S_2O_3)_2^{3-}$$
$$\text{Thiosulfate}$$

S14. The following cell has a voltage of 1.018 V. Find K_a for formic acid, HCO_2H.

$$Pt(s) ∣ UO_2^{2+} (0.050 \text{ M}), U^{4+}(0.050 \text{ M}), HCO_2H(0.10 \text{ M}), HCO_2Na(0.30 \text{ M})$$
$$‖ Fe^{3+}(0.050 \text{ M}), Fe^{2+}(0.025 \text{ M}) ∣ Pt(s)$$

S15. For the following cell , the half-reactions can be written

Right half-cell: $\qquad\qquad Pb^{2+}(right) + 2e^- \rightleftarrows Pb(s)$

Left half-cell: $\qquad\qquad Pb^{2+}(left) + 2e^- \rightleftarrows Pb(s)$

(a) Given that K_{sp} for $Pb(HPO_4)(s)$ is 2.0×10^{-10}, find $[HPO_4^{2-}]$ in the left half-cell.

(b) If the measured cell voltage is 0.097 V, calculate K_{sp} for $PbF_2(s)$.

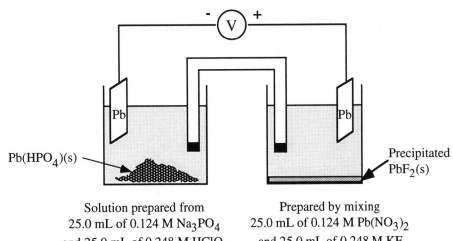

Solution prepared from
25.0 mL of 0.124 M Na_3PO_4
and 25.0 mL of 0.248 M $HClO_4$

Prepared by mixing
25.0 mL of 0.124 M $Pb(NO_3)_2$
and 25.0 mL of 0.248 M KF

S16. The monstrous cell below was set up. Then 50.0 mL of 0.044 4 M Na$_2$EDTA was added to the right-hand compartment and 50.0 mL of 0.070 0 M NaOH was added to the left-hand compartment. The cell voltage leveled off at +0.418 V. Find the formation constant for CuY^{2-} (where Y = EDTA).

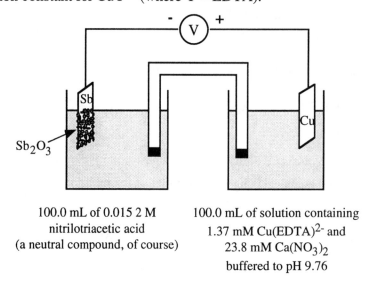

100.0 mL of 0.015 2 M nitrilotriacetic acid (a neutral compound, of course)	100.0 mL of solution containing 1.37 mM Cu(EDTA)$^{2-}$ and 23.8 mM Ca(NO$_3$)$_2$ buffered to pH 9.76

S17. Evaluate E°' for the half-reaction CO$_2$(g) + 2H$^+$ + 2e$^-$ $\rightleftarrows$ CO(g) + H$_2$O.

S18. Evaluate E°' for the half-reaction CO$_2$(g) + 2H$^+$ + 2e$^-$ $\rightleftarrows$ HCO$_2$H(aq).

S19. The standard reduction potential of anthraquinone,-2,6-disulfonate is 0.229 V.

Anthraquinone-2,6-disulfonate

$+ 2H^+ + 2e^- \rightleftharpoons$

Acidic protons

The reduced product is a diprotic acid with pK$_1$ = 8.10 and pK$_2$ = 10.52. Calculate E°' (pH = 7) for anthraquinone-2,6-disulfonate.

S1. Convert the potentials listed below. The Ag|AgCl and calomel reference electrodes are saturated with KCl.

(a) -0.222 vs S.H.E. = ? vs Ag|AgCl

(b) 0.523 vs Ag|AgCl = ? vs S.H.E.

(c) -0.523 vs S.C.E. = ? vs Ag|AgCl

(d) 0.035 vs Ag|AgCl = ? vs S.C.E.

(e) -0.035 vs calomel = ? vs Ag|AgCl

S2. What is the cell voltage when saturated calomel and Pt electrodes are dipped into a solution containing 0.002 17 M Br_2 (aq) and 0.234 M Br^-?

S3. A 50.0 mL solution of 0.100 M NaSCN was titrated with 0.200 M $AgNO_3$ in the cell:

S.C.E. ‖ titration solution | Ag(s)

Find the cell voltage for these volumes of titrant: 0.1, 10.0, 25.0 and 30.0 mL.

S4. The right-hand compartment of the cell below was titrated with ligand while monitoring the cell voltage:

Ag(s) | Ag^+(aq, 0.100 M, 25.0 mL) ‖ Ag^+(aq, 0.100 M, 75.0 mL) | Ag(s)

With 1.52 M ammonia titrant, the end point was observed at 9.8 mL, while 1.72 M ethylenediamine (Section 13-1) gave an end point of 5.0 mL. In both cases, the measured cell voltage was near 0.15 V at the end point [H. Meyer, *J. Chem. Ed.* **1992**, *69*, 499].

(a) Write the half-reactions, net reaction and Nernst equation for the cell.

(b) Write the stoichiometry of the observed reactions with the titrants.

(c) From the observed voltage, find the concentration of uncomplexed Ag^+ at the end points.

S5. The titration solution in the cell below had a total volume of 50.0 mL and contained 0.100 M Mg^{2+} and 1.00×10^{-5} M $Zn(EDTA)^{2-}$ at a pH of 10.00

S.C.E. ‖ Zn(s) | titration solution

What will be the cell voltage when 10.0 mL of 0.100 M EDTA has been added?

(*Hint*: See Exercise B in Chapter 15.)

S6. Which side of the liquid junction 0.1 M HCl | 0.1 M LiBr will be negative?

S7. How many seconds will it take for (a) OH^- and (b) F^- to migrate a distance of 1.00 cm in a field of 1.00 kV/m?

S8. (a) Write an expression analogous to Equation 15-8 to describe the response of a La^{3+}-selective electrode to La^{3+} ion.

   (b) If $\beta \approx 1.00$, by how many mV will the potential change when the electrode is removed from 1.00×10^{-4} M $LaClO_4$ and placed in 1.00×10^{-3} M $LaClO_4$?

   (c) By how many millivolts will the potential of the electrode change if the electrode is removed from 2.36×10^{-4} M $LaClO_4$ and placed in 4.44×10^{-3} M $LaClO_4$?

   (d) The electrode potential is +100 mV in 1.00×10^{-4} M $LaClO_4$ and the selectivity coefficient $k_{La^{3+},Fe^{3+}}$ is $\frac{1}{1\,200}$. What will be the potential when 0.010 M Fe^{3+} is added?

S9. A metal ion buffer was prepared from 0.050 M ML and 0.50 M L, where ML is a metal-ligand complex and L is free ligand.

$$M + L \rightleftharpoons ML \qquad K_f = 3.6 \times 10^{10}$$

Calculate the concentration of free metal ion, M, in this buffer.

S10. A calcium ion-selective electrode obeys Equation 15-8, in which $\beta = 0.970$ and $n_{Ca^{2+}} = 2$. The selectivity coefficients for several ions are listed.

Interfering ion, Y	$k_{Ca^{2+},Y}$
Mg^{2+}	0.040
Ba^{2+}	0.021
Zn^{2+}	0.081
K^+	6.6×10^{-5}
Na^+	1.7×10^{-4}

In a pure 1.00×10^{-3} M calcium solution, the reading was +300.0 mV. What would be the voltage if the solution had the same calcium concentration plus $[Mg^{2+}]$ = 1.00×10^{-3} M, $[Ba^{2+}] = 1.00 \times 10^{-3}$ M, $[Zn^{2+}] = 5.00 \times 10^{-4}$ M, $[K^+] = 0.100$ M, and $[Na^+] = 0.500$ M? (Use concentrations instead of activities to answer this question.) If they are present at *equal* concentrations, which ion in the table above interferes the most with Ca^{2+} the electrode?

S11. An ion-selective electrode used to measure the cation M^{2+} obeys the equation

$$E = \text{constant} + \frac{0.057\ 1}{2} \log \{[M^{2+}] + 0.002\ 0\ [Li^+]^2\}$$

When the electrode was immersed in 100.0 mL of unknown containing M^{2+} in 0.200 M LiCl, the reading was 194.6 mV (versus S.C.E.). When 1.00 mL of 1.07 $\times 10^{-2}$ M M^{2+} (in 0.200 M LiCl) was added to the unknown, the reading increased to 200.7 mV. Find the concentration of M^{2+} in the original unknown.

S12. A magnesium ion buffer was made by mixing 10.0 mL of 1.00 mM $MgSO_4$, 10.0 mL of 1.30 mM EDTA, and 5.00 mL of buffer, pH 10.00. What is the concentration of free metal ion in this solution? Answer the same question for $MnSO_4$ used instead of $MgSO_4$.

CHAPTER 16: SUPPLEMENTARY PROBLEMS
REDOX TITRATIONS

S1. Consider the titration of 50.0 mL of 0.020 0 M In^+ by 0.050 0 M Cu^{2+} using Pt and saturated calomel electrodes to find the end point.

    (a) Write a balanced titration reaction.

    (b) Write two different half-reactions for the indicator electrode.

    (c) Write two different Nernst equations for the net cell reaction.

    (d) Calculate E at the following volumes of Cu^{2+}: 1.00, 20.0, 39.0, 40.0, 41.0, and 80.0 mL. Sketch the titration curve.

S2. Consider the titration of 50.0 mL of 0.025 0 M Fe^{3+} by 0.050 0 M Cu^+ to give Fe^{2+} and Cu^{2+}, using Pt and saturated Ag | AgCl electrodes to find the end point.

    (a) Write a balanced titration reaction.

    (b) Write two different half-reactions for the indicator electrode.

    (c) Write two different Nernst equations for the net cell reaction.

    (d) Calculate E at the following volumes of Cu^+: 1.0, 12.5, 24.5, 25.0, 25.5, and 30.0 mL. Sketch the titration curve.

S3. Use a spreadsheet to prepare a titration curve (potential referenced to saturated Ag | AgCl electrode vs. volume of titrant) for each case below. In addition, compute the potential at the following specific points: 0.01 V_e, 0.5 V_e, 0.99 V_e, 0.999 V_e, V_e, 1.01 V_e, 1.1 V_e and 2 V_e.

    (a) Titration of 25.00 mL of 0.020 0 M Fe^{3+} with 0.010 00 M Cr^{2+} in 1.00 M $HClO_4$.

    (b) Titration of 10.0 mL of 0.050 0 M MnO_4^- with 0.050 0 M Fe^{2+} at pH -0.30 in H_2SO_4.

    (c) Titration of 50 mL of 0.020 8 M Fe^{3+} with 0.017 3 M ascorbic acid at pH 0.00 in HCl.

    (d) Titration of 50.0 mL of 0.050 M Sn^{2+} in 1 M HCl with 0.100 M UO_2^{2+} to give Sn^{4+} and U^{4+}.

S4. *Derivation of Spreadsheet Equation for Dichromate Titrations.* Consider the titration of 120.0 mL of 0.010 0 M Fe^{2+} by 0.020 0 M dichromate at pH = 1.00:

$$Cr_2O_7^{2-} + 6Fe^{2+} + 14H^+ \rightarrow 2Cr^{3+} + 6Fe^{3+} + 7H_2O$$

    Titrant   Analyte

The half-reaction for analyte is $Fe^{3+} + e^- \rightleftharpoons Fe^{2+}$ $E° = 0.771$ V

$$\Rightarrow \alpha = \frac{[Fe^{2+}]}{[Fe^{3+}]} = 10^{(0.771 - E)/0.059\ 16}$$

Following the procedure of Section 16-2 we find $[Fe^{3+}] = Fe_{total}/(1+\alpha)$.

The half-reaction for titrant is more complicated:

$$Cr_2O_7^{2-} + 14H^+ + 6e^- \rightleftharpoons 2Cr^{3+} + 7H_2O \qquad\qquad E° = 1.36\ V$$

$$E = E° - \frac{0.059\ 16}{6} \log \frac{[Cr^{3+}]^2}{[Cr_2O_7^{2-}][H^+]^{14}}$$

$$\Rightarrow \tau \equiv \frac{[Cr^{3+}]^2}{[Cr_2O_7^{2-}]} = 10^{\{6(1.36 - E)/0.059\ 16 - 14pH\}} \Rightarrow [Cr_2O_7^{2-}] = \frac{1}{\tau}[Cr^{3+}]^2$$

The mass balance for chromium is

$$Cr_{total} = [Cr^{3+}] + 2[Cr_2O_7^{2-}] = [Cr^{3+}] + \frac{2}{\tau}[Cr^{3+}]^2 \qquad\qquad (1)$$

$$\Rightarrow Cr_{total} = [Cr^{3+}]\left(\frac{\tau + 2[Cr^{3+}]}{\tau}\right)$$

$$\Rightarrow [Cr^{3+}] = Cr_{total}\left(\frac{\tau}{\tau + 2[Cr^{3+}]}\right) \qquad\qquad (2)$$

Quadratic equation 1 can be solved for $[Cr^{3+}]$ and we find the positive root to be

$$[Cr^{3+}] = \frac{-\frac{\tau}{2} + \sqrt{\frac{\tau^2}{4} + 2\tau Cr_{total}}}{2} \qquad\qquad (3)$$

Since Fe^{3+} and Cr^{3+} are created in 6:2 proportions, we can say that

$$3[Cr^{3+}] = [Fe^{3+}]$$

Substituting $[Cr^{3+}]$ from equation 2 and $[Fe^{3+}] = Fe_{total}/(1+\alpha)$ gives

$$3\ Cr_{total}\left(\frac{\tau}{\tau + 2[Cr^{3+}]}\right) = \frac{Fe_{total}}{(1 + \alpha)}$$

Now the fraction of titration for the reaction is $\phi = 3Cr_{total}/Fe_{total}$. The factor of 3 guarantees that $\phi = 1$ at the equivalence point. Rearranging the equation above

gives $$\phi = \frac{3\ Cr_{total}}{Fe_{total}} = \frac{\tau + 2[Cr^{3+}]}{\tau\ (1 + \alpha)}$$

Replacing $[Cr^{3+}]$ by its value from equation 3 gives what we seek:

$$\phi = \frac{3\ Cr_{total}}{Fe_{total}} = \frac{\frac{1}{2} + \sqrt{\frac{1}{4} + \frac{2Cr_{total}}{\tau}}}{1 + \alpha} \qquad\qquad (4)$$

But the value of ϕ depends on the total concentration of chromium at each point. We can express the total concentration of chromium in terms of ϕ as follows:

$$Cr_{total} = \frac{mmol\ of\ Cr}{mL\ of\ solution} = \frac{V_{Cr}\ C_{Cr}}{V_{Cr} + V_{Fe}^{\circ}} = \frac{\phi\ V_e\ C_{Cr}}{\phi\ V_e + V_{Fe}^{\circ}} \tag{5}$$

where C_{Cr} is the concentration of chromium ($= 2[Cr_2O_7^{2-}]$) in titrant and V_{Fe}° is the initial concentration of Fe^{2+} analyte. Substituting Cr_{total} from equation (5) into equation (4) gives the master equation for the titration curve of Fe^{2+} titrated with $Cr_2O_7^{2-}$:

$$\phi = \frac{3\ Cr_{total}}{Fe_{total}} = \frac{\frac{1}{2} + \sqrt{\frac{1}{4} + \frac{2\ \phi\ V_e\ C_{Cr}}{\tau\ \phi\ V_e + \tau\ V_{Fe}^{\circ}}}}{1 + \alpha} \tag{6}$$

To implement equation (6) on a spreadsheet, the constants C_{Cr} and V_{Fe}° are added to column A. A value of potential (vs S.H.E.) is inserted in column B and values of τ and α are computed in columns C and D. Equation (6) is used to compute ϕ in column E. In this formula, ϕ is defined in terms of itself. The spreadsheet uses automatic successive approximations to find a solution, which is painless to an anxious chemist.

Consider the titration of 120.0 mL of 0.010 0 M Fe^{2+} (buffered to pH 1.00) with 0.020 M $Cr_2O_7^{2-}$ monitored by Pt and saturated Ag|AgCl electrodes. Use a spreadsheet to preapre a graph of the titration curve. Report the potential at the following volumes of $Cr_2O_7^{2-}$: 0.100, 2.00, 4.00, 6.00, 8.00, 9.00, 9.90, 10.00, 10.10, 11.00, 12.00 mL.

S5. Prepare a spreadsheet for the titration curve in Figure 16-6 if the concentrations are changed to 5.00 mM Tl^+ and 15.0 mM Sn^{2+}. Find the potential at the following volumes: 20.0, 37.5, 74.0, 75.0, 76.0, 87.5, 100.0 and 110.0 mL.

S6. When 50.00 mL of unknown was passed through a Jones reductor, molybdate ion (MoO_4^{2-}) was converted to Mo^{3+}. The filtrate required 22.11 mL of 0.012 34 M $KMnO_4$ to reach the purple end point.

$$MnO_4^- + Mo^{3+} \rightarrow Mn^{2+} + MoO_2^{2+}$$

A blank required 0.07 mL. Find the molarity of molybdate in the unknown.

S7. A 25.00-mL sample containing La^{3+} was treated with sodium oxalate to precipitate $La_2(C_2O_4)_3$, which was washed, dissolved in acid, and titrated with 12.34 mL of 0.004 321 M $KMnO_4$. Calculate the molarity of La^{3+} in the unknown.

S8. An aqueous glycerol solution weighing 153.2 mg was treated with 50.0 mL of 0.089 9 M Ce^{4+} in 4 M $HClO_4$ at 60°C for 15 minutes to oxidize the glycerol to formic acid:

$$CH_2 - CH - CH_2$$
$$\;\;|\;\;\;\;\;\;|\;\;\;\;\;\;|$$
$$OH\;\;\;OH\;\;\;OH$$

Glycerol
MW 92.095

HCO_2H

Formic acid

The excess Ce^{4+} required 10.05 mL of 0.043 7 M Fe^{2+} to reach a ferroin end point. What is the weight percent of glycerol in the unknown?

S9. A mixture of nitrobenzene and nitrosobenzene weighing 24.43 mg was titrated with Cr^{2+} to give aniline:

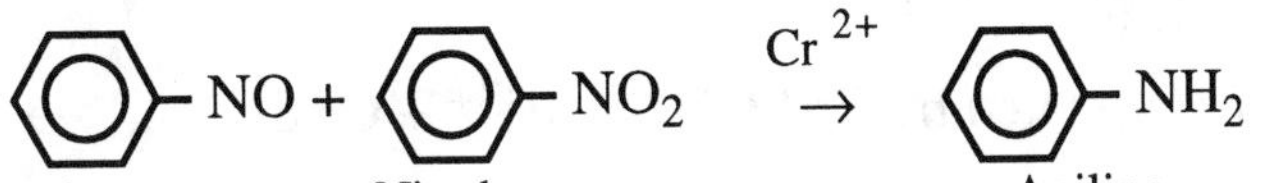

Nitrosobenzene
FW 107.112

Nitrobenzene
FW 123.111

Aniline

The titration required 21.57 mL of 0.050 00 M Cr^{2+} to reach a potentiometric end point. Find the weight percent of nitrosobenzene in the mixture.

S10. A solution containing Be and several other metals was treated with excess EDTA to *mask* the other metals. Then excess acetoacetanilide was added at 50°C at pH 7.5 to precipitate beryllium ion:

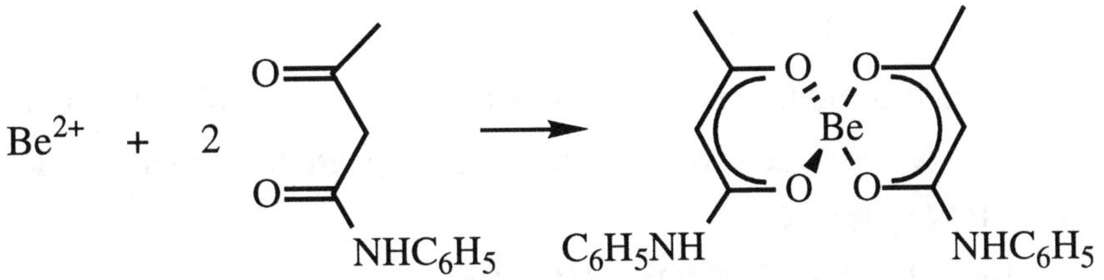

precipitate

The precipitate was dissolved in 6 M HCl and treated with 50.0 mL of solution containing 0.139 2 g of $KBrO_3$ (FW 167.00) plus 0.5 g of KBr. Bromine produced by these reagents reacts with acetoacetanilide as follows:

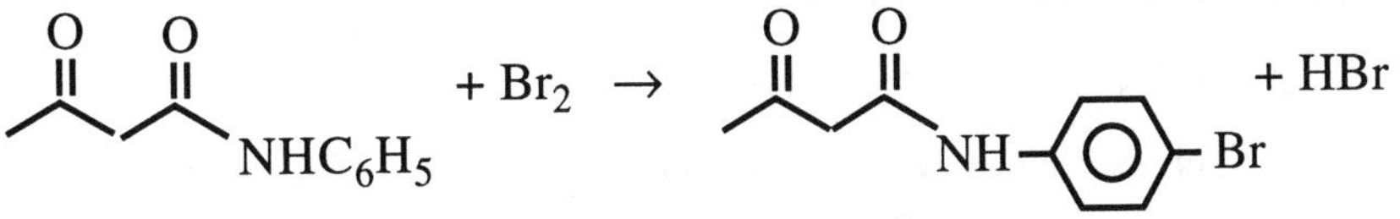

After five minutes, the excess Br_2 was destroyed by adding 2 g of KI (FW 166.00). The I_2 released by the Br_2 required 19.18 mL of 0.050 00 M $Na_2S_2O_3$. Calculate the number of milligrams of Be in the original solution.

S11. As in Problem 35 in Chapter 16, a sample of $Bi_2Sr_2(Ca_{0.8}Y_{0.2})Cu_2O_x$ (FW 770.14 + 15.999 4x) was analyzed to find the oxidation states of Bi and Cu. In Experiment A, a sample of $Bi_2Sr_2(Ca_{0.8}Y_{0.2})Cu_2O_x$ weighing 110.6 mg was dissolved in 50.0 mL of 1 M HCl containing 2.000 mM CuCl. After reaction with the superconductor, coulometry detected 0.052 2 mmol of unreacted Cu^+ in the solution. In Experiment B, 143.9 mg of superconductor was dissolved in 50.0 mL of 1 M HCl containing 1.000 mM $FeCl_2 \cdot 4H_2O$. After reaction with the superconductor, coulometry detected 0.021 3 mmol of unreacted Fe^{2+}. Find the average oxidation numbers of Bi and Cu in the superconductor, and the oxygen stoichiometry coefficient, x.

S12. A sensitive titration of Bi^{3+} is based on the following sequence [G.A. Parker, J. Chem. Ed. **1980**, *57*,721]:

1. Bi^{3+} is precipitated by $Cr(SCN)_6^{3-}$: $Bi^{3+} + Cr(SCN)_6^{3-} \rightarrow Bi[Cr(SCN)_6](s)$ (a)

2. The precipitate is filtered, washed, and treated with bicarbonate to release $Cr(SCN)_6^{3-}$:

$$Bi[Cr(SCN)_6] + HCO_3^- + H_2O \rightarrow (BiO)_2CO_3(s) + Cr(SCN)_6^{3-} + H^+ \quad \text{(b)}$$

3. I_2 is added to the filtrate after removal of the $(BiO)_2CO_3(s)$:

$$Cr(SCN)_6^{3-} + I_2 + H_2O \rightarrow SO_4^{2-} + ICN + I^- + H^+ + Cr^{3+} \quad \text{(c)}$$

4. Upon acidification to pH 2.5, HCN is created:

$$ICN + I^- + H^+ \rightarrow I_2 + HCN \quad \text{(d)}$$

5. The I_2 in this mixture is removed by extraction with chloroform. Excess bromine water is then added to the aqueous phase to convert iodide to iodate and HCN to BrCN:

$$Br_2 + I^- + H_2O \rightarrow IO_3^- + H^+ + Br^- \quad \text{(e)}$$
$$Br_2 + HCN \rightarrow BrCN + H^+ + Br^- \quad \text{(f)}$$

6. Excess Br_2 is destroyed with formic acid:

$$Br_2 + HCO_2H \rightarrow Br^- + CO_2 + H^+ \quad \text{(g)}$$

7. Addition of excess I^- produces I_2:

$$IO_3^- + I^- + H^+ \rightarrow I_2 + H_2O \quad \text{(h)}$$
$$BrCN + I^- + H^+ \rightarrow I_2 + HCN + Br^- \quad \text{(i)}$$

8. Finally, the iodine is titrated with a standard solution of sodium thiosulfate. Show that 228 mol of thiosulfate will be required for each mole of Bi^{3+} that is analyzed.

S13. *Analysis of Organic Compounds with Periodic Acid.* Periodic acid is a powerful oxidizing agent that exists under different conditions as *paraperiodic acid* (H_5IO_6), *metaperiodic acid* (HIO_4), and various deprotonated forms of the acids. Sodium metaperiodate ($NaIO_4$) is usually used to prepare a solution that is standardized by addition to excess KI in bicarbonate solution at pH 8-9:

$$IO_4^- + 3I^- + H_2O \rightleftharpoons IO_3^- + I_3^- + 2OH^- \qquad\qquad (A)$$

The I_3^- released is titrated with thiosulfate to complete the standardization. (In acidic solution, Reaction A goes further: $H_5IO_6 + 11I^- + 7H^+ \rightleftharpoons 4I_3^- + 6H_2O$.)

Periodate is especially useful for the analysis of organic compounds (such as carbohydrates) containing hydroxyl, carbonyl or amino groups adjacent to each other. In this oxidation, known as the *Malaprade reaction*, the carbon-carbon bond between the two functional groups is broken and the following changes occur:

1. A hydroxyl group is oxidized to an aldehyde or a ketone.
2. A carbonyl group is oxidized to a carboxylic acid.
3. An amine is converted to an aldehyde plus ammonia (or a substituted amine if the original compound was a secondary amine).

When there are three or more adjacent functional groups, oxidation begins near one end of the molecule.

The reactions are performed at room temperature for about one hour with a known excess of periodate. At higher temperatures, further nonspecific oxidations occur. Solvents such as methanol, ethanol, dioxane, or acetic acid may be added to the aqueous solution to increase the solubility of the organic reactant. After the reaction is complete, the unreacted periodate is analyzed by using Reaction A, followed by thiosulfate titration of the liberated I_3^-.

Some examples of the Malaprade reactions follow:

(1)
$$\underset{\text{2,3-Dihydroxybutane}}{\overset{\overset{\displaystyle OH\quad\ \ OH}{\displaystyle |\quad\ \ \ |}}{CH_3CH\text{-}\!\!\!\not\;\!\!\text{-}CHCH_3}} + IO_4^- \rightarrow CH_3\overset{O}{\overset{||}{C}}H + H\overset{O}{\overset{||}{C}}CH_3 + IO_3^- + H_2O$$

(2)
$$\underset{\text{Glycerol}}{\overset{\overset{\displaystyle OH\quad\ \ OH\quad\ \ OH}{\displaystyle |\quad\ \ \ \ |\quad\ \ \ \ |}}{CH_2\text{-}\!\!\!\not\;\!\!\text{-}CH-CH_2}} + IO_4^- \rightarrow \overset{O}{\overset{||}{C}}H_2 + H\overset{O}{\overset{||}{C}}-\overset{OH}{\overset{|}{C}}H_2 + IO_3^- + H_2O$$

$$H\overset{O}{\overset{||}{C}}\cdot\!\!\not\;\!\!\text{-}\overset{OH}{\overset{|}{C}}H_2 + IO_4^- \rightarrow H\overset{O}{\overset{||}{C}}OH + \overset{O}{\overset{||}{C}}H_2 + IO_3^-$$

(3)

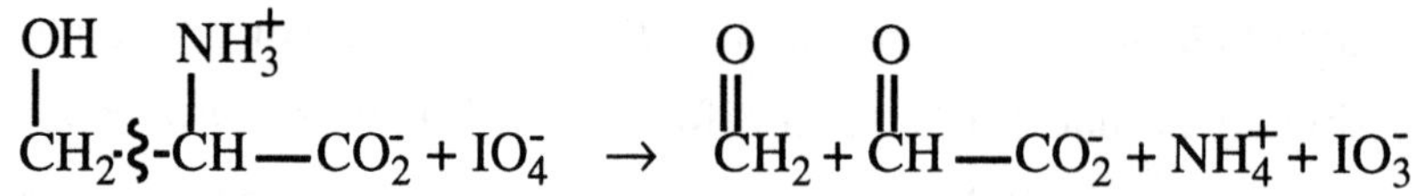

$$\underset{\text{Serine}}{\overset{\overset{\displaystyle OH \quad \overset{+}{N}H_3}{|\qquad\quad|}}{CH_2\text{-}CH\text{---}CO_2^- + IO_4^-}} \quad \rightarrow \quad \overset{\overset{\displaystyle O \quad\quad O}{||\qquad\quad||}}{CH_2 + CH\text{---}CO_2^- + NH_4^+ + IO_3^-}$$

Finally, here is your problem. Write a balanced equation for the reaction of periodate (IO_4^-) with:

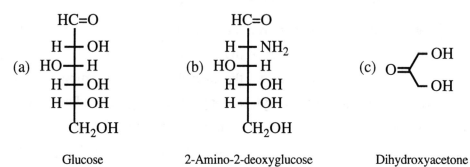

(a) Glucose

(b) 2-Amino-2-deoxyglucose

(c) Dihydroxyacetone

S1. How many amperes of current are required for 0.100 0 mol of electrons to flow through a circuit in 1.000 h?

S2. Consider the electrolysis reactions

Cathode: $H_2O(l) + e^- \rightleftarrows \frac{1}{2} H_2(g, 0.033 \text{ atm}) + OH^-(aq, 1.0 \text{ M})$

Anode: $Br^-(aq, 0.010 \text{ M}) \rightleftarrows \frac{1}{2} Br_2(l) + e^-$

(a) Calculate the equilibrium voltage needed to drive the net reaction.

(b) Suppose the cell has a resistance of 4.3 Ω and a current of 52 mA is flowing. How much voltage is needed to overcome the cell resistance? This is the ohmic potential.

(c) Suppose that the anode reaction has an overpotential (activation energy) of 0.30 V and that the cathode overpotential is 0.08 V. What voltage is necessary to overcome these effects combined with those of parts a and b?

(d) Suppose that concentration polarization occurs. The concentration of OH^- at the cathode surface increases to 2.0 M and the concentration of Br^- at the anode surface decreases to 0.002 0 M. What voltage is necessary to overcome these effects combined with those of parts b and c?

S3. Consider the cell below, whose resistance is 2.12 Ω.,

$Hg(l) \mid Hg_2Cl_2(s) \mid KCl(aq, \text{saturated}) \parallel KCl(0.080 \text{ M}) \mid Cl_2(g, 0.10 \text{ atm}) \mid Pt(s)$

Suppose that there is no concentration polarization or overpotential.

(a) Write the spontaneous galvanic cell reaction.

(b) Calculate the voltage of the galvanic cell if it produces 25.0 mA.

(c) Calculate the voltage that must be applied to run the reaction in reverse, as an electrolysis, at 25.0 mA.

(d) Calculate the electrolysis voltage for part c if concentration polarization changes the concentrations to $[Cl_2(g)]_s = 0.20 \text{ atm}$ and $[Cl^-]_s = 0.040 \text{ M}$ in the right half-cell.

(e) Find the electrolysis voltage for part d if there is an overpotential of 0.15 V at the Pt electrode.

S4. A 0.531 4 g unknown containing lead tartrate, $Pb(O_2CCHOHCHOHCO_2)$ (FW 355.3), plus inert material was electrolyzed to produce 0.122 1 g of PbO_2 (FW 239.2). Was the PbO_2 deposited at the anode or at the cathode? Find the weight percent of lead tartrate in the unknown.

S5. An acidic solution is to be electrolyzed to reduce Fe^{2+} to $Fe(s)$.

(a) Calculate the cathode potential (versus S.H.E.) needed to reduce the Fe^{2+} concentration to 0.10 μM if no concentration polarization occurs.

(b) What would be the potential versus saturated Ag|AgCl instead of S.H.E?

(c) Would the potential be more positive or more negative if concentration polarization occurred?

S6.   What equilibrium cathode potential (versus S.H.E.) is required to reduce 99.99% of Hg(II) from a solution containing 0.10 M Hg(II) in 1.0 M ammonia? Consider the reactions $Hg^{2+} + 4NH_3 \rightleftarrows Hg(NH_3)_4^{2+}$ $\beta_4 = 2 \times 10^{19}$

$Hg^{2+} + 2e^- \rightleftarrows Hg(l)$ $E° = 0.852$ V

S7.   Is it possible to remove 99% of a 1.0 μM FeY^{2-} impurity (by reduction to solid Fe) from a 10 mM CoY^{2-} solution at pH 4 without reducing any cobalt? Here, Y is EDTA and the total concentration of free EDTA is 10 mM.

S8.   The experiment in Figure 17-12 required 4.11 mA for 834 s for complete reaction of a 3.00-mL aliquot of unknown cyclohexene solution.

(a) How many moles of electrons passed through the cell?

(b) How many moles of cyclohexene reacted?

(c) What was the molarity of cyclohexene in the unknown?

S9.   Electrolytic reduction of $TiCl_4$ to Ti(s) in molten NaCl solvent is complicated by formation of soluble $TiCl_3$ and $TiCl_2$ intermediates. The average oxidation state of Ti in such a mixture is measured in two steps [C. E. Baumgartner, *Anal. Chem.* **1992**, *64*, 2001]:

1. Total Ti is determined by inductively coupled plasma atomic emission.

2. Reducing equivalents of Ti are determined by dissolving the solidified $NaCl/TiCl_4/TiCl_3/TiCl_2$ mixture in a solution containing 4.mM Fe^{3+} in 1 M KCl and 0.12 M HCl under N_2. Prior to adding unknown, the Fe^{3+} solution is exhaustively electrolyzed at +0.6 V (versus saturated Ag|AgCl) to be sure that Fe^{2+} and other reducing impurities are ozidized. Addition of the unknown reduces Fe^{3+} to Fe^{2+},

$Ti^{3+} + Fe^{3+} \rightarrow Ti^{4+} + Fe^{2+}$ $Ti^{2+} + 2Fe^{3+} \rightarrow Ti^{4+} + 2Fe^{2+}$

The resulting Fe^{2+} is coulometrically oxidized back to Fe^{3+} to measure the reducing equivalents in the unknown.

In step 1, 42.37 mg of unknown was found to contain 2.03 mg of total titanium. In step 2, 51.36 mg of unknown was dissolved in 100.0 mL of 4.00 mM Fe^{3+} solution. Controlled-potential coulometry required 9.27 C for complete oxidation of Fe^{2+} back to Fe^{3+}. Find the average oxidation number of Ti in the unknown.

S1. Polarographic data for the reduction of Al^{3+} in 0.2 M sodium acetate, pH 4.7, are given below. Construct a standard curve and determine the best straight line by the method of least squares. Calculate the standard deviation for the slope and the intercept. If an unknown solution gives $I_d = 0.904$ μA, calculate the concentration of Al^{3+} and estimate the uncertainty in concentration.

$[Al^{3+}]$ (mM)	I_d (corrected for residual current) (μA)
0.009 25	0.115
0.018 5	0.216
0.037 0	0.445
0.055 0	0.610
0.074 0	0.842
0.111	1.34
0.148	1.77
0.185	2.16
0.222	2.59
0.259	3.12

S2. The differential pulse polarogram of 3.00 mL of solution containing the antibiotic tetracycline in 0.1 M acetate, pH 4, gives a maximum current of 152 nA at a half-wave potential of -1.05 V (versus S.C.E.). When 0.500 mL containing 2.65 ppm of tetracycline was added, the current increased to 206 nA. Calculate the parts per million of tetracycline in the original solution.

S3. Chloroform can be used as an internal standard in the polarographic measurement of the pesticide DDT. A mixture containing 1.00 mM $CHCl_3$ and 1.00 mM DDT gave polarographic signals in the proportion

$$\frac{\text{Wave height of CHCl}_3}{\text{Wave height of DDT}} = 1.40$$

An unknown solution of DDT was treated with a tiny amount of pure $CHCl_3$ to give a concentration of 0.500 mM $CHCl_3$, without significantly changing the concentration of unknown. Now the relative signals are found to be

$$\frac{\text{Wave height of CHCl}_3}{\text{Wave height of DDT}} = 0.86$$

Find the concentration of DDT in the unknown.

S4. A mixture containing Tl^+, Cd^{2+}, and Zn^{2+} exhibited the following diffusion currents in two different experiments, A and B, run with the same electrolyte on different occasions:

		Concentration (mM)	I_d (μA)
Tl^+:	A	1.15	6.38
	B	1.21	6.11
Cd^{2+}:	A	1.02	6.48
	B	?	4.76
Zn^{2+}:	A	1.23	6.93
	B	?	8.54

Calculate the Cd^{2+} and Zn^{2+} concentrations in experiment B.

S5. Based on the standard addition in Fig 18-18, find the concentration of Cd^{2+} in the unknown.

S6. In a standard addition experiment similar to the one in Figure 18-18, the following signals were observed for Cu^{2+}. Dilution of the unknown by the standards was negligible. Prepare a graph similar to Figure 6-10 to find the concentration of Cu^{2+} in the unknown.

Unknown: 8.6 s   Unknown + 0.50 μg/L Cu^{2+}: 12.5 s

Unknown + 1.00 μg/L Cu^{2+}: 16.6 s

S7. Peak current (i_p) and scan rate (v) are listed below for cyclic voltammetry of a water-soluble ferrocene derivative in 0.1 M NaCl , using a Nafion-coated Pt electrode [M. E. Gomez and A. E. Kaifer, *J. Chem. Ed.* **1992**, *69*, 502]. Nafion is a polymer with fixed negative charges and mobile, exchangeable counteranions. Its structure was shown in Problem 11 in Chapter 17. Prepare graphs of i_p vs. v and i_p vs. $\sqrt{v}$ and state whether the ferrocene derivative is free in solution or confined to the electrode surface.

Scan rate (V/s)	Peak anodic current (μA)
0.050	0.73
0.100	1.32
0.150	1.89
0.200	2.22
0.250	2.89
0.300	3.39
0.350	3.87

S8. The half-wave potential ($E_{1/2}^{comp}$) for reduction of the metal-ligand complex (ML_p^+) is given by

$$ML_p^+ + e^- \rightarrow ML_q + (p\text{-}q)L$$

$$E_{1/2} = E_{1/2}^{free} - 0.059 \log \frac{\beta_p^{ox}}{\beta_q^{red}} + 0.059(q\text{-}p)\log[L]$$

where $E_{1/2}^{free}$ is the half-wave potential for reduction in the absence of ligand, β is a formation constant, and [L] is the concentration of ligand.

$$\beta_p^{ox} = \frac{[ML_p^+]}{[M^+][L]^p} \qquad\qquad \beta_q^{red} = \frac{[ML_q]}{][M][L]^q}$$

A graph of $E_{1/2}$ versus log[L] should have a slope of 0.059(q-p) and an intercept of $\{E_{1/2}^{free} - 0.059 \log (\beta_p^{ox}/\beta_q^{red})\}$. Experimental data for the reduction of an Fe^{3+} complex (M^+) to the Fe^{2+} complex (M) in the presence of the ligand imidazole (L) is shown below.

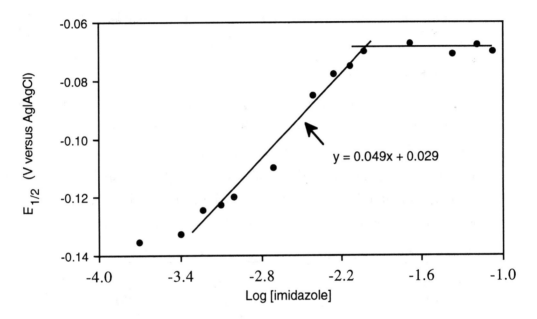

Dependence of half-wave potential (measured by cyclic voltammetry) on ligand concentration for reduction of Fe^{3+} complex to Fe^{2+} complex. [Data from D. K. Geiger. E. J. Pavlak and L. T. Kass, *J. Chem. Ed.* **1991**, *68*, 337. This article describes a student experiment for measuring binding constants.]

The species M is actually an iron-porphyrin complex (Problem 21 in Chapter 18) that can bind zero, one or two imidazole ligands in axial coordination positions:

(a) For the region -3.4 < log [L] < -2.1, the slope in the graph is +0.049 V. Write a reduction reaction with appropriate numbers of ligands for reactants and products to describe the chemistry in this region. Given that $E_{1/2}^{free} = -0.18$ V, use the intercept (+0.029) of the line segment to estimate the value of log $(\beta_p^{ox}/\beta_q^{red})$.

(b) For log [L] > -2.1, the slope in the graph is zero. Write a reduction reaction with appropriate numbers of ligands for reactants and products to describe the chemistry in this region.

S9. Ammonia can be titrated with hypobromite, but the reaction is somewhat slow:

$$2NH_3 + 3OBr^- \rightleftharpoons N_2 + 3Br^- + 3H_2O$$

The titration can be performed with a rotating Pt electrode held at +0.20 V (versus S.C.E.) to monitor the concentration of OBr^-:

Cathode: $OBr^- + 2H^+ + 2e^- \rightleftharpoons Br^- + H_2O$

The current would be near zero before the equivalence point if the OBr^- were consumed quickly in the titration. However, the sluggish titration reaction does not consume all the OBr^- after each addition, and some current is observed. The increasing current beyond the equivalence point can be extrapolated back to the residual current to find the endpoint. A solution containing 30.0 mL of 4.43×10^{-5} M NH_4Cl was titrated with NaOBr in 0.2 M $NaHCO_3$, with the following results:

OBr^- (mL)	I (μA)	OBr^- (mL)	I (μA)
0.000	0.03	0.700	1.63
0.100	1.42	0.720	2.89
0.200	2.61	0.740	4.17
0.300	3.26	0.760	5.53
0.400	3.74	0.780	6.84
0.500	3.79	0.800	8.08
0.600	3.20	0.820	9.36
0.650	2.09	0.840	10.75

Prepare a graph of current versus volume of OBr^-, and find the molarity of the NaOBr solution.

S10. Karl Fischer reagent containing I_2, SO_2, base and alcohol may be delivered from a buret, instead of generating the I_2 coulometrically. The reagent may be standardized by titration with H_2O dissolved in methanol. A 25.00-mL aliquot of Karl Fischer reagent reacted with 34.61 mL of methanol to which was added 4.163 mg of H_2O per mL. When pure "dry" methanol was titrated, 25.00 mL of methanol reacted with 3.18 mL of the same Karl Fischer reagent. A suspension of 1.000 g of a hydrated crystalline salt in 25.00 mL of methanol consumed a total of

38.12 mL of Karl Fischer reagent. Calculate the weight percent of water in the crystal.

S11. Weighted least squares. Problem 31 in Chapter 18 gave an example in which the uncertainties associated with different data points is different. We now show how to compute the least squares best straight line when errors in x_i are negligible, but each value of y_i has a standard deviation s_i. These are found by measuring y several times for each value of x. We associate a weighting factor, w_i, with each value of y_i:

$$\text{Weighting factor} = w_i = \frac{1}{s_i^2}$$

The smaller the standard deviation of y_i, the greater will be the weighting factor for y_i. That is, we weight the more certain values of y greater than the less certain values of y when we compute the least squares straight line. Using the weighting factor $w_i = 1/s_i^2$, the parameters for the straight line are

$$\text{Denominator} = D = (\Sigma w_i)(\Sigma w_i x_i^2) - (\Sigma w_i x_i)^2$$

$$\text{Slope} = m = \frac{(\Sigma w_i)(\Sigma w_i x_i y_i) - (\Sigma w_i x_i)(\Sigma w_i y_i)}{D}$$

$$\text{Intercept} = b = \frac{(\Sigma w_i y_i)(\Sigma w_i x_i^2) - (\Sigma w_i x_i)(\Sigma w_i x_i y_i)}{D}$$

$$\text{Variance of y} = s_y^2 = \frac{(\Sigma w_i)d_i^2}{n - 2}$$
$$\text{(where } d_i = y_i - mx_i - b \text{ and there are n data points)}$$

$$\text{Variance of slope} = s_m^2 = \frac{(\Sigma w_i)s_y^2}{D}$$

$$\text{Variance of intercept} = s_b^2 = \frac{(\Sigma w_i x_i^2)s_y^2}{D}$$

Prepare a spreadsheet that uses the data in Problem 31 to find the slope and intercept (and their uncertainties) for a graph of current versus $[H_2O_2]$. Compare your parameters to those found by the unweighted treatment in Problem 31. (Alternatively, you could have solved this same problem with the curve fitting program at the beginning of the *Solutions Manual*.)

S12. A spreadsheet for fitting the equation $y = A + Bx + Cx^2$. At the end of this problem is a spreadsheet that you can copy onto your computer to fit quadratic equations and to estimate the uncertainties in the parameters A, B and C. [The formulas in this spreadsheet come from R. T. O'Neill and D. C. Flaspohler (*J. Chem. Ed.* **1990**, *67*, 40), which provides a general recipe for fitting equations of

the form y = A*f(x, z, w....) + B*g(x, z, w....) + C*h(x, z, w....) + ..., where A, B, C ... are constants and f, g, h ....are functions of the variables x, z, w....]
In column A, N is the number of data points and T is the number of terms to be fit. Since we are looking for three coefficients (A, B and C), T = 3. The data are entered in columns B and C and their weights (discussed in the previous problem) are in column D. If weights are unknown, set all weights to unity. The parameters A, B and C are found in cells A15, A17 and A19 and their estimated uncertainties are in cells D15, D17 and D19. The other quantities in the table are used in the computations, but are not part of the final answer.

α-Tocopherol (Vitamin E) is a lipid-soluble anti-oxidant that can be oxidized at 0.67 V (versus S. C. E.) at a glassy carbon electrode in 0.12 M H_2SO_4 in a 1:1 ethanol:benzene mixture. (Note: In recipes calling for the carcinogen, benzene, try substituting toluene.)

α-Tocopherol (Vitamin E)

Calibration data for voltammety of standard solutions is given below.

[α-Tocopherol] (μg/mL)	Peak current (μA)	[α-Tocopherol] (μg/mL)	Peak current (μA)
13.0	18.2	62.1	82.4
26.1	36.0	91.0	115.3
37.9	52.7	118.6	148.4
50.5	68.3	169.2	200.0

(a) Copy the spreadsheet onto your computer to find the values of A, B and C for the equation

$$\text{Current} = A + B[\alpha\text{-tocopherol}] + C[\alpha\text{-tocopherol}]^2$$

(b) Prepare a graph of the calibration data, showing the curve computed in part a.

(c) Calculate the concentration (and uncertainty) of α-tocopherol in an unknown that gave a peak current of 170 ±1 μA. For the uncertainty, compute the concentrations corresponding to 170+1 = 171 μA and to 170-1 = 169 μA.

(Alternatively, you could have solved this same problem with the curve fitting program at the beginning of the *Solutions Manual* .)

	A	B	C	D	E	F	G	H	I	J	K	L
1	N (points)=	x	y	weight=w	w*y*y	w*y	w*y*x	w*y*x*x	w*x	w*x*x	w*x^3	w*x^4
2	8	13.0	18.2	1	3.3E+02	1.8E+01	2.4E+02	3.1E+03	1.3E+01	1.7E+02	2.2E+03	2.9E+04
3	C (terms)=	26.1	36.0	1	1.3E+03	3.6E+01	9.4E+02	2.5E+04	2.6E+01	6.8E+02	1.8E+04	4.6E+05
4	3	37.9	52.7	1	2.8E+03	5.3E+01	2.0E+03	7.6E+04	3.8E+01	1.4E+03	5.4E+04	2.1E+06
5		50.5	68.3	1	4.7E+03	6.8E+01	3.4E+03	1.7E+05	5.0E+01	2.6E+03	1.3E+05	6.5E+06
6		62.1	82.4	1	6.8E+03	8.2E+01	5.1E+03	3.2E+05	6.2E+01	3.9E+03	2.4E+05	1.5E+07
7		91.0	115.3	1	1.3E+04	1.2E+02	1.0E+04	9.5E+05	9.1E+01	8.3E+03	7.5E+05	6.9E+07
8		118.6	148.4	1	2.2E+04	1.5E+02	1.8E+04	2.1E+06	1.2E+02	1.4E+04	1.7E+06	2.0E+08
9		169.2	200.0	1	4.0E+04	2.0E+02	3.4E+04	5.7E+06	1.7E+02	2.9E+04	4.8E+06	8.2E+08
10		Column sums:		GG	FF	FG	FH	FK	GH	GK, HH	HK	KK
11				8	91175.83	721.3	73672.26	9363164	568.4	59668.88	7708447	1.110E+9
12	Denominator=											
13	6.307E+12											
14	A =			Standard deviation (A) =		Q' =						
15	0.9666344				0.954980		102742.6					
16	B =			Standard deviation (B) =		Q =						
17	1.3851265				0.026690		91171.61					
18	C =			Standard deviation (C) =		V =						
19	-0.001235				1.437E-4		0.844674					
20	Formulas:											

21 A13 = (F11*(J11*L11-K11*K11)+G11*(J11*K11-I11*L11)+H11*(I11*K11-J11*J11))/A13

22 A15 = (F11*(J11*L11-K11*K11)+G11*(J11*K11-I11*L11)+H11*(I11*K11-J11*J11))/A13

23 A17 = (F11*(J11*K11-I11*L11)+G11*(D11*L11-J11*J11)+H11*(I11*J11-D11*K11))/A13

24 A19 = (F11*(I11*K11-J11*J11)+G11*(I11*J11-D11*K11)+H11*(D11*J11-I11*I11))/A13

25 D15 = Sqrt(G19*(J11*L11-K11*K11)/A13)

26 D17 = Sqrt(G19*(D11*L11-J11*J11)/A13)

27 D19 = Sqrt(G19*(D11*J11-I11*I11)/A13)

28 G15 = A15*(A15*D11+A17*I11+A19*J11)+A17*(A15*I11+A17*J11+A19*K11)

29 G17 = G15+A19*(A15*J11+A17*K11+A19*L11)

30 G19 = (E11-G17)/(A2-A4)

CHAPTER 19: SUPPLEMENTARY PROBLEMS
APPLICATIONS OF SPECTROPHOTOMETRY

S1. Consider compounds X and Y in the example in Section 19-1 labeled "Analysis of a Mixture Using Equations 19-7." A mixture of X and Y in a 0.100-cm cell had an absorbance of 0.282 at 272 nm and 0.303 at 327 nm. Find the concentrations of X and Y in the mixture.

S2. Spectrophotometric data for three compounds was given in Problem 7 in Chapter 19. A solution containing X, Y and Z in a 2.000-cm cuvet had absorbances of 0.666 at 246 nm, 0.498 at 298 nm and 0.360 at 360 nm. Using a spreadsheet for three simultaneous linear equations, find the concentrations of X, Y and Z in the mixture.

S3. Compound P, which absorbs light at 517 nm, was titrated with X, which does not absorb at this wavelength. The product, PX, also absorbs at 517 nm. A series of solutions containing a fixed concentration of P (0.001 00 M) was prepared with variable concentrations of X. The absorbance of each solution was measured in a 1.000-cm cell, and the concentration of free X was determined by an independent method. The results are shown below. Prepare a Scatchard plot to find the equilibrium constant and $\Delta\varepsilon$ for the reaction $X + P \rightleftharpoons PX$.

[X] M	Absorbance	[X] M	Absorbance
0.0	0.213	0.050 9	0.493
0.005 09	0.243	0.065 0	0.563
0.008 52	0.263	0.077 9	0.633
0.017 3	0.313	0.093 2	0.703
0.029 5	0.383	0.106 2	0.763
0.038 7	0.433		

S4. Solutions of metal ions and the ligand ammonium 1-pyrrolidinecarbodithioate were prepared in aqueous solution and the resulting complexes were extracted into chloroform. Spectrophotometric results for Zn^{2+} and Ga^{3+} are given below. The reference solution for each measurement was a reagent blank prepared with ligand but no metal ion. From the absorption data, find the ligand:metal stoichiometry in the complexes.

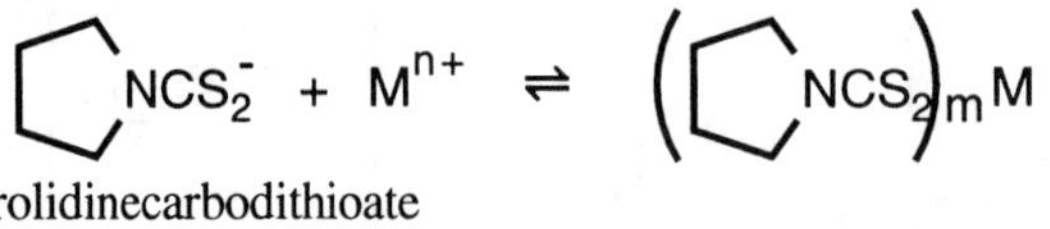

NCS2-
+ Mn+
⇌
NCS2
m
M

1-pyrrolidinecarbodithioate

Mole fraction of metal ion	Zn²⁺ Relative absorbance at 315 nm at pH 5.20	Ga³⁺ Relative absorbance at 308 nm at pH 2.30
0.05	0.145	0.204
0.10	0.298	0.406
0.15	0.440	0.594
0.20	0.589	0.750
0.25	0.720	0.778
0.30	0.818	0.759
0.33	0.836	---
0.40	0.795	0.665
0.50	0.689	0.551
0.60	0.548	0.442
0.70	0.403	0.321
0.80	0.274	0.204
0.90	0.143	0.088

Data from W. Likussar and D. F. Boltz, *Anal. Chem.* **1971**, *43*, 1273.

S5. In a time-resolved fluorescence immunoassay, a solution was irradiated at 340 nm and Eu^{3+} emission was observed at 613 nm. What is the energy difference (kJ/mol) between these two wavelengths? This difference is converted to heat in the solution.

S6. Consider a fluorescence experiment in which $\varepsilon_{ex} = 2\ 120\ M^{-1} \cdot cm^{-1}$, $\varepsilon_{em} = 810\ M^{-1} \cdot cm^{-1}$, $b_1 = 0.300$ cm, $b_2 = 0.400$ cm, and $b_3 = 0.500$ cm in Equation 19-27. Make a graph of relative fluorescence intensity versus concentration for the following concentrations of solute: 1.00×10^{-7}, 1.00×10^{-6}, 1.00×10^{-5}, 1.00×10^{-4}, 1.00×10^{-3}, and 1.00×10^{-2} M.

S7. A sensitive assay for ATP is based on its participation in the light-producing reaction of the firefly. The reaction catalyzed by the enzyme luciferase is

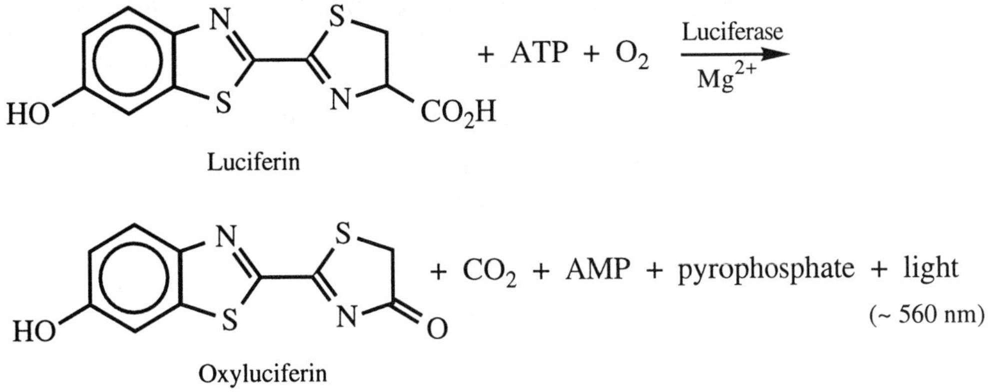

Luciferin

Oxyluciferin

When the reactants are mixed, the solution gives off light. The light intensity decays slowly due to product inhibition of the reaction. Otherwise, the light would have a steady intensity because *the rate at which reactants are consumed is negligible*. That is, ATP and luciferin maintain their original concentrations throughout the few minutes that the reaction might be monitored. Some typical experimental results are shown below.

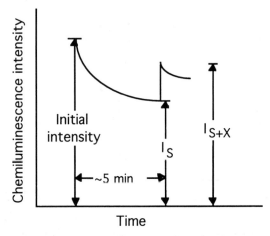

Let the initial concentration of ATP in the reaction be [S]. Suppose that additional ATP is added, increasing the concentration in the reaction to [S] + [X]. The kinetic description of the reaction predicts that the increase in light intensity after the addition will be given by

$$\frac{I_S}{I_{S+X}} = \frac{1}{[S] + [X]}\left(\frac{K[S]}{K + [S]}\right) + \frac{[S]}{K + [S]}$$

where K is constant.

(a) Suppose that [S] = 250 μM and after 5 minutes I_S = 58.7 arbitrary intensity units. Then a standard addition of [X] = 200 μM is made, and I_{S+X} is found to be 74.5 units. Use these data to find the value of K in the equation above.

(b) When the intensity had decayed to 63.5 units, an unknown aliquot of ATP was added to the reaction, and the intensity increased to 74.6 units. How much was the increase in concentration caused by the unknown aliquot?

[Problem from J. J. Lemasters and C. R. Hackenbrock, *Methods of Enzymology*, **1978**, *57*, 36. A student experiment using luciferase for the assay of ATP or reduced nicotine adenine dinucleotide (NADH) is described by T. C. Selig, K. A. Drozda, and J. A. Evans, *J. Chem. Ed.*, **1984**, *61*, 918.]

S1. A sample that scatters negligible light has an external transmittance of 70% and a reflectance of 8%. Find the absorptance.

S2. Find the internal transmittance (P_2/P_1 in Figure 20-1) of a 2.00-mm-thick plate with an absorption coefficient of $\alpha = 1.40$ cm^{-1}.

S3. Sapphire windows are sometimes used for spectroscopy of aqueous solutions or systems under high pressure. The refractive index of sapphire at a wavelength of 5 μm is 1.62 and the absorption coefficient is 0.85 cm^{-1}.
  (a) Neglecting absorption, calculate the external transmittance of a 5.00-mm-thick sapphire window at 5 μm wavelength.
  (b) Use the equation from Exercise B in Chapter 20 to find the transmittance, accounting for both reflection and absorption.

S4. (a) Find the critical value of θ_i in Figure 20-4 beyond which there is total internal reflection in an As$_2$Se$_3$-based infrared optical fiber whose core refractive index is 2.7 and whose cladding refractive index is 2.0.
  (b) Referring to Problem 20-9, calculate the quotient power out/power in for a 0.50-m-long fiber with a loss of 0.012 dB/m.

S5. The ability of a prism to spread apart (disperse) neighboring wavelengths (λ) increases as the slope $dn/d\lambda$ increases, where n is the refractive index. Use Figure 20-2 to decide whether NaCl or KBr has greater dispersion at a wavelength of 13 μm.

S6. Light passes from quartz (medium 1) to carbon disulfide (medium 2) in Figure 20-3 at (a) $\theta_1 = 30°$ or (b) $\theta_1 = 0°$. Find the angle θ_2 in each case.

S7. (a) For the 60° prism below, use Snell's law to show that light traveling through the prism parallel to the base enters and exits at the same angle, θ.
  (b) The index of refraction of the prism is 1.500 and the index of refraction of air is 1.000. Find the angle θ in the diagram.

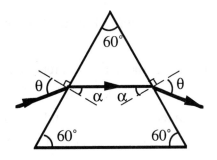

S8.   What is the ratio of power per unit area (the exitance, W/m^2) radiating from a blackbody at 900 K compared to one at 300K? Calculate the exitance at 900 K.

S9.   A long iron cylinder with a diameter of 0.64 cm is heated yellow hot (1 100°C) and then allowed to begin cooling. The density of iron is 7.86 g/cm^3 and the heat capacity at this temperature is 0.606 J/(g·K).

   (a) Consider a 1-cm length of the cylinder far from either end. Calculate the rate of cooling of this 1-cm section by blackbody radiation (ignoring the surface area at the ends of the cylinder).

   (b) Calculate the rate of cooling for a thinner cylinder with a 0.064 cm diameter at 1 100°C.

S10. Consider a reflection grating operating with an incident angle of 20° in Figure 20-12.

   (a) How many lines per centimeter should be etched in the grating if the first-order diffraction angle for 400 nm (visible) light is to be 10°?

   (b) Answer the same question for 1 000 cm^{-1} (infrared) light.

S11. (a) What resolution is required for a diffraction grating to resolve adjacent lines in the spectrum of calcium ions at wavelengths of 443.495 and 443.567 nm? What is the difference between these two spectral lines in wavenumbers (cm^{-1}) and frequency (Hz)?

   (b) With a resolution of 10^4, how close in nm is the closest line to 443.495 nm that can barely be resolved?

   (c) Calculate the second order (n=2) resolution of a 6.00-cm-long grating ruled at 212 lines/mm.

   (d) Find the angular dispersion ($\Delta\phi$) between light rays with wavelengths of 443.495 and 443.567 nm for second order diffraction (n = 2) and 20th order diffraction from a grating with 200 lines/mm and $\phi = 10.0°$.

S12. The true absorbance of a sample is 1.26, but 0.4 % stray light reaches the detector. Find the apparent transmittance and apparent absorbance of the sample.

S13. The interferometer mirror of a Fourier transform infrared spectrophotometer travels ±2 cm.

   (a) How many centimeters is the maximum retardation, Δ?

   (b) What is the approximate resolution (cm^{-1}) of the instrument?

   (c) At what retardation interval, δ, must the interferogram be sampled (converted to digital form) to cover a spectral range of 0–4 000 cm^{-1}?

S14. A spectrum has a signal-to-noise ratio of 3/1. How many spectra must be averaged to increase the signal-to-noise ratio to 9/1?

S15. A shipboard flow injection analytical procedure for measuring Fe^{2+} in sea water is based on chemiluminescence from the dye brilliant sulfoflavin in the presence of Fe^{2+} and H_2O_2. Data from a series of standard additions is shown below. Use all 12 points to construct a graph similar to Figure 6-10 and find $[Fe^{2+}]$ in the unknown.

Sample	Detector signal		
Unknown	12.0	12.0	11.0
Unknown + 5.25 nM Fe^{2+}	27.2	26.5	26.5
Unknown + 7.88 nM Fe^{2+}	39.9	41.7	39.1
Unknown + 10.5 nM Fe^{2+}	53.9	56.2	55.3

V. A. Elrod, K. S. Johnson and K. H. Coale, *Anal. Chem.* **1991**, *63*, 893

CHAPTER 21: SUPPLEMENTARY PROBLEMS
ATOMIC SPECTROSCOPY

S1. Calculate the Doppler linewidths in Hz ($= s^{-1}$) for the 248-nm line of Fe and for the 254-nm line of Hg, each at 3 000 K and 6 000 K.

S2. The first excited state of Cu is reached by absorption of 327 nm light.

(a) What is the energy difference (kJ/mol) between the ground and excited states?

(b) The relative degeneracies are $g^*/g_0 = 3$ for Cu. What is the ratio N^*/N_0 at 2 400 K?

(c) By what percentage will the fraction in part b be changed by a 15 K rise in temperature?

(d) What will be the ratio N^*/N_0 at 6 000 K?

S3. A sample containing 23.6 µg Mo/mL gave an absorbance of 0.025. Estimate the sensitivity for Mo.

S4. A series of Ca and Cu samples was run to determine the atomic absorbance of each element.

Ca(µg/mL)	$A_{422.7}$	Cu (µg/mL)	$A_{324.7}$
1.00	0.086	1.00	0.142
2.00	0.177	2.00	0.292
3.00	0.259	3.00	0.438
4.00	0.350	4.00	0.576

(a) Find the average relative absorbance ($A_{422.7}/A_{324.7}$) produced by equal concentrations (µg/mL) of Ca and Cu.

(b) Copper was used as an internal standard in a Ca determination. A sample known to contain 2.47 µg Cu/mL gave $A_{422.7} = 0.218$ and $A_{324.7} = 0.269$. Calculate the concentration of Ca in micrograms per milliliter.

S5. The following nonlinear calibration data were obtained in an atomic absorption experiment.

(a) Use the quadratic equation spreadsheet in the Supplementary Problems for Chapter 18 to find the best fit to the data in the form $y = A + Bx + Cx^2$.

(b) Use the method of Box 6-3 to estimate the concentration (and its uncertainty) of an unknown with a signal intensity of 344.0 ± 2.2.

Concentration of standard (μg/mL)	Signal intensity (arbitrary meter units)
0	0.1
1.00	49.0
2.00	96.5
3.00	142.4
4.00	183.3
5.00	222.1
6.00	258.9
7.00	293.0
8.00	321.2
9.00	350.8
10.0	377.8

CHAPTER 22: SUPPLEMENTARY PROBLEMS
INTRODUCTION TO ANALYTICAL SEPARATIONS

S1. (a) What volume of chloroform is needed to extract 99.5% of a solute from 100 mL of water, if the partition coefficient is $C_{CHCl_3}/C_{H_2O} = 610$?

(b) What will be the total volume of chloroform needed to remove 99.5% of the solute in four equal extractions instead?

S2. The acid HA ($K_a = 4.2 \times 10^{-4}$) equilibrates between water and ether.

(a) Define the distribution coefficient (D) and the partition coefficient (K) for this system.

(b) Calculate D at pH 4.00 if K = 92.

(c) Will D be greater or less at pH 3.50 than at pH 4.00? Explain why.

(d) At what pH will D be unity?

S3. The chromatogram in Problem 42 in Chapter 22 has a peak for isooctane at 13.81 min. The elution time for unretained solute is 1.06 min and the column is 30.0 m long.

(a) Find the adjusted retention time and capacity factor for isooctane.

(b) Measuring $w_{1/2}$, find the number of theoretical plates for this peak.

(c) Measure the width (w) at the base of the peak to find the number of plates.

(d) Using the answer from part c, find the plate height.

(e) The open tubular column has an inner diameter of 524 μm. Given that the linear flow rate is 30.0 m/1.06 min, find the volume flow rate.

S4. (a) How many theoretical plates produce a chromatography peak eluting at 12.83 min with a width at the base of 18.4 s?

(b) If the length of the column is 15.8 cm, find the plate height.

S5. (a) A chromatography column with a length of 50.6 cm and inner diameter of 19.2 mm is packed with a stationary phase that occupies 57.4% of the volume. Find the linear flow rate (cm/min) if the volume flow rate is 2.22 mL/min.

(b) Find the retention time for a solute with a capacity factor of 8.04.

S6. Two compounds with capacity factors of 5.00 and 5.10 are separated on a column with a plate height of 155 μm. What length of column is required to give a resolution of 1.00? Of 2.00?

S7. Consider the peaks for *n*-heptane and *p*-difluorobenzene in the chromatogram in Problem 42 in Chapter 22. The elution time for unretained solute 1.06 min. The open tubular column is 30.0 m in length and 0.530 mm in diameter, with a 3.0 μm thick layer of stationary phase on the inner wall.

(a) Find the adjusted retention times and capacity factors for both compounds.

(b) Find the relative retention.

(c) Measuring $w_{1/2}$ on the chromatogram, find the number of plates (N_1 and N_2) and the plate height for these two compounds.

(d) Measuring the width (w) at the baseline on the chromatogram, find the number of plates for these two compounds.

(e) Use your answer to part d to find the resolution between the two peaks.

(f) Using the number of plates ($N = \sqrt{N_1 N_2}$, with values from part d), the relative retention, and the capacity factors, calculate what the resolution should be and compare your answer to the the measured resolution in part e.

(g) What fraction of the time does heptane spend in the stationary phase?

S8. A 3.00-cm-diameter column with a length of 32.6 cm gives adequate resolution of a 72.4 mg mixture of unknowns, initially dissolved in 0.500 mL. If you wish to scale down to 10.0 mg of the same mixture with minimum use of chromatographic stationary phase and solvent, what length and diameter column would you use? In what volume would you dissolve the sample?

S9. Potassium ion diffuses down a steady concentration gradient of $dc/dx = 3.4$ M/cm. If the diffusion coefficient is 2.0×10^{-9} m^2/s, find the number of ions crossing a 1 cm^2 area in 5.5 s.

CHAPTER 23: SUPPLEMENTARY PROBLEMS
GAS AND LIQUID CHROMATOGRAPHY

S1. Retention times in a gas chromatogram are 1.21 min for air, 7.33 min for hexane, 7.66 min for unknown, and 8.41 min for heptane. Find the Kovats retention index for the unknown.

S2. Referring to Table 23-3, what is the order of elution of the hexane, heptane, octane, nonane, benzene, nitropropane and pyridine from

(a) $(diphenyl)_{0.05}(dimethyl)_{0.95}polysiloxane$

(b) $(cyanopropylphenyl)_{0.14}(dimethyl)_{0.86}polysiloxane$

(c) $(biscyanopropyl)_{0.9}(cyanopropylphenyl)_{0.1}$ polysiloxane

S3. Compounds C and D gave the following HPLC results:

Compound	Concentration (µg/mL) in mixture	Peak area (cm²)
C	236	4.42
D	337	5.52

A solution was prepared by mixing 1.23 mg of D in 5.00 mL with 10.00 mL of unknown containing just C, and diluting to 25.00 mL. Peak areas of 3.33 and 2.22 cm^2 were observed for C and D, respectively. Find the concentration of C (µg/mL) in the unknown.

S4. Polar solutes were separated by HPLC using a bonded phase containing polar diol substituents [—$CH(OH)CH_2OH$]. How would the retention times be affected if the eluent were changed from 40 vol % to 60 vol % acetonitrile in water?

S5. Toluene exhibited a capacity factor of 3.50 and a plate height of 10.6 µm on a 15.0-cm HPLC column. If unretained solute is eluted in 1.33 min, find the retention time and width ($w_{1/2}$) at half-height for toluene. If you want to use a faster linear flow rate, but get the same separation efficiency, should you raise or lower the column temperature?

CHROMATOGRAPHIC METHODS AND CAPILLARY ELECTROPHORESIS

S1. Measure the width of the I^- peak at half-height in Figure 24-22 and calculate the number of theoretical plates. The capillary was 60 cm long. Find the plate height.

S2. A gel-filtration column has a radius (r) of 1.27 cm and a length (l) of 28.4 cm. Blue Dextran was eluted at a void volume of 43.4 mL and a solute was eluted at 98.6 mL. Find K_{av} for the solute.

S3. Polystyrene standards of known molecular weight gave the following calibration data in a size-exclusion column [E. Meehan and G. M. Cowell, *Am. Lab.* August 1992, p. 24]. Find the molecular weight of an unknown with a retention time of 13.00 min.

Molecular weight	Retention time (min)
8.50×10^6	9.28
3.04×10^6	10.07
1.03×10^6	10.88
3.30×10^5	11.67
1.56×10^5	12.14
6.60×10^4	12.74
2.85×10^4	13.38
9.20×10^3	14.20
3.25×10^3	14.96
5.80×10^2	16.04

S4. An Ohm's law plot was constructed to find the maximum voltage that could be used in capillary electrophoresis without serious overheating. The measured points in the graph are fit by the equation $I = 3V + 0.01V^2 + 0.000\,2\ V^3$, where I is in μA and V is in kV. The straight line extrapolated from the low voltage points is $I = 3V$. At what voltage do these differ by 5%? This will be chosen as the maximum operating voltage.

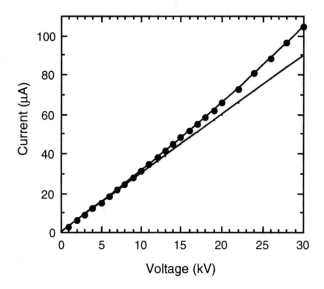

S5. Rationalize why the mobility of 13-*cis*-retinoate anion is slightly greater than that of the all-*trans* isomer:

all-*trans*-retinoate anion
$\mu_{ep} = -1.84 \times 10^{-8}$ m^2/(V·s)

13-*cis*-retinoate anion
$\mu_{ep} = -1.91 \times 10^{-8}$ m^2/(V·s)

(Solvent is 50 vol % CH$_3$CN in H$_2$O with borate buffer at pH 8.5)

S6. (a) A protein required 6.8 min to travel 82 cm to the detector in a 96-cm-long capillary tube with 25.4 kV between the ends. Find the apparent electrophoretic mobility.

   (b) How many femtomoles of 2.4 μM protein are injected electrokinetically into a 50-μm-diameter capillary at 5.0 kV for 3.0 s if the sample has half of the conductivity of the background electrolyte?

S7. To get the best separation of acetate from formate in capillary electrophoresis, it makes sense to operate at the pH at which their charge difference is greatest. What is this pH?

S8. (a) Hexanoic acid and 1-aminohexane, adjusted to pH 12 with NaOH, were passed through a cation-exchange column loaded with NaOH at pH 12. State the principal species that will be eluted.

   (b) Hexanoic acid and 1-aminohexane, adjusted to pH 2 with HCl, were passed through a cation-exchange column loaded with HCl at pH 2. State the principal species that will be eluted.

S1. A 0.123 4-g sample of impure piperazine contained 79.33 wt % piperazine (FW 86.137). How many grams of product (FW 206.243) will be formed when this sample is analyzed by Reaction 25-6?

S2. A 1.127-g sample of unknown gave 2.188 g of bis(dimethylglyoximate)nickel(II) (FW 288.91) when analyzed by Reaction 25-7. Find the mass of Ni in the unknown.

S3. A solution containing 1.263 g of unknown potassium compound was dissolved in water and treated with excess sodium tetraphenylborate, $Na^+B(C_6H_5)_4^-$ solution to precipitate 1.003 g of insoluble $K^+B(C_6H_5)_4^-$ (FW 358.33). Find the wt % of K in the unknown.

S4. Combustion of 8.732 mg of an unknown organic compound gave 16.432 mg of CO_2 and 2.840 mg of H_2O.

(a) Find the wt % of C and H in the substance.

(b) Find the smallest reasonable integer ratio of C:H in the compound.

S5. A 2.000-g sample of a solid mixture containing only $PbCl_2$ (FW 278.1), $CuCl_2$ (FW 134.45), and KCl (FW 74.55) was dissolved in water to give 100.0 mL of solution. First 50.00 mL of the unknown solution was treated with sodium piperidine dithiocarbamate to precipitate 0.726 8 g of lead piperidine dithiocarbamate:

Lead piperidine dithiocarbamate
F.W. 527.77

Then 25.00 mL of the unknown solution was treated with iodic acid to precipitate 0.838 8 g of $Pb(IO_3)_2$ and $Cu(IO_3)_2$.

$$Cu^{2+} + 2IO_3^- \rightarrow Cu(IO_3)_2$$
FW 413.35

$$Pb^{2+} + 2IO_3^- \rightarrow Pb(IO_3)_2$$
FW 557.0

Calculate the wt % of Cu in the unknown mixture.

S6. $NaHCO_3$ (FW 84.007) decomposes to Na_2CO_3 (FW 105.989) near 100°C.
$KHCO_3$ (FW 100.115) decomposes to K_2CO_3 (FW 138.206) near 150°C.

(a) Write balanced thermal decomposition reactions for the two compounds.

(b) What mass of Na_2CO_3 would be produced by decomposition of 50.0 g of
$NaHCO_3$? Answer the same for 50.0 g of $KHCO_3$.

(c) Sketch the expected thermogravimetric curve for a mixture of 50.0 mg of
$NaHCO_3$ and 50.0 mg of $KHCO_3$.

S7. Use the uncertainties from Instrument 2 in Table 25-4 to estimate the uncertainties
in the stoichiometry coefficients in the formula $C_8H_{h\pm x}N_{n\pm y}$.

S8. When the *high temperature superconductor* yttrium barium copper oxide is heated
under flowing H_2, the solid remaining at 1 000°C is a mixture of Y_2O_3, BaO and
Cu. The starting material has the formula $YBa_2Cu_3O_{7-x}$, in which the oxygen
stoichiometry varies between 7 and 6.5 (x = 0 to 0.5).

$$YBa_2Cu_3O_{7-x}(s) + (3.5-x)H_2(g) \xrightarrow{\text{1 000°C}}$$

FW
666.19-16.00x

$$\tfrac{1}{2}Y_2O_3(s) + 2BaO(s) + 3Cu(s) + (3.5-x)H_2O(g)$$

$$\underbrace{\phantom{\tfrac{1}{2}Y_2O_3(s) + 2BaO(s) + 3Cu(s)}}_{YBa_2Cu_3O_{3.5}}$$

Starting with 28.19 mg of $YBa_2Cu_3O_{7-x}$, 25.85 mg of solid product remained at
1 000°C. Find the value of x.

S9. *Pregl halogen analysis.* Combustion of organic halogen (X) compounds over a hot
Pt catalyst gives HX and X_2. The gas stream is bubbled through a carbonate
solution containing sodium sulfite to reduce X_2:

$$SO_3^{2-} + X_2 + H_2O \rightarrow SO_4^{2-} + 2X^- + 2H^+$$

Sulfite   Sulfate

After neutralization with HNO_3, the halide can be titrated with Ag^+ to precipitate
AgX. An 8.463-mg sample of unknown containing C, H, Br and Cl required
0.405 g of $AgNO_3$ titrant (containing 0.093 84 mol $AgNO_3$/kg solution) to reach
the first end point and an additional 0.787 g to reach the second end point. Find the
weight percent of Br and Cl in the compound and find the Br/Cl atomic ratio.

S1. In analyzing a lot with random sample variation, there is a sampling standard deviation of ±6%. Assuming negligible error in the analytical procedure, how many samples must be analyzed to give 95% confidence that the error in the mean is within ±3% of the true value? Answer the same question for a confidence level of 90%.

S2. In an experiment analogous to that in Figure 26-2, the following results were seen:

Sample mass (g)	Relative standard deviation (%)
0.54	6.33
1.78	3.21
4.22	2.00
8.63	1.55

(a) Calculate the average sampling constant, K_s.

(a) What mass of sample is required for a ±1% sampling standard deviation?

(b) How many samples of the size in part a are required to produce 90% confidence that the mean is known to within 0.75%?

S3. Consider a random mixture containing 87.0 wt % diamond (density 3.51 g/mL) and 13.0 wt % SiC (density 3.23 g/mL) with a uniform spherical particle radius of 0.100 mm.

(a) Calculate the mass of a single particle of diamond and the number of particles of diamond in 1.00 g of the mixture. Do the same for SiC.

(b) Calculate the relative sampling standard deviation in the number of particles of each type in a (i) 1.00-g and a (ii) 100.0-mg sample.

S1. (a) 2×10^{12} J   (c) 3.7×10^7 m   (e) 8.42×10^{-10} F

(b) 3.7×10^{-2} m   (d) 4×10^{-1} K   (f) 1.84×10^4 Pa

S2. (a) 80 µmol   (b) 10 GW   (c) 400 nL   (d) 3 cm   (e) 180 THz   (f) 5.37 TΩ

S3. $\dfrac{800 \times 10^{-12} \; \cancel{K}}{43 \times 10^{-6} \; \cancel{K}} = 1.9 \times 10^{-5}$

S4. 1 atm $=$ 101 325 N/m^2 and 1 torr $=$ 133.322 N/m^2

760 $\times$ 1 torr $=$ 760 $\times$ 133.322 N/m^2 $=$ 101 325 N/m^2 $=$ 1 atm

S5. mass of solution $=$ (250 $\cancel{mL}$) (1.00 g/$\cancel{mL}$) $=$ 250 g

$ppm = \dfrac{13.7 \times 10^{-6} \; \cancel{g}}{250 \; \cancel{g}} \times 10^6 = 0.054\ 8$ ppm

$ppb = \dfrac{13.7 \times 10^{-6} \; \cancel{g}}{250 \; \cancel{g}} \times 10^9 = 54.8$ ppb

S6. Molarity $= \dfrac{(5.00 \; \cancel{g}) \, / \, (79.101 \; \cancel{g}/mol)}{0.457 \; L} = 0.138$ M

S7. (a) Mass of solution $= 0.804 \dfrac{g}{\cancel{mL}} \times \dfrac{1\ 000 \; \cancel{mL}}{L} = 804 \dfrac{g}{L}$

Mass of ethanol $= \dfrac{0.950 \text{ g of ethanol}}{\cancel{g\ of\ solution}} \times \dfrac{804 \; \cancel{g\ of\ solution}}{L}$

$= 764 \dfrac{\text{g of ethanol}}{L}$

(b) $\dfrac{764 \dfrac{\cancel{g}}{L}}{46.07 \dfrac{\cancel{g}}{mol}} = 16.6$ M

(c) 100.0 mL of solution contains 95.0 g of ethanol and 5.0 g of water.

(95.0 g of ethanol) / (46.07 g/mol) $=$ 2.06 mol of ethanol.

Molality $= \dfrac{2.06 \text{ mol of ethanol}}{5.0 \times 10^{-3} \text{ kg of } H_2O} = 412$ m

S8. (a) 10.0 g of 10.2 wt% solution contains

$0.102 \dfrac{g \; NiSO_4 \cdot 6H_2O}{\cancel{g\ solution}} \times 10.0 \; \cancel{g\ solution} = 1.02$ g $NiSO_4 \cdot 6H_2O$

$= 3.88 \times 10^{-3}$ mol $NiSO_4 \cdot 6H_2O$

$\Rightarrow (3.88 \times 10^{-3} \; \cancel{mol\ Ni}) \times (58.693\ 4 \dfrac{g \; Ni}{\cancel{mol\ Ni}}) = 0.228$ g Ni

(b) There are 0.412 mol of $NiSO_4 \cdot 6H_2O$ $=$ 108.3 g of $NiSO_4 \cdot 6H_2O$ per L of solution. From the 10.2 wt%, we can say

$\dfrac{108.3 \; \cancel{g\ NiSO_4 \cdot 6H_2O}/L \text{ solution}}{0.102 \; \cancel{g\ NiSO_4 \cdot 6H_2O}/g \text{ solution}} = 1.06 \times 10^3 \dfrac{g \text{ solution}}{L \text{ solution}}$

$\Rightarrow$ density $= 1.06 \dfrac{g}{mL}$

S9. $M_{conc} \cdot V_{conc} = M_{dil} \cdot V_{dil}$

$12.1 \, \dfrac{mol}{L} \times V = 1.00 \, \dfrac{mol}{L} \times 0.100 \, L \Rightarrow V = 8.26 \, mL$

Dilute 8.26 mL of 12.1 M HCl to 100.0 mL in a volumetric flask.

S10.   We will use 100 g = 0.100 kg H_2O. Weigh out 0.100 kg $\times$ 0.082 mol/kg =

0.008 2 mol = 1.00 g $NaClO_4$ and dissolve in 0.100 kg H_2O.

S11.   (a)   40.0 wt% solution:

1 L = 1 430 g of solution. 40.0% of this is 572 g CsCl = 3.40 mol

Molar concentration = 3.40 M

20.0 wt% solution: 1 L = 1 180 g solution = 236 g CsCl $\Rightarrow$ 1.40 M

(b)   40 wt% solution:

1 g of solution contains 0.400 g CsCl + 0.600 g H_2O

$molality = \dfrac{mol \, CsCl}{kg \, H_2O} = \dfrac{(0.400 \, g) / (168.37 \, g/mol)}{0.000 \, 600 \, kg} = 3.96 \, \dfrac{mol}{kg}$

20.0 wt% solution:

$molality = \dfrac{(0.200 \, g) / (168.37 \, g/mol)}{0.000 \, 800 \, kg} = 1.48 \, \dfrac{mol}{kg}$

(c)   40.0 wt% solution has a concentration of 3.40 M

$M_{con} \cdot V_{con} = M_{dil} \cdot V_{dil}$

$(3.40 \, M) \, V = (0.100 \, M) \, (0.500 \, L) \Rightarrow V = 14.7 \, mL$

20.0 wt% solution has a concentration of 1.40 M $\Rightarrow$ V = 35.7 mL

It requires more than twice as much of the 20% solution because the 20%

solution is less dense than the 40% solution.

CHAPTER 2: TOOLS OF THE TRADE

S1. $m = \dfrac{(9.947 \, g) \, (1 - \dfrac{0.001 \, 2 \, g/mL}{8.0 \, g/mL})}{(1 - \dfrac{0.001 \, 2 \, g/mL}{0.88 \, g/mL})} = 9.959 \, g$

S2. $\dfrac{c' \, at \, 35°}{0.994 \, 03 \, g/mL} = \dfrac{0.027 \, 64 \, M}{0.998 \, 41 \, g/mL} = 0.027 \, 52$

S3.   14.974 g - 9.974 g = 5.000 g. Using column 3 of Table 2-6 tells us that the true

volume at 26°C is (5.000 g)(1.004 3 mL/g) = 5.022 mL. Using column 4, the true

volume at 20°C is (5.000 g)(1.004 2 mL/g) = 5.021 mL.

S1. (a) 4 (b) 4 (c) 4

S2. (a) 5.125 (b) 5.124 (c) 5.124 (d) 0.135 2 (e) 1.52 (f) 1.53

S3. (a) 12.01 (e) -17.66

(b) 10.9 (f) 5.97×10^{-3}

(c) 14 (g) 2.79×10^{-5}

(d) 14.3

S4. 95.980

S5. (a) 3.4 ± 0.2 $\quad e = \sqrt{0.2^2 + 0.1^2} = 0.224$

$\underline{+ 2.6 \pm 0.1}$

$6.0 \pm e \quad = 6.0 \pm 0.2 \ (\pm 3._7\%)$

(b) $\dfrac{3.4 \pm 0.2}{2.6 \pm 0.1} = \dfrac{3.4 \pm 5.88\%}{2.6 \pm 3.85\%} = 1.308 \pm e$

$\%e = \sqrt{5.88^2 + 3.85^2} = 7.03\%$

Answer: $1.30_8 \pm 0.09_2 \ (\pm 7._0\%)$

(c) $\dfrac{3.4 \ (\pm 0.2) \times 10^{-8}}{2.6 \ (\pm 0.1) \times 10^{3}} = \dfrac{3.4 \ (\pm 5.88\%) \times 10^{-8}}{2.6 \ (\pm 3.85\%) \times 10^{3}}$

$= 1.30_8 \ (\pm 0.09_2) \times 10^{-11} \ (\pm 7._0\%)$

(d) $3.4 \ (\pm 0.2) - 2.6 \ (\pm 0.1) = 0.8 \pm 0.2_{24} = 0.8 \pm 28.0\%$

$0.8 \ (\pm 28.0\%) \times 3.4 \ (\pm 5.88\%) = 2.72 \pm 28.6\%$

Answer: $2.7_2 \pm 0.7_8 \ (\pm 29\%)$

S6. 6C: $6 \times 12.011 \pm 0.001 = 72.066 \pm 0.006$

6H: $6 \times 1.007\ 94 \pm 0.000\ 07 = 6.047\ 64 \pm 0.000\ 42$

uncertainty $= \sqrt{0.006^2 + 0.000\ 42^2} = 0.006$ $\qquad$ MW $= 78.114 \pm 0.006$

S7. (a) Molarity $= \dfrac{0.222\ 2 \ (\pm 0.090\%) \ \text{g}}{214.001\ 0 \ (\pm 0.000\ 42\%) \ \frac{\text{g}}{\text{mol}} \times 0.050\ 00 \ (\pm 0.10\%) \ \text{L}}$

$\%e = \sqrt{0.090^2 + 0.000\ 42^2 + 0.10^2} = 0.135\%$

molarity $= 0.020\ 76_6 \pm 0.000\ 02_8$ M

(b) The uncertainty in the analysis is ~0.1%, so 0.1% uncertainty in reagent purity is significant.

S8. (a) $y = x^{1/2} \Rightarrow \%e_y = \frac{1}{2}\left(\frac{0.2}{3.4} \times 100\right) = 2.94\%$

Answer: $1.84_4 \pm 0.05_4 \ (\pm 2._9\%)$

(b) $y = x^2 \Rightarrow \%e_y = 2\left(\frac{0.2}{3.4} \times 100\right) = 11.76\%$

Answer: $11._6 \pm 1._4 \ (\pm 12\%)$

(c) $y = 10^x \Rightarrow e_y = (10^{3.4})(2.302\ 6)(0.2) = 1.16 \times 10^3$

Answer: $2._{51} \pm 1._{16} \times 10^3 \ (\pm 46\%)$

(d) $y = e^x \Rightarrow e_y = (e^{3.4})(0.2) = 5.99$

Answer: $30._0 \pm 6._0 \ (\pm 20\%)$

(e) $y = \log x \quad e_y = 0.434\ 29\left(\frac{0.2}{3.4}\right) = 0.025\ 5$

Answer: $0.53_1 \pm 0.02_6 \ (\pm 4.8\%)$

(f) $y = \ln x \Rightarrow e_y = \frac{0.2}{3.4} = 0.058\ 8$

Answer: $1.22_4 \pm 0.05_9 \ (\pm 4.8\%)$

S9. $k = \frac{R}{N} \Rightarrow \%e_k^2 = \%e_R^2 + \%e_N^2$

$\Rightarrow \%e_k^2 = \left(\frac{100 \times 0.000\ 070}{8.314\ 510}\right)^2 + \left(\frac{100 \times 0.000\ 003\ 6}{6.022\ 136\ 7}\right)^2$

$\Rightarrow \%e_k = 0.000\ 844\%$

$\Rightarrow e_k = (0.000\ 008\ 44)(1.380\ 658) = 0.000\ 012$

S1. (a) Mean $= \frac{1}{7}(2.310\ 17 + \cdots + 2.310\ 28) = 2.310\ 11$

   (b) Standard deviation $= \sigma = \left(\dfrac{\Sigma(x_i - \bar{x})^2}{6}\right)^{1/2} = 0.000\ 14_3$

   (c) Variance $= \sigma^2 = 2.0_3 \times 10^{-8}$

S2. (a) $z > 0 \Rightarrow 50\%$ (c) $z = -3$ to $z = +3 \Rightarrow 99.73\%$

   (b) $z = -1$ to $z = +1 \Rightarrow 68.26\%$ (d) $z < -2 \Rightarrow 2.27\%$

   (e) $z = -1.4$ to $z = +0.6 \Rightarrow$ area $= 0.419\ 2 + 0.225\ 8 = 64.50\%$

   (f) $z = -1.76$ to $z = -0.18 \Rightarrow$ area $= 0.460\ 6 - 0.071\ 4 = 38.92\%$

   Interpolations: $\left(\dfrac{1.76-1.70}{1.80-1.70}\right)(0.464\ 1 - 0.455\ 4) + 0.455\ 4 = 0.460\ 6$

   $\left(\dfrac{0.18-0.10}{0.20-0.10}\right)(0.079\ 3 - 0.039\ 8) + 0.039\ 8 = 0.071\ 4$

S3. $\bar{x} = 2.299\ 47$ g, $s = 0.001\ 38$ g, $n = 7$ degrees of freedom

   95% confidence: $\mu = \bar{x} \pm \dfrac{(2.365)(0.001\ 38)}{\sqrt{8}} = 2.299\ 47 \pm 0.001\ 15$

   99% confidence: $\mu = \bar{x} \pm \dfrac{(3.500)(0.001\ 38)}{\sqrt{8}} = 2.299\ 47 \pm 0.001\ 71$

S4. $\bar{x}_1 = 147._8$, $\bar{x}_2 = 157._2$, $s_{pooled} = 8._{90}$ $t = \dfrac{157._2 - 147._8}{8._{90}}\sqrt{\dfrac{5 \cdot 5}{5+5}} =$

   $1.67 < 2.306$ (Student's t for 8 degrees of freedom) The difference is <u>not</u> significant.

S5. $\bar{x} = 0.131_{17}$, $s = 0.002_{93}$

   95% confidence interval: $\bar{x} \pm \dfrac{(2.571)(0.002\ 93)}{\sqrt{6}} = 0.131_{17} \pm 0.003_{07}$

   interval $= 0.128_1$ to 0.134_2
   This does not include the known value of 0.137 wt %. Results are different.

S6. $Q = (0.217 - 0.195)/(0.224 - 0.195) = 0.76 > 0.56$. Discard 0.195.

S7. (a) $y(\pm 0.4_{08}) = -0.75_0(\pm 0.14_4)x + 3.9_{17}(\pm 0.4_{93})$

   (b) $x = \dfrac{y-b}{m} = \dfrac{1.00(\pm 0.408) - 3.917(\pm 0.493)}{-0.750(\pm 0.144)} = \dfrac{-2.917\ (\pm 0.640)}{-0.750\ (\pm 0.144)} = 3.8_9 \pm 1.1_3$

   (c) Uncertainty in $x = \dfrac{0.408}{0.75}\left[1 + (3.89)^2\left(\dfrac{3}{24}\right) + \dfrac{35}{24} - 2(3.89)\dfrac{9}{24}\right]^{1/2} = 0.65$

   In general, the greater the scatter in the data, the poorer the agreement between Equations 4-23 and 4-24.

CHAPTER 5: SUPPLEMENTARY SOLUTIONS
CHEMICAL EQUILIBRIUM

S1. (a) $K = [Cl^-] [OCl^-] / [OH^-]^2 P_{Cl_2}$ (b) $K = 1/P_{I_2}$

S2. Remain unchanged.

S3.

$$Cu^+ + N_3^- \rightleftharpoons CuN_3(s) \qquad K_1 = \dfrac{1}{4.9 \times 10^{-9}}$$

$$HN_3 \rightleftharpoons H^+ + N_3^- \qquad K_2 = 2.2 \times 10^{-5}$$

$$Cu^+ + HN_3 \rightleftharpoons CuN_3(s) + H^+ \qquad K_3 = K_1 K_2 = 4.5 \times 10^3$$

S4. $Q = [H^+][OH^-] = 6.0 \times 10^{-12} > K \implies$ reaction goes to the left.

S5. $K_3 = e^{-\Delta G_3^\circ/RT} = K_1 \cdot K_2 = e^{-\Delta G_1^\circ/RT} \, e^{-\Delta G_2^\circ/RT}$

$e^{-\Delta G_3^\circ/RT} = e^{-(\Delta G_1^\circ + \Delta G_2^\circ)/RT}$

$\therefore \quad \Delta G_3^\circ = \Delta G_1^\circ + \Delta G_2^\circ$

S6. $K = e^{-\Delta G^\circ/RT} \implies \ln K = -\Delta G^\circ/RT \implies \Delta G^\circ = -RT \ln K$

(a) $\Delta G^\circ = - (8.314\,51 \dfrac{J}{mol \cdot K}) (298.15\ K) \ln (6.5 \times 10^{-5}) = 23.9\ kJ/mol$

(b) $\Delta G^\circ = 63.6\ kJ/mol$

S7. (a) Positive.

(b) $\Delta G^\circ = (+) = \Delta H^\circ - T \Delta S^\circ$. Since $- T \Delta S^\circ$ is negative, ΔH° must be positive. The reaction is endothermic.

S8. (a) $Ag_2CrO_4(s) \overset{K_{sp}}{\rightleftharpoons} 2Ag^+ + CrO_4^{2-}$
FW 331.73 $\qquad\quad$ 2x $\quad$ x

$(2x)^2(x) = 1.2 \times 10^{-12} \implies x = 6.6_9 \times 10^{-5}\ M$

(b) $6.69 \times 10^{-5}\ M = 0.022\,2\ g/L = 0.002\,2_2\ g/100\ mL$

(c) $[Ag^+] = 13.4 \times 10^{-5}\ M = 0.014\,4\ g/L = 0.014\,4\ mg/mL = 14.4\ \mu g/mL$
$= 14.4\ ppm$

S9.

$$Cu(s) + Cu^{2+} \rightleftharpoons 2\,Cu^+ \qquad K_1 = 9.6 \times 10^{-7}$$

$$2\,Cu^+ + 2\,Cl^- \rightleftharpoons 2\,CuCl(s) \qquad K_2 = 1/K_{sp}^2 = 1/(1.9 \times 10^{-7})^2$$

$$Cu(s) + Cu^{2+} + 2\,Cl^- \rightleftharpoons 2\,CuCl(s) \qquad K = K_1 K_2 = 2.7 \times 10^7$$

S10. (a) $PbI_2(s) \overset{K_{sp}}{\rightleftharpoons} Pb^{2+} + 2I^-$
FW 461.0   x   2x

$x(2x)^2 = 7.9 \times 10^{-9} \Rightarrow x = 1.2_5 \times 10^{-3} M = 0.57_8 \ g/L$

(b) $[Pb^{2+}] (0.063\ 4)2 = 7.9 \times 10^{-9} \Rightarrow x = 1.9_7 \times 10^{-6} M = 9.0_6 \times 10^{-4} \ g/L$

S11. $Ca(C_2O_4)(s) \overset{K_{sp} = 1.3 \times 10^{-8}}{\rightleftharpoons} Ca^{2+} + C_2O_4^{2-}$
We want to reduce $[C_2O_4^{2-}]$ to $1.0 \times 10^{-4} M$:

$[Ca^{2+}] [1.0 \times 10^{-4}] = 1.3 \times 10^{-8} \Rightarrow [Ca^{2+}] = 1.3 \times 10^{-4} M$

S12. $Ca(OH)_2$: $K_{sp} = 6.5 \times 10^{-6}$ $Mg(OH)_2$: $K_{sp} = 7.1 \times 10^{-12}$
We want to reduce $[Mg^{2+}]$ to $2.0 \times 10^{-5} M$

$[Mg^{2+}] [OH^-]^2 = [2.0 \times 10^{-5}] [OH^-]^2 = 7.1 \times 10^{-12} \Rightarrow [OH^-] = 5.96 \times 10^{-4} M$

Will this precipitate $0.10 M$ Ca^{2+}?

$Q = [Ca^{2+}] [OH^-]^2 = (0.10)(5.96 \times 10^{-4})^2 = 3.55 \times 10^{-8} < K_{sp}$

$\Rightarrow Ca(OH)_2$ will not precipitate.

S13. $[Pb^{2+}] = K_{sp}/[I^-]^2 = (7.9 \times 10^{-9}) / (0.050)^2 = 3.1_6 \times 10^{-6} M$

$[PbI^+] = K_1[Pb^{2+}] [I^-] = 1.5_8 \times 10^{-5} M$

$[PbI_2(aq)] = \beta_2[Pb^{2+}] [I^-]^2 = 1.1_1 \times 10^{-5} M$

$[PbI_3^-] = \beta_3[Pb^{2+}] [I^-]^3 = 3.2_8 \times 10^{-6} M$

$[PbI_4^{2-}] = \beta_4[Pb^{2+}] [I^-]^4 = 5.9_2 \times 10^{-7} M$

S14. $[Ag^+] = K_{sp}/[Cl^-]$

$[AgCl_2^-] = K_2[Cl^-]$

$[AgCl_3^{2-}] = K_3[AgCl_2^-] [Cl^-]$

$[Ag]_{total} = [Ag^+] + [AgCl_2^-] + [AgCl_3^{2-}]$

	(a) 0.010 M Cl^-	(b) 0.20 M Cl^-	(c) 2.0 M Cl^-
$[Ag^+]$	$1.8_0 \times 10^{-8}$ M	$9.0_0 \times 10^{-10}$ M	$9.0_0 \times 10^{-11}$ M
$[AgCl_2^-]$	$1.5_0 \times 10^{-4}$ M	$0.003\ 0_0$ M	0.030_0 M
$[AgCl_3^{2-}]$	$7.3_5 \times 10^{-7}$ M	$0.000\ 29_4$ M	0.029_4 M
$[Ag]_{total}$	$1.5_0 \times 10^{-4}$ M	$0.003\ 2_9$ M	0.059_4 M

S15. HSO_3^- , H_2O

S16.

acid	base
H_2O	OH^-
$H_3\overset{+}{N}CH_2CH_2NH_2$	$H_2NCH_2CH_2NH_2$

S17. (a) $[H^+] = 1.0 \times 10^{-3}$ M $\Rightarrow$ pH = -log $[H^+]$ = 3.00

(b) $[H^+] = 0.050$ M $\Rightarrow$ pH = 1.30

(c) $[OH^-] = 0.050$ M $\Rightarrow [H^+] = K_w / [OH^-] = 2.0 \times 10^{-13}$ M $\Rightarrow$ pH = 12.70

(d) $[OH^-] = 3.0$ M $\Rightarrow [H^+] = 3.3 \times 10^{-15}$ M $\Rightarrow$ pH = 14.48

(e) $[OH^-] = 0.005\ 0$ M $\Rightarrow [H^+] = 2.0 \times 10^{-12}$ M $\Rightarrow$ pH = 11.70

S18. $HCO_2H \underset{}{\overset{K_a}{\rightleftharpoons}} HCO_2^- + H^+$ $CH_3NH_3^+ \underset{}{\overset{K_a}{\rightleftharpoons}} CH_3NH_2 + H^+$

S19.

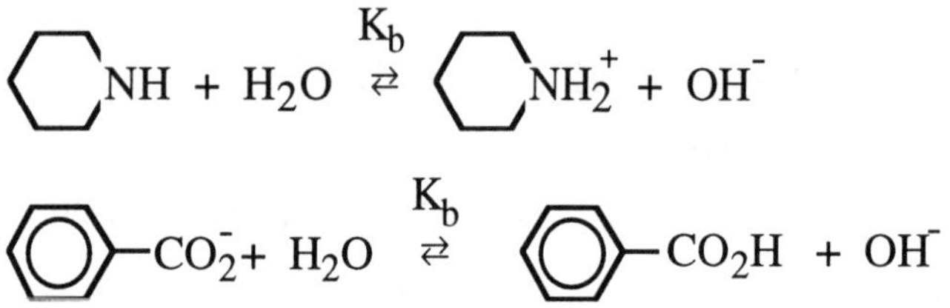

$\underset{}{\overset{K_b}{\rightleftharpoons}}$

(piperidine) $NH + H_2O \rightleftharpoons NH_2^+ + OH^-$

$\underset{}{\overset{K_b}{\rightleftharpoons}}$

(benzoate) $CO_2^- + H_2O \rightleftharpoons CO_2H + OH^-$

S20. $HPO_4^{2-} \underset{}{\overset{K_a}{\rightleftharpoons}} H^+ + PO_4^{3-}$ $HPO_4^{2-} + H_2O \underset{}{\overset{K_b}{\rightleftharpoons}} H_2PO_4^- + OH^-$

S21.

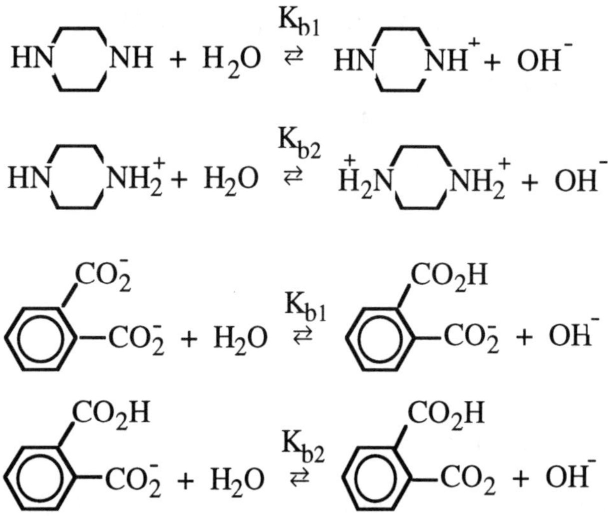

$HN\ \ NH + H_2O \underset{}{\overset{K_{b1}}{\rightleftharpoons}} HN\ \ NH^+ + OH^-$

$HN\ \ NH_2^+ + H_2O \underset{}{\overset{K_{b2}}{\rightleftharpoons}} \overset{+}{H_2N}\ \ NH_2^+ + OH^-$

$CO_2^- / CO_2^- + H_2O \underset{}{\overset{K_{b1}}{\rightleftharpoons}} CO_2H / CO_2^- + OH^-$

$CO_2H / CO_2^- + H_2O \underset{}{\overset{K_{b2}}{\rightleftharpoons}} CO_2H / CO_2 + OH^-$

S22. 4-nitrophenol has the larger K_a.

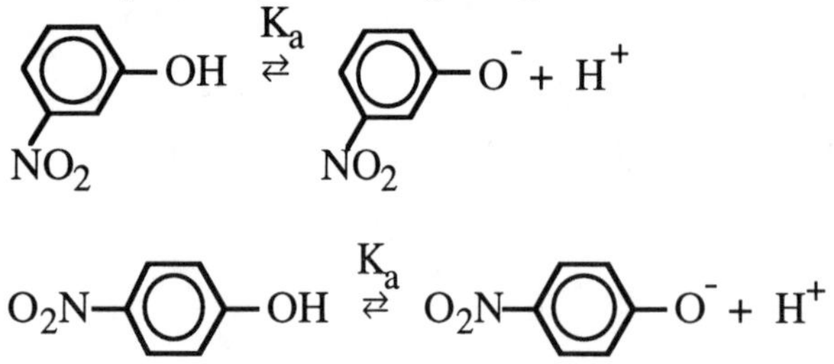

(3-nitrophenol) $OH \underset{}{\overset{K_a}{\rightleftharpoons}} O^- + H^+$

$O_2N-\ -OH \underset{}{\overset{K_a}{\rightleftharpoons}} O_2N-\ -O^- + H^+$

S23. Cyclohexylamine has the larger K_b.

S24. $OCl^- + H_2O \rightleftharpoons HOCl + OH^-$ $K_b = K_w/K_a = 3.3 \times 10^{-7}$

S25. $HSO_4^- \overset{Ka2}{\rightleftharpoons} H^+ + SO_4^{2-}$

S26. $K_{b1} = \dfrac{K_w}{K_{a3}} = 2.49 \times 10^{-8}$

$K_{b2} = \dfrac{K_w}{K_{a2}} = 5.78 \times 10^{-10}$

$K_{b3} = \dfrac{K_w}{K_{a1}} = 1.34 \times 10^{-11}$

CHAPTER 6: SUPPLEMENTARY SOLUTIONS
A FIRST LOOK AT SPECTROPHOTOMETRY

S1. Using $\nu = c/\lambda$, $\tilde{\nu} = 1/\lambda$, and $E = h\nu$, we find

	250 nm	10 μm
ν (hz)	1.20×10^{15}	3.00×10^{13}
$\tilde{\nu}$ (cm^{-1})	4.00×10^{4}	$1\,000$
J/photon	7.95×10^{-19}	1.99×10^{-20}
kJ/mol	479	12.0

S2. (a) $\dfrac{15.0 \times 10^{-3} \text{ g}}{(384.63 \text{ g/mol})(5 \times 10^{-3} \text{ L})} = 7.80 \times 10^{-3}$ M

(b) One tenth dilution $\Rightarrow 7.80 \times 10^{-4}$ M

(c) $\varepsilon = A/bc = 0.634/[(0.500 \text{ cm})(7.80 \times 10^{-4} \text{ M})] = 1.63 \times 10^{3}$ M^{-1} cm^{-1}

S3. Original concentration $= \dfrac{(0.267 \text{ g})(337.69 \text{ g/mol})}{(0.100\,0 \text{ L})} = 7.91 \times 10^{-3}$ M

Diluted concentration $= \left(\dfrac{2.000}{100.0}\right)(7.91 \times 10^{-3} \text{ M}) = 1.58 \times 10^{-4}$ M

$\varepsilon = \dfrac{A}{bc} = \dfrac{0.728}{(1.58 \times 10^{-4} \text{ M})(2.00 \text{ cm})} = 2.30 \times 10^{3}$ M^{-1} cm^{-1}

S4. $A = -\log T = 0.085\,1$ for $T = 82.2\%$ and 0.295 for $T = 50.7\%$.
Ratio $= 0.085\,1/0.295 = 0.288$.

S5. (a) $\varepsilon = \dfrac{A}{cb} = \dfrac{0.494 - 0.053}{(3.73 \times 10^{-5} \text{ M})(1.000 \text{ cm})} = 1.18_2 \times 10^{-4}$ M^{-1} cm^{-1}

(b) $c = \dfrac{A}{\varepsilon b} = \dfrac{0.777 - 0.053}{(1.182 \times 10^{4} \text{ M}^{-1} \text{ cm}^{-1})(1.000 \text{ cm})} = 6.12_5 \times 10^{-5}$ M^{-1}

Original concentration was $\dfrac{250.0}{5.00}$ times as great $= 3.06$ mM

S6. (a) The concentration of phosphorus in solution A is 1.196×10^{-3} M. When 0.140 mL of A is diluted to 5.00 mL, $[P] = 3.349 \times 10^{-5}$ M.
$\varepsilon = A/bc = (0.829 - 0.017)/[(1.00 \text{ cm})(3.349 \times 10^{-5} \text{ M})] = 2.42_5 \times 10^{4}$ M^{-1} cm^{-1}.

(b) [P] in analyte $= \dfrac{A}{\varepsilon b} = \dfrac{(0.836 - 0.038)}{(2.425 \times 10^{4} \text{ M}^{-1} \text{ cm}^{-1})(1.00 \text{ cm})} = 3.291 \times 10^{-5}$ M

$$[P] \text{ in } 1.00 \text{ mL of undiluted analyte} = \left(\frac{5.00}{0.300}\right)(3.291 \times 10^{-5} \text{ M}) =$$

5.485×10^{-4} M.

1.00 mL contains 5.485×10^{-7} mol P = 1.699×10^{-5} g = 1.26% phosphorus.

S7. (a) Subtract 0.002 from each area, and plot (µg/mL) vs area:

⇒ slope = 0.008 72 and intercept = 0.002 6

(b) Corrected area = 0.156 = 0.008 72 [Ag] + 0.002 6 ⇒ [Ag] = 17.6 µg/mL.

S8. First, note that s_y = 0.003 2.

0.156 (±0.003 2) = 0.008 72 (±0.000 14) [Ag] + 0.002 6 (±0.002 5)

$$[Ag] = \frac{0.156 \ (\pm 0.003 \ 2) - 0.002 \ 6 \ (\pm 0.002 \ 5)}{0.008 \ 72 \ (\pm 0.000 \ 14)} = 17.6 \ (\pm 0.5) \text{ µg/mL.}$$

S9. (a) $[\text{dopamine}]_f = [\text{dopamine}]_i \left(\dfrac{90.0}{100.0}\right)$

(b) $[S]_f = (0.015 \ 6 \text{ M}) \left(\dfrac{2.00 \text{ mL}}{100.0 \text{ mL}}\right) = 0.312$ mM

(c) $\dfrac{[\text{dopamine}]_i}{0.312 \text{ mM} + 0.900[\text{dopamine}]_i} = \dfrac{34.6 \text{ nA}}{58.4 \text{ nA}} \Rightarrow [\text{dopamine}]_i = 0.396$ mM

S10. $\dfrac{[A]/[B] \text{ in unknown}}{[A]/[B] \text{ in standard}} = \dfrac{\text{signal ratio in unknown}}{\text{signal ratio in standard}}$

$\dfrac{[A]/(930 \text{ nM})}{(80.0 \text{ nM}/64.0 \text{ nM})} = \dfrac{1.21}{0.71} \Rightarrow [A] = 1.98 \times 10^3 \text{ nM} = 1.98 \text{ µM}$

S11. (a)-(b) Corrected signal (mV) = 94.4 (±3.7) [C_2H_2, vol%] + 10.05 (±0.7$_3$)

Setting the left side to zero gives

$$[C_2H_2, \text{ vol\%}] = \left| \frac{-10.05 \ (\pm 0.73)}{94.4 \ (\pm 3.7)} \right| = 0.106 \ (\pm 0.009) \text{ vol\%}$$

S12. $\dfrac{[X]_i}{[S]_f + [X]_f} = \dfrac{I_X}{I_{S+X}} \Rightarrow [X]_i I_{S+X} = [S]_f I_X + [X]_f I_X$

$$I_{S+X} = \underbrace{\left(\frac{I_X}{[X]_i}\right)}_{\text{slope}} [S]_f + \underbrace{\frac{[X]_f}{[X]_i} I_X}_{\text{intercept}}$$

Setting $I_{S+X} = 0$ gives $\dfrac{I_X}{[X]_i} [S]_f = -\dfrac{[X]_f}{[X]_i} I_X$, which is true when $[S]_f = -[X]_f$

S1. 1.00 mL of 0.027 3 M Ce^{4+} = 0.027 3 mmol of Ce^{4+}. This will react with half as many mol of oxalic acid = 0.013 65 mmol of $H_2C_2O_4 \cdot 2H_2O$ = 1.72 mg.

S2. Let x = mg $FeCl_2$ and (27.73 - x) = mg KCl

mmol Ag^+ = 2 mmol $FeCl_2$ + mmol KCl

$$(18.49 \text{ mL})(0.022\ 37 \text{ M}) = \frac{2\ x\ mg}{126.75\ mg/mmol} + \frac{(27.73 - x)\ mg}{74.55\ mg/mmol}$$

$$\Rightarrow x = 17.61 \text{ mg } FeCl_2 = 7.76 \text{ mg Fe} = \frac{7.76\ mg}{27.73\ mg} \times 100 = 28.0 \text{ wt\% Fe.}$$

S3. (a) $mol\ Hg^{2+} = \frac{1}{2} mol\ Cl^- = \frac{1}{2} \dfrac{0.147\ 6\ g}{58.442\ 5\ g/mol} = 1.263 \times 10^{-3}$ mol

$$[Hg^{2+}] = \frac{1.263 \times 10^{-3}\ mol}{0.028\ 06\ L} = 0.045\ 00 \text{ M}$$

(b) $(0.022\ 83 \text{ L } Hg(NO_3)_2)\ (0.045\ 00 \text{ mol/L}) = 1.027 \text{ mmol } Hg^{2+}$

= 2.055 mmol Cl^- = 72.85 mg Cl^- in 2.000 mL = 36.42 mg Cl^-/mL

S4. $Br^- + Ag^+ \rightarrow AgBr(s)$

Equivalence point $= \dfrac{(20.00 \text{ mL})(0.053\ 20 \text{ M})}{(0.051\ 10 \text{ M})} = 20.82 \text{ mL}$

(a) $[Br^-] = \left(\dfrac{20.82 - 20.00}{20.82}\right)(0.053\ 20 \text{ M})\left(\dfrac{20.00}{40.82}\right) = 1.027 \times 10^{-3} \text{ M}$

$$\underbrace{\hphantom{\dfrac{20.82 - 20.00}{20.82}}}_{\substack{\text{Fraction remaining}}} \quad \underbrace{\hphantom{(0.053)}}_{\substack{\text{Initial}\\\text{concentration}}} \quad \underbrace{\hphantom{\dfrac{20.00}{40.82}}}_{\substack{\text{Dilution}\\\text{factor}}}$$

$$[Ag^+] = \frac{K_{sp}}{[Br^-]} = \frac{5.0 \times 10^{-13}}{1.027 \times 10^{-3}} = 4.9 \times 10^{-10} \text{ M}$$

$$pAg^+ = -\log(4.9 \times 10^{-10}) = 9.31$$

(b) At the equivalence point, $[Ag^+] = [Br^-] = \sqrt{K_{sp}} = 7.07 \times 10^{-7} \text{ M}$.

$$pAg^+ = -\log[Ag^+] = 6.15$$

(c) $[Ag^+] = (0.051\ 1 \text{ M})\left(\dfrac{22.60 - 20.82}{42.60}\right) = 2.14 \times 10^{-3} \text{ M} \Rightarrow pAg^+ = 2.67$

$$\underbrace{\hphantom{(0.051)}}_{\substack{\text{Initial}\\\text{concentration}}} \quad \underbrace{\hphantom{\dfrac{22.60}{42.60}}}_{\substack{\text{Dilution}\\\text{factor}}}$$

S5. $Hg^{2+} + 2SCN^- \rightarrow Hg(SCN)_2(s)$ $K_{sp} = 2.8 \times 10^{-20}$

Equivalence point $= 2\dfrac{(50.00 \text{ mL})(0.024\ 6 \text{ M})}{(0.104 \text{ M})} = 23.65 \text{ mL}$

$$0.25 \; V_e = 5.91 \; mL : [Hg^{2+}] = \frac{3}{4} \quad (0.024\ 6 \; M) \quad \left(\frac{50.00}{55.91}\right) = 0.016\ 5 \; M$$

Fraction remaining   Initial concentration   Dilution factor

$$pHg^{2+} = -\log [Hg^{2+}] = 1.78$$

$$0.50 \; V_e = 11.83 \; mL : [Hg^{2+}] = \frac{1}{2} (0.024\ 6 \; M) \left(\frac{50.00}{61.83}\right) = 0.009\ 95 \; M$$

$$pHg^{2+} = 2.00$$

$$0.75 \; V_e = 17.74 \; mL : [Hg^{2+}] = \frac{1}{4} (0.024\ 6 \; M) \left(\frac{50.00}{67.74}\right) = 0.004\ 54 \; M$$

$$pHg^{2+} = 2.34$$

$$V_e = [Hg^{2+}][SCN^-]^2 = (x)(2x)^2 = K_{sp} \Rightarrow x = [Hg^{2+}] = 1.9 \times 10^{-7} \; M$$

$$pHg^{2+} = 6.72$$

$$1.05 \; V_e = 24.84 \; mL : [SCN^-] = (0.104 \; M)\left(\frac{24.84 - 23.65}{74.84}\right) = 0.001\ 65 \; M$$

Initial concentration   Dilution factor

$$[Hg^{2+}] = \frac{K_{sp}}{[SCN^-]^2} = \frac{2.8 \times 10^{-20}}{(0.001\ 65)^2} = 1.02 \times 10^{-14} \; M \Rightarrow pHg^{2+} = 13.99$$

$$1.25 \; V_e = 29.57 \; mL : [SCN^-] = (0.104 \; M)\left(\frac{29.57 - 23.65}{79.57}\right) = 0.007\ 74 \; M$$

$$[Hg^{2+}] = \frac{K_{sp}}{[SCN^-]^2} = 4.68 \times 10^{-16} \; M \Rightarrow pHg^{2+} = 15.33$$

S6. First reaction: $SO_4^{2-} + Ra^{2+} \rightarrow RaSO_4(s) \; K_{sp} = 4.3 \times 10^{-11}$

$$V_{e1} = (100 \; mL)\left(\frac{0.050\ 0 \; M}{0.250 \; M}\right) = 20.00 \; mL$$

Second reaction: $SO_4^{2-} + Sr^{2+} \rightarrow SrSO_4(s) \qquad K_{sp} = 3.2 \times 10^{-7}$

$$V_{e2} = 20.00 + 20.00 = 40.00 \; mL$$

(a) 10.00 mL: Half of the Ra^{2+} has reacted.

$$[Ra^{2+}] = \frac{1}{2} \quad (0.050\ 0 \; M) \left(\frac{100.00}{110.00}\right) = 0.022\ 7 \; M$$

Fraction remaining   Initial concentration   Dilution factor

$$[SO_4^{2-}] = \frac{K_{sp}(RaSO_4)}{[Ra^{2+}]} = 1.89 \times 10^{-9} \; M \Rightarrow pSO_4^{2-} = 8.72$$

(b) 19.00 mL: 19/20 of the Ra^{2+} has reacted.

$$[Ra^{2+}] = \left(\frac{1.00}{20.00}\right)(0.050\ 0 \; M)\left(\frac{100.00}{119.00}\right) = 0.002\ 10 \; M$$

$$[SO_4^{2-}] = \frac{K_{sp}(RaSO_4)}{[Ra^{2+}]} = 2.05 \times 10^{-8} \text{ M} \implies pSO_4^{2-} = 7.69$$

(c) 21.00 mL: 1/20 of the Sr^{2+} has reacted.

$$[Sr^{2+}] = \left(\frac{19.00}{20.00}\right)(0.050\ 0 \text{ M})\left(\frac{100.00}{121.00}\right) = 0.039\ 3 \text{ M}$$

$$[SO_4^{2-}] = \frac{K_{sp}(SrSO_4)}{[Sr^{2+}]} = 8.14 \times 10^{-6} \text{ M} \implies pSO_4^{2-} = 5.09$$

(d) 30.00 mL: Half of the Sr^{2+} has reacted.

$$[Sr^{2+}] = \frac{1}{2}(0.050\ 0 \text{ M})\left(\frac{100.00}{130.00}\right) = 0.019\ 2 \text{ M}$$

$$[SO_4^{2-}] = \frac{K_{sp}(SrSO_4)}{[Sr^{2+}]} = 1.67 \times 10^{-5} \text{ M} \implies pSO_4^{2-} = 4.78$$

(e) 40.00 mL: Second equivalence point.

$$[Sr^{2+}][SO_4^{2-}] = x^2 = 3.2 \times 10^{-7} \implies [SO_4^{2-}] = 5.7 \times 10^{-4} \implies pSO_4^{2-} = 3.25$$

(f) 50.00 mL: There is 10.00 mL of excess SO_4^{2-}.

$$[SO_4^{2-}] = (0.250 \text{ M})\left(\frac{10.00}{150.00}\right) = 0.016\ 7 \text{ M} \implies pSO_4^{2-} = 1.78$$

S7.

	A	B	C	D	E
1	Ksp =	pM	[M]	[X]	Vm
2	1E-25	11	1.00E-11	1.00E-01	0.000
3	Cm =	10.8	1.58E-11	7.36E-02	11.825
4	0.1	10.6	2.51E-11	5.41E-02	22.479
5	Cx =	10.3	5.01E-11	3.41E-02	35.762
6	0.1	10	1.00E-10	2.15E-02	45.735
7	Vx =	9	1.00E-09	4.64E-03	61.664
8	100	7	1.00E-07	2.15E-04	66.428
9		5	1.00E-05	1.00E-05	66.672
10		3	1.00E-03	4.64E-07	68.350
11		2	1.00E-02	1.00E-07	85.185
12	C2 = 10^-B2				
13	D2 = ($A$2/(C2*C2))^(1/3)				
14	E2 = $A$8*(2*$A$6+3*C2-2*D2)/(3*$A$4-3*C2+2*D2)				

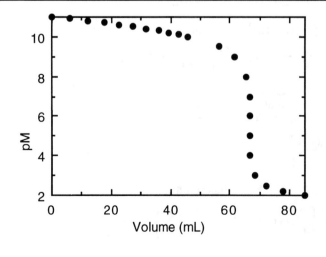

S1. (a) 0.008 7 M   (b) 12 mM   (c) 10 mM   (d) $13 + 8 = 21$ mM

S2. (a) 0.868   (b) 0.725   (c) 0.10   (d) 0.902

S3. The ionic strength is half way between 0.01 and 0.05 M $\Rightarrow \gamma = \frac{1}{2} (0.900 + 0.81)$

   $= 0.85_5$

S4. (a) $\log \gamma = \dfrac{-0.51 \cdot 1^2 \cdot \sqrt{0.038}}{1 + (350 \sqrt{0.038} / 305)} = -0.081\ 2 \Rightarrow \gamma = 10^{-0.081\ 2} = 0.82_9$

   (b) $\gamma = (\dfrac{0.038 - 0.01}{0.05 - 0.01}) (0.81 - 0.900) + 0.900 = 0.83_7$

S5. $[Ag^+]^2 \gamma_{Ag^+}^2 [CrO_4^{2-}] \gamma_{CrO_4^{2-}} = 1.2 \times 10^{-12}$

   (a) For 0.05 M $KClO_4$, $\mu = 0.05$, $\gamma_{Ag^+} = 0.80$, $\gamma_{CrO_4^{2-}} = 0.445$

   $(2x)^2 (0.80)^2 (x) (0.445) = 1.2 \times 10^{-12} \Rightarrow x = 1.017 \times 10^{-4}$ M

   $[Ag^+] = 2x = 2.0_3 \times 10^{-4}$ M

   (b) For 0.001 67 M K_2CrO_4, $\mu = 0.005$ M , $\gamma_{Ag^+} = 0.924$, $\gamma_{CrO_4^{2-}} = 0.740$

   $[Ag^+]^2 (0.924)^2 (0.001\ 67) (0.740) = 1.2 \times 10^{-12} \Rightarrow [Ag^+] = 3.3_7 \times 10^{-5}$ M

S6. Since we don't know the ionic strength, we begin by neglecting activity
coefficients:
   $[Tl^+][Br^-] = x^2 = K_{sp} = 3.6 \times 10^{-6} \Rightarrow x = 1.90 \times 10^{-3}$ M
   $\Rightarrow \mu = 1.90 \times 10^{-3}$ M $\Rightarrow \gamma_{Tl^+} = 0.955$ and $\gamma_{Br^-} = 0.955$

   For a second approximation, we add activity coefficients from the first calculation:
   $[Tl^+]\gamma_{Tl^+} [Br^-]\gamma_{Br^-} = (x)(0.955)(x)(0.955) = K_{sp} \Rightarrow x = 1.99 \times 10^{-3}$ M
   $\Rightarrow \mu = 1.99 \times 10^{-3}$ M $\Rightarrow \gamma_{Tl^+} = 0.954$ and $\gamma_{Br^-} = 0.954$

   The third approximation is $(x)(0.954)(x)(0.954) = K_{sp} \Rightarrow x = 1.99 \times 10^{-3}$ M

S7. (a) $\mu = 0.050$ M $\Rightarrow \gamma_{H^+} = 0.86$
   $\mathcal{A}_{H^+} = (0.050)(0.86) = 0.043$ $pH = -\log \mathcal{A}_{H^+} = 1.37$
   (b) $\mu = 0.10$ M $\Rightarrow \gamma_{H^+} = 0.83$
   $\mathcal{A}_{H^+} = (0.10)(0.83) = 0.083$ $pH = -\log \mathcal{A}_{H^+} = 1.08$

S1. $[H^+] + [^+H_3NCH_2CO_2H] = [OH^-] + [H_2NCH_2CO_2^-]$

S2. $[H^+] + 3[Al^{3+}] + 2[AlOH^{2+}] + [Al(OH)_2^+] + [K^+] = [OH^-] + [Al(OH)_4^-]$

S3. $[H_2NCH_2CO_2^-] + [^+H_3NCH_2CO_2^-] + [^+H_3NCH_2CO_2H] = 0.05$ M

S4. $[Al^{3+}] + [AlOH^{2+}] + [Al(OH)_2^+] + [Al(OH)_3(aq)] + [Al(OH)_4^-] = 0.005\ 0$ M

$[K^+] = 0.45$ M

S5.   1. Pertinent reaction: $(Hg_2)_3[Co(CN)_6]_2$ (s) $\rightleftarrows$ $3Hg_2^{2+} + 2Co(CN)_6^{3-}$

2. Charge balance: $2[Hg_2^{2+}] = 3[Co(CN)_6^{3-}]$

3. Mass balance: $2[Hg_2^{2+}] = 3[Co(CN)_6^{3-}]$

4. Equilibrium constant: $K_{sp} = 1.9 \times 10^{-37} = [Hg_2^{2+}]^3[Co(CN)_6^{3-}]^2$

Solve: If $[Hg_2^{2+}] = x$, then $[Co(CN)_6^{3-}] = \frac{2}{3}x$. Putting these values into the

solubility product gives $(x)^3 (\frac{2}{3}x)^2 = K_{sp} \Rightarrow x = [Hg_2^{2+}] = 5.3 \times 10^{-8}$ M

S6.   (a) $M_t = [M^{2+}] + [ML^+]$

(b) $L_t = [L^-] + [ML^+] + [HL]$

(c) $2[M^{2+}] + [H^+] + [ML^+] = [L^-] + [OH^-]$

(d) $K = 1.0 \times 10^8 = \dfrac{[ML^+]}{[M^{2+}][L^-]}$ (1)

$K_a = 1.0 \times 10^{-5} = \dfrac{[H^+][L^-]}{[HL]}$ (2)

Equation (2) gives $[L^-] = [HL]$ when $[H^+] = 1.0 \times 10^{-5}$ M .

Putting $[L^-] = [HL]$ into the L_t equation gives

$L_t = 0.1$ M $= 2[L^-] + [ML^+] \Rightarrow [L^-] = 0.05 - \frac{1}{2}[ML^+]$.

Substituting this last expression for $[L^-]$ and the expression
$[M^{2+}] = M_t - [ML^+] = 0.1 - [ML^+]$ into Equation (1) gives

$$1.0 \times 10^8 = \frac{[ML^+]}{(0.10 - [ML^+])(0.050 - \frac{1}{2}[ML^+])} \quad \begin{matrix}\text{Solve quadratic}\\\text{equation}\\\Rightarrow\end{matrix} \quad [ML^+] \approx 0.10 \text{ M}$$

Equation (1) gives $[M^{2+}] = \dfrac{[ML^+]}{K\ [L^-]} = \dfrac{0.10}{(1.0 \times 10^8)(0.050)} = 2.0 \times 10^{-8}$ M

S7. $\dfrac{[HF][OH^-]}{[F^-]} = 1.5 \times 10^{-11}$ $\overset{pH\ 3.00}{\Rightarrow}$ $[OH^-] = 10^{-11}$ M $[HF] = 1.5\,[F^-]$

Mass balance: $[F^-] + [HF] = 2.5\,[F^-] = 2[Mg^{2+}] \Rightarrow [F^-] = 0.80[Mg^{2+}]$

Substitution into solubility product gives

$[Mg^{2+}][F^-] = [Mg^{2+}](0.80[Mg^{2+}])^2 = 6.6 \times 10^{-9} \Rightarrow [Mg^{2+}] = 2.1_8 \times 10^{-3}$ M

$[F^-] = 0.80[Mg^{2+}] = 1.7_4 \times 10^{-3}$ M

$[HF] = 1.5\,[F^-] = 2.6_1 \times 10^{-3}$ M

S8. $\dfrac{[A^-]}{[HA(aq)][OH^-]} = 6.3 \times 10^5.$

Putting in $[HA(aq)] = 0.008\,5$ M and $[OH^-] = 1.0 \times 10^{-4}$ M $\Rightarrow [A^-] = 0.536$ M

$[HA] + [A^-] = 0.54$ M.

S9. (a) Mass balance: $[Sr^{2+}] = [SO_4^{2-}] + [HSO_4^-]$ (1)

Charge balance: $2[Sr^{2+}] + [H^+] = 2[SO_4^{2-}] + [HSO_4^-] + [OH^-]$ (2)

(b) If pH = 2.50, $[OH^-] = 3.1_6 \times 10^{-12}$ M

$\dfrac{[HSO_4^-][OH^-]}{[SO_4^{2-}]} = 9.8 \times 10^{-13} \Rightarrow [HSO_4^-] = 0.31_0[SO_4^{2-}]$

Putting this expression for $[HSO_4^-]$ into the mass balance gives

$[Sr^{2+}] = [SO_4^{2-}] + 0.31_0[SO_4^{2-}] = 1.31_0[SO_4^{2-}]$

Putting this expression for $[SO_4^{2-}]$ into the solubility product gives

$[Sr^{2+}][SO_4^{2-}] = [Sr^{2+}]\dfrac{[Sr^{2+}]}{1.31_0} = K_{sp} \Rightarrow [Sr^{2+}] = 6.5 \times 10^{-4}$ M.

S10. (a) Mass balance: $[Ca^{2+}] = [C_2O_4^{2-}] + [HC_2O_4^-] + [H_2C_2O_4]$

Charge balance: $2[Ca^{2+}] + [H^+] = 2[C_2O_4^{2-}] + [HC_2O_4^-] + [OH^-]$

(b) If pH = 2.30, $[OH^-] = 2.0_0 \times 10^{-12}$ M

$\dfrac{[HC_2O_4^-][OH^-]}{[C_2O_4^{2-}]} = K_{b1} \Rightarrow [HC_2O_4^-] = 90._0[C_2O_4^{2-}]$

$\dfrac{[H_2C_2O_4][OH^-]}{[HC_2O_4^-]} = K_{b2} \Rightarrow [H_2C_2O_4] = 0.090[HC_2O_4^-] = 8.1[C_2O_4^{2-}]$

Substituting into the mass balance gives

$[Ca^{2+}] = [C_2O_4^{2-}] + 90._0[C_2O_4^{2-}] + 8.1[C_2O_4^{2-}] = 99.1[C_2O_4^{2-}]$

Substituting into the solubility product gives

$[Ca^{2+}][C_2O_4^{2-}] = [Ca^{2+}]\dfrac{[Ca^{2+}]}{99.1} = 1.3 \times 10^{-8} \Rightarrow [Ca^{2+}] = 1.1_4 \times 10^{-3}$ M

S11.   (a) Mass balance: $2[Zn^{2+}] = 3\{[AsO_4^{3-}] + [HAsO_4^{2-}] + [H_2AsO_4^-] + [H_3AsO_4]\}$

Charge balance: $2[Zn^{2+}] + [H^+] = 3[AsO_4^{3-}] + 2[HAsO_4^{2-}] + [H_2AsO_4^-] + [OH^-]$

(b) pH = 6.00 $\Rightarrow$ $[OH^-] = 1.0 \times 10^{-8}$

$$\frac{[HAsO_4^{2-}][OH^-]}{[AsO_4^{3-}]} = K_{b1} \Rightarrow [HAsO_4^{2-}] = 3.1 \times 10^5[AsO_4^{3-}]$$

$$\frac{[H_2AsO_4^-][OH^-]}{[HAsO_4^{2-}]} = K_{b2} \Rightarrow [H_2AsO_4^-] = 9.1[HAsO_4^{2-}] = 2.82 \times 10^6[AsO_4^{3-}]$$

$$\frac{[H_3AsO_4][OH^-]}{[H_2AsO_4^-]} = K_{b3} \Rightarrow [H_3AsO_4] = 1.7 \times 10^{-4}[H_2AsO_4^-] = 480[AsO_4^{3-}]$$

Putting these expressions into the mass balance gives

$$2[Zn^{2+}] = 3[AsO_4^{3-}](1 + 3.1 \times 10^5 + 2.82 \times 10^6 + 480) = 9.39 \times 10^6[AsO_4^{3-}]$$

$$[AsO_4^{3-}] = 2.13 \times 10^{-7}[Zn^{2+}]$$

Substituting into the solubility product gives $[Zn^{2+}]$:

$$[Zn^{2+}]^3 \{2.13 \times 10^{-7}[Zn^{2+}]\}^2 = K_{sp} \Rightarrow [Zn^{2+}] = 0.001\,8_6 \text{ M}$$

S12.

	A	B	C	D	E	F	G	H
1	K =	pH	[H+]	b	[M2+]	[L-]	[ML+]	[HL]
2	1.0E+08	0	1E+00	1.0E+05	9.5E-03	9.5E-08	9.0E-02	9.5E-03
3	Ka =	2	1E-02	1.0E+03	1.0E-03	9.9E-07	9.9E-02	9.9E-04
4	1.0E-05	5	1E-05	2.0E+00	4.5E-05	2.2E-05	1.0E-01	2.2E-05
5	M(total) =	6	1E-06	1.1E+00	3.3E-05	3.0E-05	1.0E-01	3.0E-06
6	0.1	8	1E-08	1.0E+00	3.2E-05	3.2E-05	1.0E-01	3.2E-08
7	L(total) =	10	1E-10	1.0E+00	3.2E-05	3.2E-05	1.0E-01	3.2E-10
8	0.1	12	1E-12	1.0E+00	3.2E-05	3.2E-05	1.0E-01	3.2E-12
9		14	1E-14	1.0E+00	3.2E-05	3.2E-05	1.0E-01	3.2E-14
10	C2 = 10^-B2							
11	D2 = 1+$A$2*$A$8-$A$2*$A$6+(C2/$A$4)							
12	E2=(-D2+Sqrt(D2^2+4*$A$2*$A$6*(1+C2/$A$4)))/(2*$A$2))							
13	F2 = ($A$6/E2-1)/$A$2							
14	G2 = $A$2*E2*F2							
15	H2 = F2*C2/$A$4							

CHAPTER 10: SUPPLEMENTARY SOLUTIONS
MONOPROTIC ACID-BASE EQUILIBRIA

S1. (a) $pH = -\log [H^+] = -\log (5.0 \times 10^{-4}) = 3.30$

(b) $pH = -\log (K_w/[OH^-]) = -\log (1.0 \times 10^{-14}) / (5.0 \times 10^{-4}) = 10.70$

S2. $[OH^-] = [H^+] + [(CH_3)_4N^+] = [H^+] + 2.0 \times 10^{-7}$

$[H^+] [OH^-] = K_w = [H^+] ([H^+] + 2.0 \times 10^{-7}) \Rightarrow [H^+] = 4.14 \times 10^{-8}$ M

$pH = -\log [H^+] = 7.38$

$[OH^-]_{total} = K_w/[H^+] = 2.41 \times 10^{-7}$ M $\quad [OH^-]_{from\ H_2O} = [H^+] = 4.1 \times 10^{-8}$ M

$$\text{Fraction of } [OH^-] \text{ from } H_2O = \frac{4.1 \times 10^{-8}\ M}{2.41 \times 10^{-7}\ M} = 0.17$$

S3. (a) For 0.050 M HBr, $pH = -\log [H^+]\, \gamma_{H^+} = -\log[(0.050)(0.86)] = 1.37$

(b) For 0.050 M NaOH, $pH = -\log \mathcal{A}_{H^+} = -\log(\frac{K_w}{\mathcal{A}_{OH^-}}) = -\log \frac{K_w}{[OH^-]\,\gamma_{OH^-}}$

$$= -\log \frac{1.0 \times 10^{-14}}{(0.050)\,(0.81)} = 12.61$$

S4. 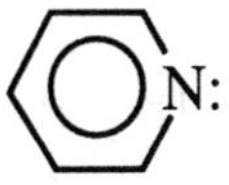 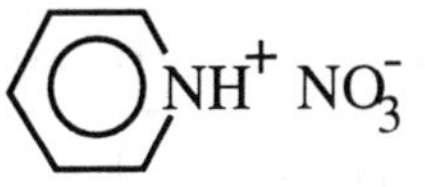

pyridine — pyridinium nitrate

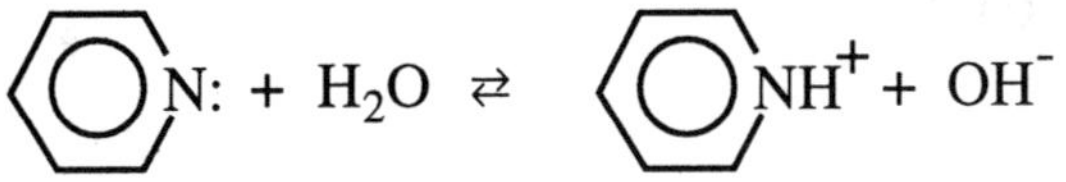

$K_b = K_w/K_a = 1.69 \times 10^{-9}$

$pK_b = 8.77$

S5. Let $x = [H^+] = [A^-]$ and $0.010\,0 - x = [HA]$.

$$\frac{x^2}{0.010\,0-x} = 1.00 \times 10^{-4} \Rightarrow x = 9.51 \times 10^{-4}\ M \Rightarrow pH = -\log x = 3.02$$

$$\alpha = \frac{x}{F} = \frac{9.51 \times 10^{-4}}{0.010\,0} = 9.51 \times 10^{-2}$$

S6. 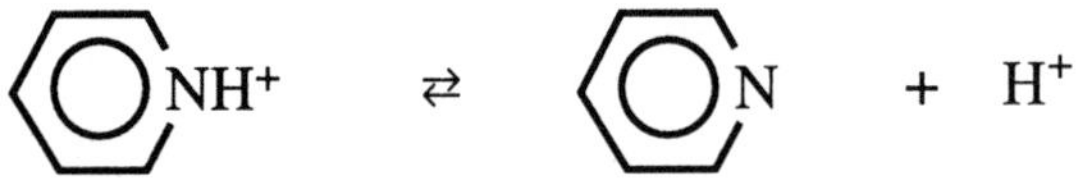

$K_a = 5.90 \times 10^{-6}$

F-x x x

$$\frac{x^2}{0.085\,0-x} = K_a \Rightarrow x = 7.05 \times 10^{-4} \Rightarrow pH = 3.15$$

S7. $\underset{0.020 - x}{BH^+}$ $\rightleftarrows$ $\underset{x}{B}$ + $\underset{x}{H^+}$ $K_a = 2.3 \times 10^{-11}$

$\dfrac{x^2}{0.020-x} = 2.3 \times 10^{-11} \Rightarrow x = 6.78 \times 10^{-7}$ M

pH $= -\log x = 6.17$ $[B] = x = 6.8 \times 10^{-7}$ M $[BH^+] = 0.020$ M

S8. $\underset{0.100 - 10^{-2.36}}{HA}$ $\rightleftarrows$ $\underset{10^{-2.36}}{H^+}$ + $\underset{10^{-2.36}}{A^-}$

$K_a = \dfrac{(10^{-2.36})(10^{-2.36})}{0.100 - 10^{-2.36}} = 1.99 \times 10^{-4}$

$pK_a = -\log K_a = 3.70$

S9.   (a) $\underset{0.010\,0 - x}{HA}$ $\rightleftarrows$ $\underset{x}{H^+}$ + $\underset{x}{A^-}$ $K_a = 4.2 \times 10^{-10}$

$\dfrac{x^2}{0.010\,0 - x} = 4.2 \times 10^{-10} \Rightarrow x = 2.05 \times 10^{-6}$ M $\Rightarrow$ pH $= 5.69$

$\alpha = \dfrac{x}{F} = 2.05 \times 10^{-4}$

(b)   For $F = 10^{-9.00}$ M , the pH must be very close to 7.00.

$K_a = 4.2 \times 10^{-10} = \dfrac{[H^+][A^-]}{F - [A^-]} = \dfrac{(10^{-7.00})[A^-]}{F - [A^-]} \Rightarrow [A^-] = 4.18 \times 10^{-12}$ M

$\alpha = \dfrac{x}{F} = \dfrac{4.18 \times 10^{-12}}{10^{-9.00}} = 0.004\,2$

S10. $\underset{F-x}{A}$ + H_2O $\rightleftarrows$ $\underset{x}{AOH^-}$ + $\underset{x}{H^+}$

$\dfrac{x^2}{F-x} = 10^{-5.4} \Rightarrow x = 2.8 \times 10^{-4}$ M $\Rightarrow$ pH $= 3.55$

S11. $\underset{0.050 - x}{B}$ + H_2O $\rightleftarrows$ $\underset{x}{BH^+}$ + $\underset{x}{OH^-}$ K_b

$\dfrac{x^2}{0.050 - x} = 1.00 \times 10^{-4} \Rightarrow x = 2.19 \times 10^{-3}$

pH $= -\log (K_w/x) = 11.34$ $\alpha = \dfrac{x}{F} = 0.044$

S12. $\underset{0.030 - x}{B}$ + H_2O $\rightleftarrows$ $\underset{x}{BH^+}$ + $\underset{x}{OH^-}$ K_b

$\dfrac{x^2}{0.030 - x} = 4.7 \times 10^{-10} \Rightarrow x = 3.75 \times 10^{-6}$

pH $= -\log (K_w/x) = 8.57$

$[B] = [(CH_3CH_2)_2NH] = 0.030$ M $[BH^+] = 3.8 \times 10^{-6}$ M

S13. $OCl^- + H_2O \rightleftharpoons HOCl + OH^-$
0.026 - x   x   x

$$K_b = K_w/K_a = K_w/(3.0 \times 10^{-8}) = 3.33 \times 10^{-7}$$

$$\frac{x^2}{0.026 - x} = K_b \Rightarrow x = [OH^-] = 9.3 \times 10^{-5} M$$

$$pH = -\log(K_w/[OH^-]) = 9.97$$

$$\alpha = \frac{x}{F} = \frac{9.3 \times 10^{-5}}{0.026} = 0.003\,6$$

S14. $HCO_2^- + H_2O \rightleftharpoons HCO_2H + OH^-$
F-x   x   x

$$K_b = K_w/K_a = 5.56 \times 10^{-11}$$

For $F = 10^{-1}$ and 10^{-2} M , we solve the equation $x^2 / (F - x) = K_b$. The fraction of association is $\alpha = \frac{x}{F}$. This gives $x = 2.36 \times 10^{-6}$ M for $F = 10^{-1}$ M
($\Rightarrow \alpha = 2.36 \times 10^{-5}$) and $x = 7.46 \times 10^{-7}$ M for $F = 10^{-2}$ ($\Rightarrow \alpha = 7.46 \times 10^{-5}$).

For $F = 10^{-12}$ M , pH = 7.00 and $[OH^-] = 10^{-7}$ M .

$$K_b = \frac{[BH^+][OH^-]}{[B]} = \frac{[BH^+]\,10^{-7}}{[B]} \Rightarrow [BH^+] = 10^7 K_b\,[B]$$

Putting this relation between $[BH^+]$ and $[B]$ into the definition of fraction of association gives

$$\alpha = \frac{[BH^+]}{[BH^+] + [B]} = \frac{10^7\,K_b\,[B]}{10^7\,K_b[B] + [B]} = \frac{10^7\,K_b}{10^7\,K_b + 1} = 5.56 \times 10^{-4}$$

F	α
10^{-1}	2.36×10^{-5}
10^{-2}	7.46×10^{-5}
10^{-12}	5.56×10^{-4}

S15. $B + H_2O \rightleftharpoons BH^+ + OH^-$
0.030 - x   x   x

$$pH = 10.50 \Rightarrow [OH^-] = 10^{-3.50} = 3.16 \times 10^{-4} M$$

$$K_b = \frac{x^2}{0.030 - x} = \frac{(3.16 \times 10^{-4})^2}{0.030 - 3.16 \times 10^{-4}} = 3.3_6 \times 10^{-6}$$

S16. $\alpha = \frac{x}{F} = 0.002\,7$. Since F = 0.030 M , $x = 8.10 \times 10^{-5}$ M .

$$K_b = \frac{x^2}{F - x} = \frac{(8.10 \times 10^{-5})^2}{0.030 - 8.10 \times 10^{-5}} = 2.2 \times 10^{-7}$$

S17.
$$B + H_2O \rightleftarrows BH^+ + OH^-$$
$$10^{-2.00} - x \qquad\qquad x \qquad x$$

$$K_b = K_w/K_a = 2.38 \times 10^{-5}$$

$$\frac{x^2}{10^{-2.00} - x} = K_b \Rightarrow x = 4.76 \times 10^{-4} \Rightarrow pH = -\log(K_w/x) = 10.68$$

$$\alpha = \frac{x}{F} = 0.048$$

S18. The pK_a values are (a) 10.774, (b) 9.244, (c) 5.96 and (d) 8.39. Since pK_a for 3-nitrophenol is closest to 8.5, buffer (d) will have the greatest buffer capacity at pH 8.5.

S19. $pH = pK_a + \log\dfrac{[A^-]}{[HA]} = 3.46 + \log\dfrac{(5.13\ g)/(112.126\ g/mol)}{(2.53\ g)/(74.036\ g/mol)} = 3.59$

S20. $pH = 10.64 + \log\dfrac{[CH_3NH_2]}{[CH_3NH_3^+]}$

pH	$[CH_3NH_2]/[CH_3NH_3^+]$
4.00	2.3×10^{-7}
10.64	1.00
12.0	23

S21. $pK_a = 14.00 - pK_b = 0.17 \qquad pH = 0.17 + \log\dfrac{[IO_3^-]}{[HIO_3]}$

pH	$[HIO_3^-]/[IO_3^-]$
7.00	1.5×10^{-7}
1.00	0.15

S22. (a) $pH = pK_a + \log\dfrac{[tris]}{[trisH^+]} = 8.075 + \log\dfrac{(10.0\ g)/(121.136\ g/mol)}{(10.0\ g)/(157.597\ g/mol)} = 8.19$

(b)
$$trisH^+ + OH^- \rightarrow tris + H_2O$$

Initial mmol:	63.45	5.25	82.55
Final mmol:	58.20	—	87.8'0

$$pH = 8.075 + \log\frac{87.80}{58.20} = 8.25$$

S23. $B \;+\; H^+ \;\rightarrow\; BH^+$

Initial mmol: 0.699 5   x   —

Final mmol: 0.699 5 - x   —   x

$$8.00 \;=\; 8.492 + \log\frac{0.699\,5 - x}{x} \;\Rightarrow\; x = 0.529\,1 \text{ mmol}$$

mL required = (0.529 1 mmol)/(0.113 mmol/mL) = 4.68 mL

S24. (a) 250.0 mL of 0.100 M buffer requires 0.025 0 mol NaOAc $\cdot$ 2H$_2$O = 2.95 g

$$OAc^- \;\;+\;\; H^+ \;\;\rightarrow\;\; HOAc$$

Initial mol: 0.025 0   —   —

Final mol: 0.025 0 - x $10^{-5.00}$ x

$$pH \;=\; pK_a + \log\frac{[OAc^-]}{[HOAc]}$$

$$5.00 \;=\; 4.757 - \log\frac{0.025\,0 - x}{x} \;\Rightarrow\; x = 9.091 \text{ mmol} = 90.9 \text{ mL HCl}$$

(b) 1. Calibrate the pH electrode and meter at 5° C.

2. Dissolve 2.95 g NaOAc $\cdot$ 2H$_2$O in ~100 mL H$_2$O at 5°C.

3. While measuring the pH, add enough HCl to bring the pH to 5.00 at 5°C.

4. Dilute to exactly 250 mL at 5°C.

S25. The solution will have a pH near pK$_a$ (= 10.566), so we can neglect [H$^+$] relative to [OH$^-$] in Equations 10-20 and 10-21:

[HA] $\approx$ F$_{HA}$ + [OH$^-$]   [A$^-$] $\approx$ F$_A$ - [OH$^-$]

B + H$_2$O $\rightleftarrows$ BH$^+$ + OH$^-$ K$_b$ = 3.68 $\times$ 10^{-4}

$$K_b \;=\; \frac{[BH^+][OH^-]}{[B]} \;=\; \frac{(F_{BH^+} + [OH^-])[OH^-]}{F_B - [OH^-]}$$

$$3.68 \times 10^{-4} \;=\; \frac{(0.000\,100 + [OH^-])[OH^-]}{0.000\,100 - [OH^-]} \;\Rightarrow\; [OH^-] = 6.86 \times 10^{-5} \text{ M}$$

[B] = 0.000 100 - [OH$^-$] = 3.14 $\times$ 10^{-5} M

[BH$^+$] = 0.000 100 + [OH$^-$] = 1.69 $\times$ 10^{-4} M

S1.

$$\underset{\substack{CH_2OH \\ |}}{H_2NCHCO_2^-} + H_2O \rightleftharpoons \underset{\substack{CH_2OH \\ +| }}{H_3NCHCO_2^-} + OH^- \qquad K_{b1} = K_w/K_{a2} = 1.62 \times 10^{-5}$$

$$\underset{\substack{+CO_2H \\ +| }}{H_3NCHCO_2^-} + H_2O \rightleftharpoons \underset{\substack{CH_2OH \\ +| }}{H_3NCHCO_2H} + OH^- \quad K_{b2} = K_w/K_{a1} = 1.54 \times 10^{-12}$$

S2. (a) $\dfrac{x^2}{0.100 - x} = K_1 \Rightarrow x = 9.95 \times 10^{-4}\,M = [H^+] = [HA^-] \Rightarrow pH = 3.00$

$[H_2A] = 0.100 - x = 0.099\,0\,M$

$[A^{2-}] = \dfrac{K_2\,[HA^-]}{[H^+]} = 1.00 \times 10^{-9}\,M$

(b) $[H^+] = \sqrt{\dfrac{K_1 K_2 F + K_1 K_w}{K_1 + F}} = 1.00 \times 10^{-7}\,M \Rightarrow pH = 7.00$

$[HA^-] \approx 0.100\,M$

$[H_2A] = \dfrac{[H^+][HA^-]}{K_1} = 1.00 \times 10^{-3}\,M$

$[A^{2-}] = \dfrac{K_2[HA^-]}{[H^+]} = 1.00 \times 10^{-3}\,M$

(c) $\dfrac{x^2}{0.100-x} = \dfrac{K_w}{K_2} \Rightarrow x = [OH^-] = [HA^-] = 9.95 \times 10^{-4}\,M \Rightarrow pH = 11.00$

$[A^{2-}] = 0.100 - x = 0.099\,0\,M$

$[H_2A] = \dfrac{[H^+][HA^-]}{K_1} = 1.00 \times 10^{-9}\,M$

	pH	$[H_2A]$	$[HA^-]$	$[A^{2-}]$
0.100 M H_2A	3.00	9.90×10^{-2}	9.95×10^{-4}	1.00×10^{-9}
0.100 M NaHA	7.00	1.00×10^{-3}	0.100	1.00×10^{-3}
0.100 M Na_2A	11.00	1.00×10^{-9}	9.95×10^{-4}	9.90×10^{-2}

S3. $N\langle\ \rangle NH^+\,Cl^-$ ($BH^+\,Cl^-$) is the intermediate form of a diprotic system, with

$F = 0.150\,M$, $K_1 = 4.65 \times 10^{-6}$ and $K_2 = 1.86 \times 10^{-10}$.

$[H^+] = \sqrt{\dfrac{K_1 K_2 F + K_1 K_w}{K_1 + F}} = 2.94 \times 10^{-8}\,M \Rightarrow pH = 7.53$

$[BH^+] \approx 0.150\,M \quad [B] = \dfrac{K_2[BH^+]}{[H^+]} = 9.49 \times 10^{-4}\,M$

$[BH_2^{2+}] = \dfrac{[BH^+][H^+]}{K_1} = 9.48 \times 10^{-4}\,M$

S4.   Charge balance: $[H^+] + [Na^+] + [BH^+] = [Cl^-] + [HA^-] + 2[A^{2-}] + [OH^-]$

Mass balances : $[Cl^-] = F_1$

$[Na^+] = 2F_2$

$F_2 = [H_2A] + [HA^-] + [A^{2-}]$

$F_3 = [B] + [BH^+]$

Equilibria : $[H^+]\gamma_{H^+} [OH^-] \gamma_{OH^-} = K_w$

$K_1 = [H^+]\gamma_{H^+} [HA^-] \gamma_{HA^-} / [H_2A] \gamma_{H_2A}$

$K_2 = [H^+] \gamma_{H^+} [A^{2-}] \gamma_{A^{2-}} / [HA^-] \gamma_{HA^-}$

$K_b = [BH^+] \gamma_{BH^+} [OH^-] \gamma_{OH^-} / [B]\gamma_B$

S5.   Tartaric acid: H_2A $pK_1 = 3.036$ $pK_2 = 4.366$

(a) At pH 3.00, there is a mixture of H_2A and HA^-.

$$H_2A \quad + \quad OH^- \quad \rightarrow \quad HA^-$$

Initial mmol: 3.331 6   x

Final mmol: 3.331 6 - x   x

$3.00 = 3.036 + \log \dfrac{x}{3.331\ 6 - x}$ $\Rightarrow$ $x = 1.597$ mmol $= 3.77$ mL KOH.

(b) At pH 4.00, there is a mixture of HA^- and A^{2-}. We must add 3.331 6 mmol ($= 7.876$ mL) of KOH to convert H_2A into HA^-. Then we need to add more KOH:

$$HA^- \quad + \quad OH^- \quad \rightarrow \quad A^{2-} \quad + \quad H_2O$$

Initial mmol: 3.331 6   x   —

Final mmol: 3.331 6 - x   —   x

$4.00 = 4.366 + \log \dfrac{x}{3.331\ 6 - x}$ $\Rightarrow$ $x = 1.003$ mmol $= 2.370$ mL KOH

Total KOH $= 7.876 + 2.370 = 10.25$ mL

S6.   Malonic acid $= H_2A$ $pK_1 = 2.847$ $pK_2 = 5.696$

(a) At pH 6.00, there is a mixture of A^{2-} and HA^-.

$$A^{2-} \quad + \quad H^+ \quad \rightarrow \quad HA^-$$

Initial mmol: 2.775   x   —

Final mmol: 2.775 - x   —   x

$6.00 = 5.696 + \log \dfrac{2.775 - x}{x}$ $\Rightarrow$ $x = 0.920\ 8$ mmol $= 2.19$ mL HCl.

(b) At pH 3.20, there is a mixture of H_2A and HA^-. We first add 2.775 mmol
(= 6.591 mL) of HCl to convert A^{2-} into HA^-. Then we need to add more HCl

$$HA^- \quad + \quad H^+ \quad \rightarrow \quad H_2A$$

Initial mmol:	2.775	x	—
Final mmol:	2.775 - x	—	x

$$3.20 = 2.847 + \log\frac{2.775 \text{ - x}}{x} \Rightarrow x = 0.852\,7 \text{ mmol} = 2.025 \text{ mL HCl}$$

Total HCl = 6.591 + 2.025 = 8.62 mL

S7. Oxalic acid = H_2A $pK_1 = 1.252$ $pK_2 = 4.266$

At pH 3.20, there is a mixture of A^{2-} and HA^-. We begin with 5.00 g = 0.030 08
mol A^{2-}. The reaction of H_2A with A^{2-} creates 2 moles of HA^-:

$$H_2A \quad + \quad A^{2-} \quad \rightarrow \quad 2HA^-$$

Initial mmol:	x	0.030 08	—
Final mmol:	—	0.030 08 - x	2x

$$3.20 = 4.266 + \log\frac{0.030\,08 \text{ - x}}{2x} \Rightarrow x = 0.025\,67 \text{ mol}$$

= 2.31 g oxalic acid.

S8.

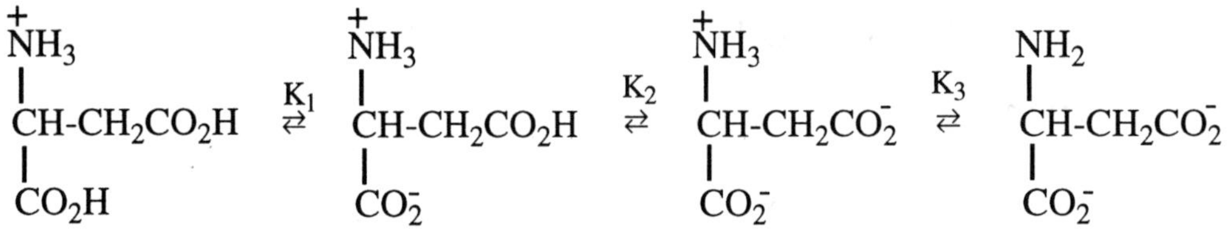

Aspartic acid

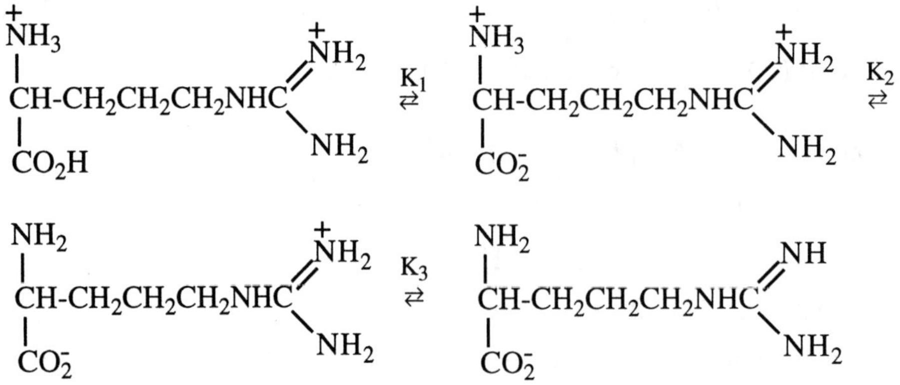

Arginine

S9.

(a) $H_3His^{2+} \quad \underset{\rightleftarrows}{\overset{pK_1 = 1.7}{}} \quad H_2His^+ \quad \underset{\rightleftarrows}{\overset{pK_2 = 6.02}{}} \quad HHis \quad \underset{\rightleftarrows}{\overset{pK_3 = 9.08}{}} \quad His^-$

Histidine

For 0.050 0 M histidine, $[H^+] = \sqrt{\dfrac{K_2K_3F + K_2K_w}{K_2 + F}} = 2.82 \times 10^{-8}$ M

$\Rightarrow$ pH = 7.55

$7.55 = 6.02 + \log \dfrac{[HHis]}{[H_2His^+]} \quad \Rightarrow \quad \dfrac{[H_2His^+]}{[HHis]} = 0.029\ 5$

(b) For 0.050 0 M $H_2His^+Cl^-$, $[H^+] = \sqrt{\dfrac{K_1K_2F + K_1K_w}{K_1 + F}} = 1.17 \times 10^{-4}$ M

$\Rightarrow$ pH = 3.93

$3.93 = 6.02 + \log \dfrac{[HHis]}{[H_2His^+]} \quad \Rightarrow \quad \dfrac{[H_2His^+]}{[HHis]} = 123$

S10.

$$\overset{+}{N}H_3$$
$$|$$
$$CH\text{-}CH_2CH_2CH_2NHC\overset{\displaystyle \nearrow \overset{+}{N}H_2}{\searrow NH_2}$$
$$|$$
$$CO_2^-$$

For H_3Arg^{2+} $pK_1 = 1.823$

$pK_2 = 8.991$

$pK_3 = 12.48$

H_2Arg^+ found in Arginine $\cdot$ HCl

$[H^+] = \sqrt{\dfrac{K_1K_2(0.012\ 0) + K_1K_w}{K_1 + (0.012\ 0)}} = 2.61 \times 10^{-6}$ M $\Rightarrow$ pH = 5.58

$\dfrac{[H_2Arg^+][H^+]}{[H_3Arg^{2+}]} = K_1 \Rightarrow [H_3Arg^{2+}] = \dfrac{(0.012\ 0)(2.61 \times 10^{-6})}{10^{-1.823}}$

$= 2.08 \times 10^{-6}$ M

$\dfrac{[HArg][H^+]}{[H_2Arg^+]} = K_2 \Rightarrow [HArg] = \dfrac{(10^{-8.991})(0.012\ 0)}{2.61 \times 10^{-6}} = 4.69 \times 10^{-6}$ M

$\dfrac{[Arg^-][H^+]}{[HArg]} = K_3 \Rightarrow [Arg^-] = \dfrac{(10^{-12.48})(4.69 \times 10^{-6})}{2.61 \times 10^{-6}} = 5.95 \times 10^{-13}$ M

S11. (a) Since pK = 11.305, BH^+ is predominant at pH 11 and B is predominant at pH 12.

(b) 11.305

(c) $12.00 = 11.305 + \log \dfrac{[B]}{[BH^+]} \quad \Rightarrow \quad \dfrac{[B]}{[BH^+]} = 5.0$

$2.00 = 11.305 + \log \dfrac{[B]}{[BH^+]} \quad \Rightarrow \quad \dfrac{[B]}{[BH^+]} = 5.0 \times 10^{-10}$

S12. (a) HSO_3^- (b) HSO_3^- (c) HSO_3^- (d) SO_3^{2-}

S13. $H_3Cit \overset{pK_1 = 3.128}{\rightleftarrows} H_2Cit^- \overset{pK_2 = 4.761}{\rightleftarrows} HCit^{2-} \overset{pK_3 = 6.396}{\rightleftarrows} Cit^{3-}$

At pH 5.00, $HCit^{2-}$ is dominant.

S14. Fraction in form HA $= \alpha_{HA} = \dfrac{[H^+]}{[H^+] + K_a} = \dfrac{10^{-8.00}}{10^{-8.00} + 10^{-7.00}} = 0.090\ 9$

Fraction in form $A^- = \alpha_{A^-} = \dfrac{K_a}{[H^+] + K_a} = 0.909\ 1.$

$\dfrac{[A^-]}{[HA]} = \dfrac{0.909\ 1}{0.090\ 9} = 10.0.$

S15. Fraction in form $BH_2^{2+} = \alpha_{BH_2^{2+}} = \dfrac{[H^+]^2}{[H^+]^2 + [H^+]K_1 + K_1K_2}$, where

$K_1 = 10^{-6.00}$ and $K_2 = 10^{-12.00}$ $\Rightarrow$ $\alpha_{BH_2^{2+}} = 9.98 \times 10^{-4}$

S16. $K_1 = 3.8 \times 10^{-5}$ $K_2 = 3.8 \times 10^{-6}$

		pH 5.00	pH 6.00
$\alpha_{H_2A} = \dfrac{[H^+]^2}{[H^+]^2 + [H^+]K_1 + K_1K_2}$		= 0.16	0.005 4
$\alpha_{HA^-} = \dfrac{[H^+]K_1}{[H^+]^2 + [H^+]K_1 + K_1K_2}$		= 0.61	0.21
$\alpha_{A^{2-}} = \dfrac{K_1K_2}{[H^+]^2 + [H^+]K_1 + K_1K_2}$		= 0.23	0.79

S17. $K_1 = 2.2 \times 10^{-11}$ $K_2 = 1 \times 10^{-12}$

			pH		
	10.00	10.66	11.00	12.00	12.50
α_{H_2A}	0.82	0.49	0.29	0.022	0.003 4
α_{HA^-}	0.18	0.49	0.64	0.49	0.24
$\alpha_{A^{2-}}$	0.001 8	0.022	0.064	0.49	0.76

S18. Isoelectric pH $= \dfrac{pK_1 + pK_2}{2} = 7.36$

Isoionic pH $= \sqrt{\dfrac{K_1K_2(0.010) + K_1K_w}{K_1 + (0.010)}}$ $\Rightarrow$ 4.38×10^{-8} M $\Rightarrow$ pH $= 7.36$

S19. $k_a + k_b = \dfrac{[H_bA^{4-}][H^+]}{[H_aH_bA^{3-}]} + \dfrac{[H_aA^{4-}][H^+]}{[H_aH_bA^{3-}]} = \dfrac{([H_aA^{4-}] + [H_bA^{4-}])\,[H^+]}{[H_aH_bA^{3-}]} = K_1$

$\dfrac{k_{ab}k_{ba}}{k_{ab} + k_{ba}} = \dfrac{1}{\dfrac{1}{k_{ba}} + \dfrac{1}{k_{ab}}} = \dfrac{1}{\dfrac{[H_aA^{4-}]}{[A^{5-}][H^+]} + \dfrac{[H_bA^{4-}]}{[A^{5-}][H^+]}} = \dfrac{[A^{5-}][H^+]}{[H_aA^{4-}] + [H_bA^{4-}]} = K_2$

$k_a\, k_{ab} = \dfrac{\cancel{[H_bA^{4-}]}[H^+]}{[H_aH_bA^{3-}]}\,\dfrac{[A^{5-}][H^+]}{\cancel{[H_bA^{4-}]}} = \dfrac{\cancel{[H_aA^{4-}]}[H^+]}{[H_aH_bA^{3-}]}\,\dfrac{[A^{5-}][H^+]}{\cancel{[H_aA^{4-}]}} = k_b\, k_{ba}$

S20. $f_a = \dfrac{[H_2A] + [H_aA^-]}{[H_2A] + [H_aA^-] + [H_bA^-] + [A^{2-}]}$

divide
everything
$=$
by
$[H_2A]$

$\dfrac{1 + \dfrac{[H_aA^-]}{[H_2A]}}{1 + \dfrac{[H_aA^-]}{[H_2A]} + \dfrac{[H_bA^-]}{[H_2A]} + \dfrac{[A^{2-}]}{[H_2A]}}$

But $[H_aA^-]/[H_2A] = k_b/[H^+]$, $[H_bA^-]/[H_2A] = K_a/[H^+]$, and $[A^{2-}]/[H_2A] = k_bk_{ba}/[H^+]^2$. Substituting these expressions into the equation above gives

$f_a = \dfrac{1 + \dfrac{k_b}{[H^+]}}{1 + \dfrac{k_b}{[H^+]} + \dfrac{k_a}{[H^+]} + \dfrac{k_bk_{ba}}{[H^+]^2}}$

multiply
everything
$=$
by
$[H^+]^2$

$\dfrac{[H^+] + k_b[H^+]}{[H^+]^2 + k_b[H^+] + k_a[H^+] + k_bk_{ba}}$

The f_b expression is derived similarly.

S21. (a) $K_1 = k_a + k_b = 4.4 \times 10^{-7}$; $K_2 = \dfrac{k_{ab}k_{ba}}{k_{ab} + k_{ba}} = 5.0 \times 10^{-8}$

(b) $f_a = \dfrac{10^{-14} + 1.5 \times 10^{-14}}{10^{-14} + 1.5 \times 10^{-14} + 2.9 \times 10^{-14} + 2.25 \times 10^{-14}} = 0.33$

$f_b = \dfrac{10^{-14} + 2.9 \times 10^{-14}}{10^{-14} + 1.5 \times 10^{-14} + 2.9 \times 10^{-14} + 2.25 \times 10^{-14}} = 0.51$

(c) $\dfrac{[A^{2-}]}{[H_2A]} = \dfrac{k_{ba}k_b}{[H^+]^2} = 2.25$

Measurements with a ruler on the graph show that $[A^{2-}]/[H_2A] \approx 2.25$ when equivalents of $OH^- \approx 1.15$.

S22. (a) $^{18}\alpha = \dfrac{K/R}{(K/R) + [H^+]} = \dfrac{K}{K + R[H^+]}$

Approximating $\bar{\alpha}$ as $^{16}\alpha$, we can write

$$\frac{\Delta\alpha}{\sqrt{\bar{\alpha}}} \approx \frac{\dfrac{K}{K + [H^+]} - \dfrac{K}{K + R[H^+]}}{\sqrt{\dfrac{K}{K + [H^+]}}}$$

$$\frac{\Delta\alpha}{\sqrt{\bar{\alpha}}} \approx \frac{(R - 1)\, K[H^+]}{(K + [H^+])(K + R[H^+])} \frac{\sqrt{K + [H^+]}}{\sqrt{K}} = \frac{(R - 1)\sqrt{K}[H^+]}{\sqrt{K + [H^+]}(K + R[H^+])}$$

(b) $\dfrac{d\left(\dfrac{\Delta\alpha}{\sqrt{\bar{\alpha}}}\right)}{d[H^+]} = 0 = \sqrt{K + [H^+]}\ (K + R[H^+])(R-1)\sqrt{K} -$

$$(R-1)\sqrt{K}[H^+]\{R\sqrt{K+[H^+]} + (K+R[H^+])\tfrac{1}{2}(K+[H^+])^{-1/2}\}$$

Solving for [H$^+$] gives [H+] $= \dfrac{K + K\sqrt{1+8R}}{2R}$

(c) Setting R = 1 gives [H$^+$] $= \dfrac{K + K\sqrt{9}}{2} = 2K$

$$-\log [H^+] = -\log K - \log 2$$
$$pH = pK - 0.30$$

S1. <u>0 mL</u>: $pH = -\log(0.050\ 0) = 1.30$

<u>1 mL</u>: $[H^+] = \left(\dfrac{11.5}{12.5}\right)(0.050\ 0\ M)\left(\dfrac{25.0}{26.0}\right) = 0.044\ 2\ M \Rightarrow pH = 1.35$

<u>5 mL</u>: 1.60

<u>10 mL</u>: 2.15

<u>12.4 mL</u>: 3.57

<u>12.5 mL</u>: Equivalence point $\Rightarrow pH = 7.00$

<u>12.6 mL</u>: $[OH^-] = \left(\dfrac{0.1}{37.6}\right)(0.100\ M) = 2.66 \times 10^{-4}\ M \Rightarrow pH = 10.43$

<u>13 mL</u>: 11.42

S2. <u>0 mL</u>: $HA \rightleftarrows H^+ + A^-$

$\dfrac{x^2}{0.050\ 0-x} = 10^{-4} \Rightarrow x = 2.19 \times 10^{-3}\ M \Rightarrow pH = 2.66$

<u>1 mL</u>: $pH = pK_a + \log\dfrac{[A^-]}{[HA]} = 4.00 + \log\dfrac{1}{4} = 3.40$

<u>2.5 mL</u>: $pH = 4.00 + \log\left(\dfrac{2.5}{2.5}\right) = 4.00$

<u>4 mL</u>: $pH = 4.00 + \log\left(\dfrac{4}{1}\right) = 4.60$

<u>4.9 mL</u>: $pH = 4.00 + \log\left(\dfrac{4.9}{0.1}\right) = 5.69$

<u>5 mL</u>: $A^- + H_2O \rightleftarrows HA + OH^-$ $\qquad\qquad pK_b = 10.00$

$\left(\dfrac{50}{55}\right)(0.050\ 0)-x \qquad\qquad x \qquad x$

$\dfrac{x^2}{0.045\ 5-x} = 1.00 \times 10^{-10} \Rightarrow x = [OH^-] = 2.13 \times 10^{-6} \Rightarrow pH = 8.33$

<u>5.1 mL</u>: $[OH^-] = \left(\dfrac{0.1}{55.1}\right)(0.500\ M) = 9.07 \times 10^{-4}\ M \Rightarrow pH = 10.96$

<u>6 mL</u>: $[OH^-] = \left(\dfrac{1}{56}\right)(0.500\ M) = 8.93 \times 10^{-3}\ M \Rightarrow pH = 11.95$

S3. $V_e = 25.0\ mL \qquad\qquad pK_a = 7.15$

<u>0 mL</u>: $HA \rightleftarrows A^- + HA$

$\dfrac{x^2}{0.050\ 0-x} = 7.1 \times 10^{-8} \Rightarrow x = 5.95 \times 10^{-5}\ M \Rightarrow pH = 4.23$

To find the volume at which $pH = pK_a - 1$, we set up a Henderson-Hasselbalch

equation: $pK_a - 1 = pK_a + \log\dfrac{[A^-]}{[HA]} \Rightarrow [A^-]/[HA] = 0.1$, or $[A^-] = 1/11$ and

$[HA] = 10/11$ of total. Volume of $OH^- = \dfrac{V_e}{11} = 2.27\ mL$

Similarly, when $pH = pK_a + 1$, volume of $OH^- = \dfrac{10V_e}{11} = 22.73\ mL$

When volume of $OH^- = \frac{1}{2}V_e = 12.5\ mL$, $pH = pK_a$

$\underline{V_e = 25.0 \text{ mL}}$: HA is converted to A⁻:

$$A^- + H_2O \rightleftarrows HA + OH^- \qquad\qquad K_b = 1.4 \times 10^{-7}$$
$$\left(\tfrac{50}{75}\right)(0.05)\text{-}x \qquad\qquad x \qquad x$$

$$\frac{x^2}{0.033\ 3\text{-}x} = 1.4 \times 10^{-7} \Rightarrow [OH^-] = 6.82 \times 10^{-5}\ M \Rightarrow pH = 9.83$$

$\underline{1.2V_e = 30.0 \text{ mL}}$: $[OH^-] = \left(\tfrac{5}{80}\right)(0.100\ M) = 6.25 \times 10^{-3}\ M \Rightarrow pH = 11.80$

S4.
$$HA \quad + \quad OH^- \quad \rightarrow \quad A^- \quad + \quad H_2O$$

Initial mmol: 2.345 0   1.044 2   —

Final mmol: 1.300 8   —   1.044 2

$$3.62 = pK_a + \log\frac{1.044\ 2}{1.300\ 8} \Rightarrow pK_a = 3.72$$

S5. $\underline{0 \text{ mL}}$: $B + H_2O \rightleftarrows BH^+ + OH^-$
$$0.050\text{-}x \qquad\quad x \qquad x$$

$$\frac{x^2}{0.050\text{-}x} = 1.00 \times 10^{-4} \Rightarrow x = [OH^-] = 2.19 \times 10^{-3}\ M \Rightarrow pH = 11.34$$

$\underline{1 \text{ mL}}$: $pH = pK_{BH^+} + \log\dfrac{[B]}{[BH^+]} = 10.00 + \log\dfrac{4}{1} = 10.60$

$\underline{2.5 \text{ mL}}$: $pH = pK_{BH^+} = 10.00$

$\underline{4 \text{ mL}}$: $pH = 10.00 + \log\dfrac{1}{4} = 9.40$

$\underline{4.9 \text{ mL}}$: $pH = 10.00 + \log\dfrac{0.1}{4.9} = 8.31$

$\underline{5 \text{ mL}}$: $BH^+ + H_2O \rightleftarrows B + H^+$
$$\left(\tfrac{50}{55}\right)(0.050)\text{-}x \qquad\qquad x \qquad x$$

$$\frac{x^2}{0.045\ 5\text{-}x} = 1.00 \times 10^{-10} \Rightarrow x = [H^+] = 2.13 \times 10^{-6}\ M \Rightarrow pH = 5.67$$

$\underline{5.1 \text{ mL}}$: $[H^+] = \left(\tfrac{0.1}{55.1}\right)(0.500\ M) = 9.07 \times 10^{-4}\ M \Rightarrow pH = 3.04$

$\underline{6 \text{ mL}}$: $[H^+] = \left(\tfrac{1}{56}\right)(0.500\ M) = 8.93 \times 10^{-3}\ M \Rightarrow pH = 2.05$

S6. Titration reaction: $CH_3CH_2CO_2^- + H^+ \rightarrow CH_3CH_2CO_2H$

At the equivalence point, moles of HCl = moles of NaA $\Rightarrow V_e = 47.79$ mL.

$\underline{0 \text{ mL}}$: $A^- + H_2O \rightleftarrows HA + OH^-$
$$0.040\ 0\text{-}x \qquad\quad x \qquad x$$

$$\frac{x^2}{0.040\ 0\text{-}x} = K_b = \frac{K_w}{K_a} \Rightarrow x = 5.47 \times 10^{-6}\ M \qquad pH = -\log\frac{K_w}{x} = 8.74$$

$\underline{1/4\ V_e}$: $pH = pK_a + \log\dfrac{[A^-]}{[HA]} = 4.874 + \log\dfrac{3}{1} = 5.35$

$\underline{1/2\ V_e}$: $pH = pK_a = 4.87$

$\underline{3/4\ V_e}$: $pH = pK_a + \log\dfrac{1}{3} = 4.40$

$\underline{V_e}$: $HA = H^+ + A^-$
      $F\text{-}x \quad x \quad x$

$\quad F_{HA} = \left(\dfrac{100.0}{147.79}\right)(0.040\ 0) = 0.027\ 1\ M$

$\quad \dfrac{x^2}{0.027\ 1\text{-}x} = K_a \Rightarrow x = 5.96 \times 10^{-4}\ M \quad \Rightarrow pH = 3.22$

$\underline{1.1\ V_e}$: $[H^+] = \left(\dfrac{V_e/10}{100.0 + V_e}\right)(0.083\ 7) = 2.71 \times 10^{-3}\ M \qquad pH = 2.57$

S7. $V_{e1} = 5\ mL \qquad V_{e2} = 10\ mL$

$\underline{0\ mL}$: $B + H_2O \rightleftarrows BH^+ + OH^-$
       $0.050\ 0\text{-}x \qquad\quad x \quad x$

$\quad \dfrac{x^2}{0.050\ 0\text{-}x} = 10^{-5.00} \Rightarrow x = [OH^-] = 7.02 \times 10^{-4}\ M \Rightarrow pH = 10.85$

$\underline{1\ mL}$: $pH = pK_{BH^+} + \log\dfrac{[B]}{[BH^+]} = -\log\dfrac{10^{-14}}{10^{-5}} + \log\dfrac{4}{1} = 9.60$

$\underline{2.5\ mL}$: $pH = pK_{BH^+} = 9.00$

$\underline{4\ mL}$: $pH = pK_{BH^+} + \log\dfrac{1}{4} = 8.40$

$\underline{4.8\ mL}$: $pH = pK_{BH^+} + \log\dfrac{0.2}{4.8} = 7.62$

$\underline{5\ mL}$: $[H^+] = \sqrt{\dfrac{K_1 K_2 F + K_1 K_w}{K_1 + F}} = \sqrt{\dfrac{10^{-5}\ 10^{-9}\ (0.045\ 5) + 10^{-5}\ 10^{-14}}{10^{-5} + 0.045\ 5}}$

$\qquad\qquad = 1.00 \times 10^{-7}\ M \Rightarrow pH = 7.00$

$\underline{5.2\ mL}$: $pH = pK_{BH_2^{2+}} + \log\dfrac{[B]}{[BH_2^{2+}]} = 5.00 + \log\dfrac{4.8}{0.2} = 6.38$

$\underline{6\ mL}$: $pH = pK_{BH_2^{2+}} + \log\dfrac{4}{1} = 5.60$

$\underline{7.5\ mL}$: $pH = pK_{BH_2^{2+}} = 5.00$

$\underline{9\ mL}$: $pH = pK_{BH_2^{2+}} + \log\dfrac{1}{4} = 4.40$

$\underline{9.8\ mL}$: $pH = pK_{BH_2^{2+}} + \log\dfrac{0.2}{4.8} = 3.62$

$\underline{10\ mL}$: $BH_2^{2+} \rightleftarrows BH^+ + H^+$
       $0.041\ 7\text{-}x \qquad x \quad x$

$\quad \dfrac{x^2}{0.041\ 7\text{-}x} = 10^{-5.00} \Rightarrow x = 6.41 \times 10^{-4}\ M \quad \Rightarrow pH = 3.19$

$\underline{10.2\ mL}$: $[H^+] = \left(\dfrac{0.2}{60.2}\right)(0.500\ M) = 1.66 \times 10^{-3}\ M \Rightarrow pH = 2.78$

$\underline{11\ mL}$: $[H^+] = \left(\dfrac{1}{61}\right)(0.500\ M) = 8.20 \times 10^{-3}\ M \Rightarrow pH = 2.09$

$\underline{12\ mL}$: $[H^+] = \left(\dfrac{2}{62}\right)(0.500\ M) = 1.61 \times 10^{-2}\ M \Rightarrow pH = 1.79$

S8. $V_{e1} = 5$ mL $V_{e2} = 10$ mL

<u>0 mL:</u> $H_2A \quad \rightleftharpoons \quad H^+ + \quad HA^-$
            $0.050\ 0{-}x \qquad x \qquad x$

$$\frac{x^2}{0.050\ 0{-}x} = 10^{-5.00} \Rightarrow x = 7.02 \times 10^{-4}\ M \ \Rightarrow\ pH = 3.15$$

<u>1 mL:</u> $pH = pK_1 + \log \dfrac{[HA^-]}{[H_2A]} = 5.00 + \log \dfrac{1}{4} = 4.40$

<u>2.5 mL:</u> $pH = pK_1 = 5.00$

<u>4 mL:</u> $pH = pK_1 + \log \dfrac{4}{1} = 5.60$

<u>4.8 mL:</u> $pH = pK_1 + \log \dfrac{4.8}{0.2} = 6.38$

<u>5 mL:</u> $[H^+] = \sqrt{\dfrac{K_1 K_2\,(0.045\ 5) +\ K_1 K_w}{K_1 + (0.045\ 5)}} \ \Rightarrow\ pH = 7.00$

<u>5.2 mL:</u> $pH = pK_2 + \log \dfrac{0.2}{4.8} = 7.62$

<u>6 mL:</u> $pH = pK_2 + \log \dfrac{1}{4} = 8.40$

<u>7.5 mL:</u> $pH = pK_2 = 9.00$

<u>9 mL:</u> $pH = pK_2 + \log \dfrac{4}{1} = 9.60$

<u>9.8 mL:</u> $pH = pK_2 + \log \dfrac{4.8}{0.2} = 10.38$

<u>10 mL:</u> $A^- + H_2O \rightleftharpoons HA^- + OH^- \qquad K_{b1} = \dfrac{K_w}{K_2} = 10^{-5.00}$
              $0.041\ 7{-}x \qquad\quad x \qquad x$

$$\frac{x^2}{0.041\ 7{-}x} = K_2 \Rightarrow x = [OH^-] = 6.41 \times 10^{-4}\ M \ \Rightarrow pH = 10.81$$

<u>10.2 mL:</u> $[OH^-] = \left(\dfrac{0.2}{60.2}\right)(0.500\ M) = 1.66 \times 10^{-3}\ M \ \Rightarrow\ pH = 11.22$

<u>11 mL:</u> $[OH^-] = \left(\dfrac{1}{61}\right)(0.500\ M) = 8.20 \times 10^{-3}\ M \ \Rightarrow\ pH = 11.91$

<u>12 mL:</u> $[OH^-] = \left(\dfrac{2}{62}\right)(0.500\ M) = 1.61 \times 10^{-2}\ M \ \Rightarrow\ pH = 12.21$

S9.   Titration reaction: ⬡$-NH_2 + H^+ \;\rightarrow\;$ ⬡$-NH_3^+ \quad$ $pK_a = 10.64$
                                                        $V_e = 10.0$ mL

<u>0 mL:</u> $B + H_2O \rightleftharpoons BH^+ + OH^- \quad K_b = \dfrac{K_w}{K_a} = 4.37 \times 10^{-4}$
            $0.100{-}x \qquad\quad x \qquad x$

$$\frac{x^2}{0.100{-}x} = K_b \Rightarrow x = [OH^-] = 6.40 \times 10^{-3}\ M \ \Rightarrow pH = 11.81$$

<u>2 mL:</u> $pH = pK_a + \log \dfrac{[B]}{[BH^+]} = 10.64 + \log \dfrac{8}{2} = 11.24$

$\underline{4\,mL}$: $\quad pH = 10.64 + \log\frac{6}{4} = 10.82$

$\underline{6\,mL}$: $\quad pH = 10.64 + \log\frac{4}{6} = 10.46$

$\underline{8\,mL}$: $\quad pH = 10.64 + \log\frac{2}{8} = 10.04$

$\underline{10\,mL}$:

$$BH^+ \rightleftarrows B + H^+ \quad K_a$$
$$\frac{25}{35}(0.100)-x \quad\quad x \quad\quad x$$

$$\frac{x^2}{0.071\,4-x} = 2.3\times10^{-11}\ M \quad\Rightarrow\quad x = 1.28\times10^{-6}\ M \quad\Rightarrow\quad pH = 5.89$$

$\underline{12\,mL}$: $\quad [H^+] = \left(\frac{2}{37}\right)(0.250\ M) = 1.35\times10^{-2}\ M \quad\Rightarrow\quad pH = 1.87$

S10. (a)

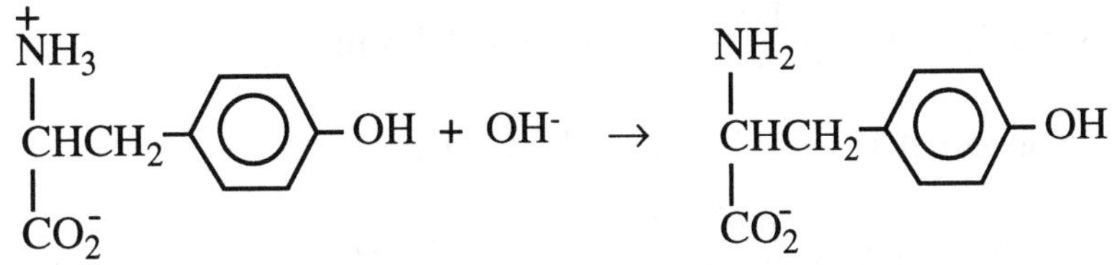

(b) V mL of tyrosine will require $\dfrac{0.010\,0}{0.004\,00}\,V = 2.50\,V$ mL of KOH. The formal concentration of product will be $\left(\dfrac{V}{V+2.50\,V}\right)(0.010\,0) = 0.002\,86\ M$.

$$[H^+] = \sqrt{\frac{K_2 K_3(0.002\,86) + K_2 K_w}{K_2 + 0.002\,86}} = 1.56\times10^{-10}\ M \quad\Rightarrow\quad pH = 9.81$$

S11. Leucine is HL. The reaction is:

$$HL + OH^- \rightarrow L^- + H_2O$$

Initial mmol: $\quad$ 7.27 $\quad\quad$ x $\quad\quad$ —

Final mmol: $\quad$ 7.27-x $\quad\quad$ — $\quad\quad$ x

$$pH = 8.00 = pK_2 + \log\frac{x}{7.27-x} \quad\Rightarrow\quad x = 0.128\ mmol$$

$$\frac{0.128\ mmol}{0.043\,1\ mmol/mL} = 2.97\ mL\ of\ NaOH$$

S12. (a)

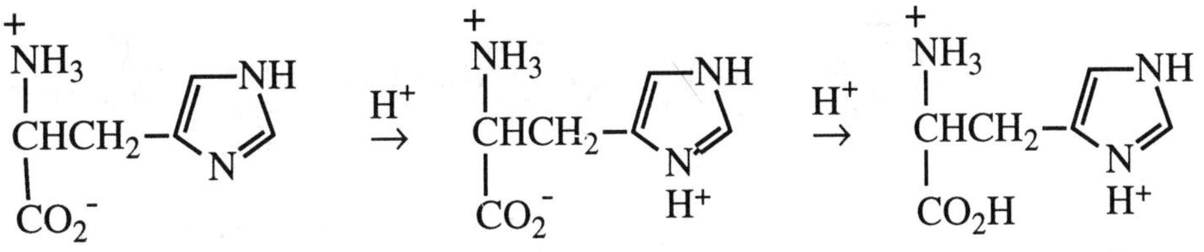

(b) $H_3A^{2+} \xrightleftharpoons{pK_1 = 1.7} H_2A^+ \xrightleftharpoons{pK_2 = 6.02} HA \xrightleftharpoons{pK_3 = 9.08} A^-$

$$\uparrow$$

$$pH \approx \frac{1}{2}(1.7 + 6.02) = 3.86$$

At pH 3.00, HA has been converted entirely to H_2A^+, and some H_2A^+ has been converted to H_3A^{2+}.

$$H_2A^+ \;+\; H^+ \;\rightarrow\; H_3A^{2+}$$

Initial mmol: 1.00   x   —

Final mmol: 1.00-x   —   x

$$pH = pK_1 + \log \frac{[H_2A^+]}{[H_3A^{2+}]}$$

$$3.00 = 1.70 + \log \frac{1-x}{x} \Rightarrow x = 0.047\ 7 \text{ mmol}$$

We need $1.000 + 0.047\ 7$ mmol $HClO_4$.

$$\text{Volume} = \frac{1.047\ 7 \text{ mmol}}{0.050\ 0 \text{ mmol/mL}} = 20.95 \text{ mL}$$

S13. The initial color will be blue. The color will change near the first equivalence point from blue to green. After the first equivalence point, the color will turn yellow and will remain yellow.

S14. (a) red   (b) yellow   (c) yellow   (d) purple

S15. There is not an abrupt change in pH at V_e. The color of the indicator would change very gradually.

S16.   ⬡$\overset{+}{N}HBr^-$ $pK_a = 5.229$ $K_a = 5.90 \times 10^{-6}$

$\underline{0.99\ V_e}$: $pH = pK_a + \log \dfrac{0.99}{0.01} = 7.22$

$\underline{V_e}$:   This is a 0.050 M pyridine solution :

⬡N + H_2O ⇄ ⬡$\overset{+}{N}H$ + OH^- $K_b = \dfrac{K_w}{5.90 \times 10^{-6}} = \dfrac{x^2}{0.050-x}$

0.050-x   x   x

$$\Rightarrow x = 9.2 \times 10^{-6} \quad [H^+] = \frac{K_w}{x} \Rightarrow pH = 8.96$$

$\underline{1.01\ V_e}$:   This solution contains $\dfrac{(0.01)(0.10\text{ M})\ V_e}{(2.01)\ V_e} = 5.0 \times 10^{-4}$ M $[OH^-]$

$$\Rightarrow pH = 10.70$$

Two possibilities are cresol red (orange → red) and phenophthalein (colorless → red).

S17. $[B] + [BH^+] = \dfrac{(6.390 \times 10^{-3} \text{ g})/(278.16 \text{ g/mol})}{0.100\ 0\ L} = 2.297 \times 10^{-4} \text{ M}$

Absorbance $= \varepsilon_B [B] + \varepsilon_{BH^+} [BH^+]$

$0.350 = (2\ 860)[B] + (937)(2.297 \times 10^{-4} - [B])$

$\Rightarrow [B] = 7.007 \times 10^{-5} \text{ M}, \ [BH^+] = 1.596 \times 10^{-4} \text{ M}$

$pH = pK_a + \log \dfrac{[B]}{[BH^+]} = (14.00 - 14.79) + \log \dfrac{7.007 \times 10^{-5}}{1.596 \times 10^{-4}} = -1.15$

S18. (a) First equivalence point : $pH \approx \frac{1}{2}(pK_1 + pK_2) = 3.46$

Second equivalence point : We need to estimate the formal concentration of A^{2-} at this point. Volume of NaOH needed to react with 1 g of acid is

$\dfrac{2(1 \text{ g}/132.073 \text{ g/mol})}{0.094\ 32 \text{ M}} = 160.6 \text{ mL}.$

$[A^{2-}] \approx \left(\dfrac{100}{260.6}\right)\left(\dfrac{10 \text{ g/L}}{132.073 \text{ g/mol}}\right) = 0.029\ 1 \text{ M}.$

$$\begin{array}{ccccccc} A^{2-} & + & H_2O & \rightleftarrows & HA^- & + & OH^- \\ 0.029\ 1-x & & & & x & & x \end{array}$$

$\dfrac{x^2}{0.029\ 1-x} = \dfrac{K_w}{10^{-4.37}} \Rightarrow x = 2.61 \times 10^{-6} \text{ M} \Rightarrow pH = 8.42$

(b) 2nd. (There will be little break in the curve near the 1st end point.)

(c) Thymolphthalein - first trace of blue.

S19. One mole of borax reacts with two moles of H^+.

$[HNO_3](0.021\ 61 \text{ L}) = 2\left(\dfrac{0.261\ 9 \text{ g}}{381.367 \text{ g/mol}}\right) \Rightarrow [HNO_3] = 0.063\ 56 \text{ M}$

S20. Derivation of equation for titration of dibasic B with strong acid (HCl):

Charge balance: $[H^+] + [BH^+] + 2[BH_2^{2+}] = [OH^-] + [Cl^-]$

$\phi = \text{fraction of titration} = \dfrac{C_a V_a}{C_b V_b}$

Substitutions: $[Cl^-] = \dfrac{C_a V_a}{V_a + V_b}$

$[BH^+] = \alpha_{BH^+} \dfrac{C_b V_b}{V_a + V_b}$

$[BH_2^{2+}] = \alpha_{BH_2^{2+}} \dfrac{C_b V_b}{V_a + V_b}$

Putting these expressions into the charge balance gives

$[H^+] + \dfrac{\alpha_{BH^+} C_b V_b}{V_a + V_b} + \dfrac{2\alpha_{BH_2^{2+}} C_b V_b}{V_a + V_b} = [OH^-] + \dfrac{C_a V_a}{V_a + V_b}$

Now multiply by $V_a + V_b$:

$$[H^+]V_a + [H^+]V_b = \alpha_{BH^+}C_bV_b + 2\alpha_{BH_2^{2+}}C_bV_b = [OH^-]V_a + [OH^-]V_b + C_aV_a$$

Collect V_a and V_b terms:

$$V_a([H^+] - [OH^-] - C_a) = V_b([OH^-] - [H^+] - \alpha_{BH^+}C_b - 2\alpha_{BH_2^{2+}}C_b)$$

$$\frac{V_a}{V_b} = \frac{(\alpha_{BH^+} + 2\alpha_{BH_2^{2+}})C_b + [H^+] - [OH^-]}{C_a - [H^+] + [OH^-]}$$

Finally, multiply both sides by $\dfrac{1/C_b}{1/C_a}$

$$\phi = \frac{C_aV_a}{C_bV_b} = \frac{\alpha_{BH^+} + 2\alpha_{BH_2^{2+}} + \dfrac{[H^+] - [OH^-]}{C_b}}{1 - \dfrac{[H^+] - [OH^-]}{C_a}}$$

S21. Derivation of fractional composition equations with activity coefficients:

$$HA \underset{\rightleftarrows}{\overset{K_a}{}} H^+ + A^- \qquad\qquad K_a = \frac{[H^+]\gamma_{H^+}[A^-]\gamma_{A^-}}{[HA]\gamma_{HA}}$$

Mass balance: $F = [HA] + [A^-] \Rightarrow [A^-] = F - [HA]$

Substitute for $[A^-]$ in the equilibrium expression:

$$K_a = \frac{[H^+]\gamma_{H^+}(F - [HA])\gamma_{A^-}}{[HA]\gamma_{HA}}$$

$$\Rightarrow [HA] = \frac{[H^+]\gamma_{H^+}\gamma_{A^-}F}{[H^+]\gamma_{H^+}\gamma_{A^-} + K_a\gamma_{HA}}$$

$$\alpha_{HA} = \frac{[HA]}{F} = \frac{[H^+]\gamma_{H^+}\gamma_{A^-}}{[H^+]\gamma_{H^+}\gamma_{A^-} + K_a\gamma_{HA}}$$

To derive an expression for α_{A^-}, substitute $[HA] = F - [A^-]$ into the equilibrium expression:

$$K_a = \frac{[H^+]\gamma_{H^+}[A^-]\gamma_{A^-}}{(F - [A^-])\gamma_{HA}}$$

$$\Rightarrow [A^-] = \frac{K_aF\gamma_{HA}}{[H^+]\gamma_{H^+}\gamma_{A^-} + K_a\gamma_{HA}}$$

$$\alpha_{A^-} = \frac{[A^-]}{F} = \frac{K_a\gamma_{HA}}{[H^+]\gamma_{H^+}\gamma_{A^-} + K_a\gamma_{HA}}$$

S1. $\alpha_{Y^{4-}} = \dfrac{10^{-0.0}10^{-1.5}...10^{-10.24}}{(10^{-6.62})^6 + (10^{-6.62})^5 10^{-0.0} + ... + 10^{-0.0}10^{-1.5}...10^{-10.24}} = 1.78 \times 10^{-4}$

S2. (a) $K_f' = \alpha_{Y^{4-}} K_f = 0.36 \times 10^{10.69} = 1.76 \times 10^{10}$

(b)
$$Ca^{2+} + EDTA \rightleftarrows CaY^{2-}$$
$$\phantom{Ca^{2+} +} x x 0.050\text{-}x$$

$\dfrac{0.050\text{-}x}{x^2} = 1.76 \times 10^{10} \Rightarrow [Ca^{2+}] = 1.7 \times 10^{-6}$ M

S3. $\dfrac{[MgY^{2-}]}{[Mg^{2+}][Y^{4-}]} \cdot \dfrac{[Na^+][Y^{4-}]}{[NaY^{3-}]} = \dfrac{K_f \text{ (for } MgY^{2-})}{K_f \text{ (for } NaY^{3-})} = 1.35 \times 10^7$

$[Mg^{2+}] = \dfrac{[Na^+]}{1.35 \times 10^7}\left(\dfrac{[MgY^{2-}]}{[NaY^{3-}]}\right)$ (1)

We know that $[Mg^{2+}] + [MgY^{2-}] = 0.050\,0$ M and $[Na^+] = [NaY^{3-}] = 0.100$ M. It seems like a good approximation to say that $[Na^+] = 0.100$ M and $[MgY^{2-}] = 0.050\,0$ M , since we expect almost all of the Na^+ to be free and almost all of the Mg^{2+} to be bound to EDTA. If the ratio $[MgY^{2-}] / [NaY^{3-}]$ is called R, we can say that $[Mg^{2+}] \approx 0.050\,0/R$. (To see this, suppose R = 1 000. Then 1/1 000 of the Mg^{2+} will not be bound to EDTA and $[Mg^{2+}] = (0.050\,0/1\,000.)$ In Equation 1 we can set $[Mg^{2+}] = 0.050\,0/R$ and $[MgY^{2-}]/[NaY^{3-}] = R$.

$$\dfrac{0.050\,0}{R} = \dfrac{[Na^+]}{1.35 \times 10^7} R \qquad (2)$$

Putting the value $[Na^+] = 0.100$ into Equation 2 gives R = 2 600. This large value of R confirms the approximations that $[MgY^{2-}] \approx 0.050\,0$ M and $[NaY^{3-}] \approx 0.100$ M.

S4. (a) $(V_e)(0.100$ M $) = (100.0$ mL$)(0.050\,0$ M $) \Rightarrow V_e = 50.0$ mL

(b) $[EDTA] = \left(\dfrac{1}{2}\right) \underset{\substack{\text{fraction}\\\text{remaining}}}{} (0.050\,0$ M $) \underset{\substack{\text{original}\\\text{concentration}}}{} \left(\dfrac{100}{125}\right) \underset{\substack{\text{dilution}\\\text{factor}}}{} = 0.020\,0$ M

(c) 0.36 (Table 13-1)

(d) $K_f' = (0.36)(10^{8.00}) = 3.6 \times 10^7$

(e) $[MY^{n-4}] = (0.100$ M $)\left(\dfrac{100}{150}\right) = 0.066\,7$ M

$\dfrac{[MY^{n-4}]}{[M^{n+}][EDTA]} = \dfrac{0.066\,7\text{-}x}{x^2} = 3.6 \times 10^7 \Rightarrow x = [M^{n+}] = 4.3 \times 10^{-5}$ M

(f) $[M^{n+}] = (0.100 \text{ M}) \left(\dfrac{5.0}{155.0}\right) = 3.23 \times 10^{-3} \text{ M}$

$[MY^{n-4}] = (0.050\ 0 \text{ M}) \left(\dfrac{100.0}{155.0}\right) = 3.23 \times 10^{-2} \text{ M}$

$\dfrac{[MY^{n-4}]}{[M^{n+}][EDTA]} = \dfrac{(3.23 \times 10^{-2})}{(3.23 \times 10^{-3})[EDTA]} = 3.6 \times 10^7 \Rightarrow [EDTA] = 2.8 \times 10^{-7} \text{ M}$

S5. $Fe^{2+} + EDTA \rightleftarrows FeY^{2-} \quad \alpha_{Y^{4-}} K_f = 4.8 \times 10^9$

$V_e = (25.00) \left(\dfrac{0.020\ 26 \text{ M}}{0.038\ 55 \text{ M}}\right) = 13.14 \text{ mL}$

(a) <u>12.00 mL</u>: $[EDTA] = \left(\dfrac{13.14 - 12.00}{13.14}\right)(0.020\ 26 \text{ M})\left(\dfrac{25.00}{37.00}\right) = 1.19 \times 10^{-3} \text{ M}$

$[FeY^{2-}] = \left(\dfrac{12.00}{37.00}\right)(0.038\ 55 \text{ M}) = 1.25 \times 10^{-2} \text{ M}$

$[Fe^{2+}] = \dfrac{[FeY^{2-}]}{[EDTA]\ K_f'} = 2.2 \times 10^{-9} \text{ M} \Rightarrow pFe^{2+} = 8.66$

(b) <u>V_e</u>: Formal concentration of FeY^{2-} is $\left(\dfrac{25.00}{38.14}\right)(0.020\ 26 \text{ M}) = 1.33 \times 10^{-2} \text{ M}$

$\begin{array}{ccccc} Fe^{2+} & + & EDTA & \rightleftarrows & FeY^{2-} \\ x & & x & & 1.33 \times 10^{-2} - x \end{array}$

$\dfrac{1.33 \times 10^{-2} - x}{x^2} = \alpha_{Y^{4-}} K_f \Rightarrow x = 1.66 \times 10^{-6} \text{ M} \Rightarrow pFe^{2+} = 5.78$

(c) <u>14.00 mL</u>: $[Fe^{2+}] = \left(\dfrac{14.0 - 13.14}{39.00}\right)(0.038\ 55 \text{ M}) = 8.50 \times 10^{-4} \text{ M}$

$\Rightarrow pFe^{2+} = 3.07$

S6. $(50.0 \text{ mL})(0.011\ 1 \text{ M}) = (V_e)(0.022\ 2 \text{ M}) \Rightarrow V_e = 25.0 \text{ mL}$

<u>0 mL</u>: $pY^{3+} = -\log(0.011\ 1) = 1.95$

<u>10.0 mL</u>: $[Y^{3+}] = \left(\dfrac{25.0 - 10.0}{25.0}\right)(0.011\ 1 \text{ M})\left(\dfrac{50.0}{60.0}\right) = 0.005\ 55 \text{ M} \Rightarrow pY^{3+} = 2.26$

<u>20.0 mL</u>: $[Y^{3+}] = \left(\dfrac{25.0 - 20.0}{25.0}\right)(0.011\ 1 \text{ M})\left(\dfrac{50.0}{70.0}\right) = 0.001\ 59 \text{ M} \Rightarrow pY^{3+} = 2.80$

<u>24.0 mL</u>: $[Y^{3+}] = \left(\dfrac{25.0 - 24.0}{25.0}\right)(0.011\ 1 \text{ M})\left(\dfrac{50.0}{74.0}\right) = 3.00 \times 10^{-4} \text{ M} \Rightarrow pY^{3+} = 3.52$

<u>24.9 mL</u>: $[Y^{3+}] = \left(\dfrac{25.0 - 24.9}{25.0}\right)(0.011\ 1 \text{ M})\left(\dfrac{50.0}{74.9}\right) = 2.96 \times 10^{-5} \text{ M} \Rightarrow pY^{3+} = 4.53$

<u>25.0 mL</u>: $\begin{array}{ccccc} Y^{3+} & + & EDTA & \rightleftarrows & Y(EDTA)^- \\ & & & & \\ x & & x & & \frac{50.0}{75.0}(0.011\ 1) - x \end{array} \qquad K_f' = 4.6 \times 10^{11}$

$\dfrac{0.007\ 40 - x}{x^2} = 4.6 \times 10^{11} \Rightarrow x = 1.3 \times 10^{-7} \Rightarrow pY^{3+} = 6.90$

$$\frac{0.007\ 40-x}{x^2} = 4.6 \times 10^{11} \Rightarrow x = 1.3 \times 10^{-7} \Rightarrow pY^{3+} = 6.90$$

<u>25.1 mL</u>: $[EDTA] = \left(\dfrac{0.1}{75.1}\right)(0.022\ 2\ M) = 2.96 \times 10^{-5}\ M$

$[Y(EDTA)^-] = \left(\dfrac{50.0}{75.1}\right)(0.011\ 1\ M) = 0.007\ 39\ M$

$[Y^{3+}] = \dfrac{[Y(EDTA)^-]}{[EDTA]\ K_f'} = 5.4 \times 10^{-10}\ M \Rightarrow pY^{3+} = 9.27$

<u>26.0 mL</u>: $[EDTA] = \left(\dfrac{1.0}{76.0}\right)(0.022\ 2\ M) = 2.92 \times 10^{-4}\ M$

$[Y(EDTA)^-] = \left(\dfrac{50.0}{76.0}\right)(0.011\ 1\ M) = 0.007\ 30\ M$

$[Y^{3+}] = 5.4 \times 10^{-11}\ M \Rightarrow pY^{3+} = 10.26$

<u>30.0 mL</u>: $[EDTA] = \left(\dfrac{5.0}{80.0}\right)(0.022\ 2\ M) = 1.39 \times 10^{-3}\ M$

$[Y(EDTA)^-] = \left(\dfrac{50.0}{80.0}\right)(0.011\ 1\ M) = 6.94 \times 10^{-3}\ M$

$[Y^{3+}] = 1.09 \times 10^{-11}\ M \Rightarrow pY^{3+} = 10.96$

S7. $Cd^{2+} + Y^{4-} \rightleftarrows CdY^{2-}$ $K_f = 10^{16.46} = 2.9 \times 10^{16}$

$\alpha_{Y4-} = 0.98$ at pH 12.00 (Table 13-1)

For Cd^{2+} and NH_3, Appendix I gives $K_1 = 10^{2.51}$, $K_2 = 10^{1.96}$, $K_3 = 10^{1.30}$ and $K_4 = 10^{0.79}$. This means $\beta_1 = K_1 = 3.2 \times 10^2$, $\beta_2 = K_1K_2 = 3.0 \times 10^4$, $\beta_3 = K_1K_2K_3 = 5.9 \times 10^5$ and $\beta_4 = K_1K_2K_3K_4 = 3.6 \times 10^6$.

$$\alpha_{Cd2+} = \frac{1}{1+\beta_1(0.200)+\beta_2(0.200)^2+\beta_3(0.200)^3+\beta_4(0.200)^4} = 8.5 \times 10^{-5}$$

$K_f' = \alpha_{Y4-}\ K_f = 2.8 \times 10^{16}$

$K_f'' = \alpha_{Y4-}\ \alpha_{Cd2+}\ K_f = 2.4 \times 10^{12}$

Equivalence point = 5.00 mL

(a)   At 0 mL, the total concentration of cadmium is $Cd^{2+} = 0.001\ 00\ M$ and

$[Cd^{2+}] = \alpha_{Cd2+}\ C_{Cd2+} = 8.5 \times 10^{-8}\ M \Rightarrow pCd^{2+} = 7.07$

(b)   At 1.00 mL, $C_{Cd2+} = \left(\dfrac{4.00}{5.00}\right)(0.001\ 00\ M)\left(\dfrac{10.00}{11.00}\right) = 7.27 \times 10^{-4}\ M$

$$\underset{\substack{\text{fraction} \\ \text{remaining}}}{}\ \underset{\substack{\text{original} \\ \text{concentration}}}{}\ \underset{\substack{\text{dilution} \\ \text{factor}}}{}$$

$[Cd^{2+}] = \alpha_{Cd2+}\ C_{Cd2+} = 6.2 \times 10^{-8}\ M \Rightarrow pCd^{2+} = 7.21$

(c)   At 4.90 mL, $C_{Cd2+} = \left(\dfrac{0.10}{5.00}\right)(0.001\ 00\ M)\left(\dfrac{10.00}{14.90}\right) = 1.34 \times 10^{-5}\ M$

$[Cd^{2+}] = \alpha_{Cd2+}\ C_{Cd2+} = 1.1 \times 10^{-9}\ M \Rightarrow pCd^{2+} = 8.94$

(d) At the equivalence point, we can write

$$C_{Cd^{2+}} \quad + \quad EDTA \quad \rightleftarrows \quad CdY^{2-}$$
$$\qquad\quad x \qquad\qquad\quad x \qquad\qquad \left(\frac{10.00}{15.00}\right)(0.001\ 00) - x$$

$$\frac{0.000\ 667 - x}{x^2} = 2.4 \times 10^{12} \Rightarrow x = C_{Cd^{2+}} = 1.7 \times 10^{-8}\ M$$

$$[Cd^{2+}] = \alpha_{Cd^{2+}} C_{Cd^{2+}} = 1.4 \times 10^{-12}\ M \Rightarrow pCd^{2+} = 11.85$$

(e) Past the equivalence point at 5.10 mL, we can say

$$[EDTA] = \left(\frac{0.10}{15.10}\right)(0.002\ 00\ M) = 1.3 \times 10^{-5}\ M$$

$$\underset{\substack{\text{dilution}\\\text{factor}}}{} \quad \underset{\substack{\text{original}\\\text{concentration}}}{}$$

$$[CdY^{2-}] = \left(\frac{10.00}{15.10}\right)(0.001\ 00\ M) = 6.6 \times 10^{-4}\ M$$

$$K_f' = \frac{[CdY^{2-}]}{[Cd^{2+}][EDTA]} = \frac{(6.6 \times 10^{-4})}{[Cd^{2+}]\,(1.3 \times 10^{-5})}$$

$$\Rightarrow [Cd^{2+}] = 1.8 \times 10^{-15}\ M \Rightarrow pCd^{2+} = 14.74$$

(f) At 6.00 mL,

$$[EDTA] = \left(\frac{1.00}{16.00}\right)(0.002\ 00\ M) = 1.2 \times 10^{-4}\ M$$

$$[CdY^{2-}] = \left(\frac{10.00}{16.00}\right)(0.001\ 00\ M) = 6.2 \times 10^{-4}\ M$$

$$K_f' = \frac{[CdY^{2-}]}{[Cd^{2+}][EDTA]} \Rightarrow [Cd^{2+}] = 1.8 \times 10^{-16}\ M \Rightarrow pCd^{2+} = 15.73$$

S8. (a) Equations 1 and 2 can be represented as

For noncooperative binding, $\square\square = \boxtimes\square = \square\boxtimes = \boxtimes\boxtimes$

at 50% saturation. Calling each concentration C gives $K_1 = \dfrac{(C+C)}{(C)[M]}$ and

$$K_2 = \frac{(C)}{(C+C)[M]}, \text{ or } \frac{K_1}{K_2} = \frac{\dfrac{2C}{CM}}{\dfrac{C}{2CM}} = 4.$$

S9. H_3In^{3-}, yellow, red, violet $\rightarrow$ red

S10. $[Fe^{3+}] = \dfrac{\text{mmol EDTA - mmol } Mg^{2+}}{25.00 \text{ mL}} = \dfrac{0.367 - 0.109}{25.00 \text{ mL}} = 0.010 \ 3 \text{ M}$

S11.   Hardness = 3.2 mM + 1.1 mM = 4.3 mM = 430 mg $CaCO_3$/L, since the formula weight of $CaCO_3$ is 100.09 g/mol.

S12.   Concentration of Ni^{2+} in the standard solution is $\left(\dfrac{39.3}{30.0}\right)(0.013 \ 0 \text{ M}) = 0.017 \ 0 \text{ M}$.

The quantity of Ni^{2+} in 25.0 mL is (25.0 mL)(0.017 0 M) = 0.425 mmol. Excess Ni equals (10.1 mL) (0.013 0 M) = 0.131 mmol.

Therefore 0.425 - 0.131 = 0.294 mmol of Ni^{2+} reacted with CN^-.

Since each mole of Ni^{2+} reacts with 4 moles of CN^-, there must have been 4(0.294) = 1.176 mmol of CN^- in 12.7 mL.

$[CN^-]$ = 1.176 mmol/12.7 mL = 0.092 6 M .

S13.

Total EDTA used	= (39.98 mL) (0.045 00 M)	=	1.799 1 mmol
- mmol of Mg^{2+}	= (10.26 mL) (0.020 65 M)	=	0.211 9 mmol
- mmol of Zn^{2+}	= (15.47 mL) (0.020 65 M)	=	0.319 5 mmol
mmol of Mn^{2+}		=	1.267 7 mmol

mg of each metal : Mn^{2+} - 69.64; Mg^{2+} - 5.150; Zn^{2+} - 20.89

S14. $[Bi^{3+}] = \dfrac{\text{mmol EDTA in step 1}}{25.00 \text{ mL}} = \dfrac{0.171 \ 5 \text{ mmol}}{25.00 \text{ mL}} = 6.861 \text{ mM}$

$[Ti^{4+}] = \dfrac{\text{mmol } Zn^{2+} \text{ in step 3}}{25.00 \text{ mL}} = \dfrac{0.053 \ 85 \text{ mmol}}{25.00 \text{ mL}} = 2.153 \text{ mM}$

$[Al^{3+}] = \dfrac{\text{mmol } Zn^{2+} \text{ in step 4}}{25.00 \text{ mL}} = \dfrac{0.033 \ 07 \text{ mmol}}{25.00 \text{ mL}} = 13.23 \text{ mM}$

Color change in steps 3 and 4: yellow $\rightarrow$ red

S1. (a) $TeO_3^{2-} + 3H_2O + 4e^- \rightleftarrows Te(s) + 6OH^-$
Oxidant

$S_2O_4^{2-} + 4OH^- \rightleftarrows 2SO_3^{2-} + 2H_2O + 2e^-$
Reductant

(b) (1.00 g Te)/(127.60 g/mol) = 7.84 mmol, which requires $4 \times 7.84 = 31.3$
mmol of electrons $(3.13 \times 10^{-2} \text{ mol e}^-)(9.649 \times 10^4 \text{ C/mol}) = 3.02 \times 10^3$ C

(c) Current = $(3.02 \times 10^3 \text{ C})/(3\ 600 \text{ s}) = 0.840$ A

S2. (a) oxidation: $Fe(CN)_6^{4-} \rightleftarrows Fe(CN)_6^{3-} + e^-$
reduction: $Ag(CN)_2^- + e^- \rightleftarrows Ag(s) + 2CN^-$

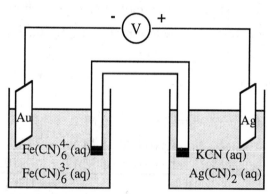

(b) oxidation: $2Hg(l) + 2Cl^- \rightleftarrows Hg_2Cl_2(s) + 2e^-$
reduction: $Zn^{2+} + 2e^- \rightleftarrows Zn(s)$

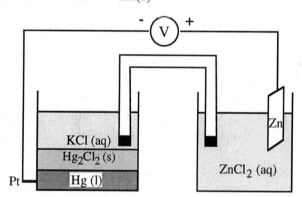

S3. The calculations below show that the hydrogen-oxygen fuel cell produces the most
electricity per kg of reactants and the lead-acid battery produces the least.

$$1 \text{ A·h} = \frac{(1 \text{ C/s})(3\ 600 \text{ s/h})(1 \text{ h})}{9.649 \times 10^4 \text{ C/mol e}^-} = 0.037\ 31 \text{ mol e}^-$$

(a) In the lead-acid battery, each mole of Pb(s) transfers 2 mol of e^- to each mol of
PbO_2 to make $PbSO_4$. Therefore 642.6 g of reactants transfers 2 mol of e^-.

$$\frac{\text{ampere·hour}}{\text{kilogram of reactants}} = \frac{(2 \text{ mol e}^-)/(0.037\ 31 \text{ mol e}^-/\text{A·h})}{0.642\ 6 \text{ kg}} = 83.42 \frac{\text{A·h}}{\text{kg}}$$

(b) One mol of Zn transfers 2 mol of e^-.

$$\frac{(2 \text{ mol } e^-)/(0.037\ 31 \text{ mol } e^-/A\cdot h)}{0.346\ 25 \text{ kg}} = 154.8 \frac{A\cdot h}{kg}$$

(c) One mol of $Cd(OH)_2$ receives 2 mol of e^-.

$$\frac{(2 \text{ mol } e^-)/(0.037\ 31 \text{ mol } e^-/A\cdot h)}{0.331\ 84 \text{ kg}} = 161.5 \frac{A\cdot h}{kg}$$

(d) Two mol of Al transfer 6 mol of e^-.

$$\frac{(6 \text{ mol } e^-)/(0.037\ 31 \text{ mol } e^-/A\cdot h)}{0.372\ 525 \text{ kg}} = 431.7 \frac{A\cdot h}{kg}$$

(e) One mol of O_2 receives 4 mol of e^-.

$$\frac{(4 \text{ mol } e^-)/(0.037\ 31 \text{ mol } e^-/A\cdot h)}{0.036\ 031 \text{ kg}} = 2\ 975 \frac{A\cdot h}{kg}$$

S4. If we organize the half-reactions as in Table 14-1, Cr^{2+} is the product in the reaction with the most negative reduction potential. Therefore it is the strongest reducing agent. The half-reaction is $Cr^{3+} + e^- \rightleftarrows Cr^{2+}$ ($E° = -0.42$ V).

S5. Become stronger : Cl_2, Al, H_2S, $S_2O_3^{2-}$
Unchanged : MnO_4^{2-}

S6. (a) right half-cell: $Hg_2Cl_2(s) + 2e^- \rightleftarrows 2Hg(l) + 2Cl^-$ $E_+° = 0.268$ V
left half-cell: $2H^+ + 2e^- \rightleftarrows H_2(g)$ $E_-° = 0$ V

right half-cell: $E_+ = \left\{ 0.269 - \dfrac{0.059\ 16}{2} \log [Cl^-]^2 \right\}$

$= \left\{ 0.269 - \dfrac{0.059\ 16}{2} \log [0.200]^2 \right\} = 0.310_4$ V

left half-cell: $E_- = \left\{ 0 - \dfrac{0.059\ 16}{2} \log \dfrac{P_{H_2}}{[H^+]^2} \right\}$

$= \left\{ -\dfrac{0.059\ 16}{2} \log \dfrac{0.100}{[10^{-2.54}]^2} \right\} = -0.120_7$ V

(b) $E = E_+ - E_- = 0.310_4 - (-0.120_7) = 0.431$ V. Since the voltage is positive, electrons flow from the left-hand electrode to the right-hand electrode. Reduction occurs at the right-hand electrode.

S7. (a) $Pt(s) \mid Cr^{2+}(aq), Cr^{3+}(aq) \parallel Tl^+(aq) \mid Tl(s)$

(b) right half-cell: $Tl^+ + e^- \rightleftarrows Tl(s)$ $E_+° = -0.336$ V
left half-cell: $Cr^{3+} + e^- \rightleftarrows Cr^{2+}$ $E_-° = -0.42$ V

right half-cell: $E_+ = \left\{ -0.336 - 0.059\ 16 \log \dfrac{1}{[Tl^+]} \right\} = -0.336$ V

left half-cell: $E_- = \left\{ -0.42 - 0.059\ 16 \log \dfrac{[Cr^{2+}]}{[Cr^{3+}]} \right\} = -0.42$ V

$E = E_+ - E_- = -0.336 - (-0.42) = 0.084$ V.

(c)　Since the voltage is positive, electrons flow from Pt to Tl and the reaction is
$$Tl^+ + Cr^{2+} \rightleftharpoons Tl(s) + Cr^{3+}$$

(d)　Pt is the anode, since Cr^{2+} is oxidized.

S8.　Balanced reaction : $HNO_2 + e^- + H^+ \rightleftharpoons NO + H_2O$

$NO_3^- \rightarrow NO_2$	$\Delta G_1^\circ = -1F(0.773)$
$NO_2 \rightarrow HNO_2$	$\Delta G_2^\circ = -1F(1.108)$
$\underline{HNO_2 \rightarrow NO}$	$\underline{\Delta G_3^\circ = -1FE_3^\circ}$
$NO_3^- \rightarrow NO$	$\Delta G_4^\circ = -3F(0.955)$

$FE_3^\circ = 3F(0.955) - F(0.773) - F(1.108)$　　　　　　　　$E_3^\circ = 0.984$ V

S9.　(a)　$Cu^{2+} + e^- \rightleftharpoons Cu^+$　　　　　　　　　　　　　$E_+^\circ = 0.161$ V

　　　　$\underline{^-Cu^+ + e^- \rightleftharpoons Cu(s)}$　　　　　　　　　　　　$\underline{E_-^\circ = 0.518 \text{ V}}$

　　　　$Cu(s) + Cu^{2+} \rightleftharpoons 2Cu^+$　　　　　　　　　$E^\circ = -0.357$ V

　　　　$\Delta G^\circ = -1FE^\circ = 34.4$ kJ　　　　$K = 10^{1E^\circ/0.059\,16} = 9.2 \times 10^{-7}$

　　(b)　$2F_2 + 4e^- \rightleftharpoons 4F^-$　　　　　　　　　　　　$E_+^\circ = 2.890$ V

　　　　$\underline{^-F_2O + 2H^+ + 4e^- \rightleftharpoons H_2O + 2F^-}$　　　　$\underline{E_-^\circ = 2.168 \text{ V}}$

　　　　$2F_2 + H_2O \rightleftharpoons F_2O + 2F^- + 2H^+$　　　$E^\circ = 0.722$ V

　　　　$\Delta G^\circ = -4FE^\circ = -279$ kJ　　　　$K = 10^{4E^\circ/0.059\,16} = 7 \times 10^{48}$

S10.　(a)　$2IO_3^- + I^- + 12H^+ + 10e^- \rightleftharpoons I_3^- + 6H_2O$　　$E_+^\circ = 1.210$ V

　　　　$\underline{^-5I_3^- + 10e^- \rightleftharpoons 15I^-}$　　　　　　　　　　$\underline{E_-^\circ = 0.535 \text{ V}}$

　　　　$2IO_3^- + 16I^- + 12H^+ \rightleftharpoons 6I_3^- + 6H_2O$　　　$E^\circ = 0.675$ V

　　(b)　$\Delta G^\circ = -10FE^\circ = -651$ kJ.　　$K = 10^{10E^\circ/0.059\,16} = 10^{114}$

　　(c)　$E = \left\{ 1.210 - \dfrac{0.059\,16}{10} \log \dfrac{[I_3^-]}{[IO_3^-]^2[I^-][H^+]^{12}} \right\} - \left\{ 0.535 - \dfrac{0.059\,16}{10} \log \dfrac{[I^-]^{15}}{[I_3^-]^5} \right\}$

　　　　$= 0.178$ V

　　(d)　$\Delta G = -10FE = -172$ kJ

　　(e)　At equilibrium, $E = 0 \Rightarrow E^\circ = \dfrac{0.059\,16}{10} \log \dfrac{[I_3^-]^6}{[IO_3^-]^2[I^-]^{16}[H^+]^{12}}$

　　　　$\Rightarrow [H^+] = 3.1 \times 10^{-9} \Rightarrow pH = 8.51$

S11.　$Mg(OH)_2(s) + 2e^- \rightleftharpoons Mg(s) + 2OH^-$　　　　　$E_+^\circ = -2.690$ V

　　　$\underline{^-Mg^{2+} + 2e^- \rightleftharpoons Mg(s)}$　　　　　　　　　　$\underline{E_-^\circ = -2.360 \text{ V}}$

　　　$Mg(OH)_2(s) \rightleftharpoons Mg^{2+} + 2OH^-$　　　　　　　$E^\circ = -0.330$ V

　　　$K_{sp} = 10^{2(-0.330)/0.059\,16} = 7 \times 10^{-12}$

S12. $Cl_2(aq) + 2e^- \rightleftarrows 2Cl^-$ $E_+^\circ = ?$

     $^-\underline{\ Cl_2(g) + 2e^- \rightleftarrows 2Cl^-\ \ \ \ \ \ \ \ \ \ \ \ \ \ \ \ \ }$ $\underline{E_-^\circ = 1.360\ V}$

       $Cl_2(aq) \rightleftarrows Cl_2(g)$ $E^\circ = E_+^\circ - 1.360$

But $E^\circ = -\Delta G^\circ / (nF) = -(-6.9 \times 10^3\ J\ mol^{-1}) / (2 \cdot 9.648\ 5 \times 10^4\ C/mol)$

$\Rightarrow E^\circ = +0.036\ V \Rightarrow E_+^\circ = E^\circ + 1.360 = 1.396\ V$

S13. (a) $Ag(S_2O_3)_2^{3-} + e^- \rightleftarrows Ag(s) + 2S_2O_3^{2-}$ $E_+^\circ = 0.017\ V$

      $^-\underline{\ H^+ + e^- \rightleftarrows \frac{1}{2}H_2(g)\ }$ $\underline{E_-^\circ = 0\ V}$

    $Ag(S_2O_3)_2^{3-} + \frac{1}{2}H_2(g) \rightleftarrows Ag(s) + 2S_2O_3^{2-} + H^+$ $E^\circ = 0.017\ V$

$$E = E^\circ - 0.059\ 16 \log \frac{[S_2O_3^{2-}]^2[H^+]}{[Ag(S_2O_3)_2^{3-}]\sqrt{P_{H_2}}}$$

$$E = 0.017 - 0.059\ 16 \log \frac{(0.050)^2(1)}{(0.010)\sqrt{1}} = 0.053\ V$$

    (b) $Ag^+ + e^- \rightleftarrows Ag(s)$ $E_+^\circ = 0.799\ V$

      $^-\underline{\ H^+ + e^- \rightleftarrows \frac{1}{2}H_2(g)\ \ \ \ \ \ \ \ \ \ \ }$ $\underline{E_-^\circ = 0\ V}$

    $Ag^- + \frac{1}{2}H_2(g) \rightleftarrows Ag(s) + H^+$ $E^\circ = 0.799\ V$

$$E = E^\circ - 0.059\ 16 \log \frac{[H^+]}{[Ag^+]\sqrt{P_{H_2}}}$$

$$0.053 = 0.799 - 0.059\ 16 \log \frac{(1)}{[Ag^+]\sqrt{1}} \Rightarrow [Ag^+] = 2._5 \times 10^{-13}\ M$$

    (c) $K_f = \dfrac{[Ag(S_2O_3)_2^{3-}]}{[Ag^+][S_2O_3^{2-}]^2} = \dfrac{0.010}{(2.5 \times 10^{-13})(0.050)^2} = 1._6 \times 10^{13}$

S14. $2Fe^{3+} + 2e^- \rightleftarrows 2Fe^{2+}$ $E_+^\circ = 0.771\ V$

     $^-\underline{\ UO_2^{2+} + 4H^+ + 2e^- \rightleftarrows U^{4+} + 2H_2O\ \ \ \ \ }$ $\underline{E_-^\circ = 0.273\ V}$

    $U^{4+} + 2Fe^{3+} + 2H_2O \rightleftarrows UO_2^{2+} + 2Fe^{2+} + 4H^+$ $E^\circ = 0.498\ V$

$$1.018 = 0.498 - \frac{0.059\ 16}{2} \log \frac{(0.050)(0.025)^2[H^+]^4}{(0.050)(0.050)^2} \Rightarrow [H^+] = 5.7 \times 10^{-5}\ M$$

$$K_a = \frac{[H^+][HCO_2^-]}{[HCO_2H]} = \frac{5.7 \times 10^{-5}\ (0.30)}{(0.10)} = 1.7 \times 10^{-4}$$

S15. (a) 25.0 mL of 0.124 M Na_3PO_4 plus 25.0 mL of 0.248 M $HClO_4$ give 50.0 mL of 0.062 0 M $H_2PO_4^-$, the first intermediate form of the triprotic acid, H_3PO_4.

$$[H^+] = \sqrt{\frac{K_1K_2(0.062\ 0) + K_1K_w}{K_1 + 0.062\ 0}} = 2.01 \times 10^{-5}\ M$$

$$[HPO_4^{2-}] = \frac{K_2[H_2PO_4^-]}{[H^+]} = \frac{(6.23 \times 10^{-8})(0.062\ 0)}{(2.01 \times 10^{-5})} = 1.95 \times 10^{-4}\ M$$

(b) In the left half-cell

$$[Pb^{2+}(left)] = \frac{K_{sp}\ for\ PbHPO_4}{[HPO_4^{2-}]} = \frac{2.0 \times 10^{-10}}{1.95 \times 10^{-4}} = 1.0 \times 10^{-6}\ M$$

The net cell reaction is $Pb^{2+}(left) \rightleftarrows Pb^{2+}(left)$, for which $E° = 0$ and

$$E = -\frac{0.059\ 16}{2} \log \frac{[Pb^{2+}(left)]}{[Pb^{2+}(right)]}$$

$$0.097 = -\frac{0.059\ 16}{2} \log \frac{1.0 \times 10^{-6}}{[Pb^{2+}(right)]} \quad \Rightarrow\ [Pb^{2+}(right)] = 1.9 \times 10^{-3}\ M$$

The right half-cell contains exactly twice as many moles of F^- as Pb^{2+}.

Therefore $PbF_2(s)$ precipitates and in the solution $[F^-] = 2[Pb^{2+}]$.

$$K_{sp} = [Pb^{2+}][F^-]^2 = (1.9 \times 10^{-3})(3.8 \times 10^{-3})^2 = 2.7 \times 10^{-8}$$

S16.
$$
\begin{array}{ll}
3[Cu^{2+} + 2e^- \rightleftarrows Cu(s)] & E_+° = 0.339\ V \\
- \underline{Sb_2O_3(s) + 6H^+ + 6e^- \rightleftarrows 2Sb(s) + 3H_2O} & \underline{E_-° = 0.147\ V} \\
3Cu^{2+} + 2Sb(s) + 3H_2O \rightleftarrows 3Cu(s) + Sb_2O_3(s) + 6H^+ & E° = 0.192\ V
\end{array}
$$

First find $[H^+]$ in left half-cell:

$$H_3A \xrightarrow{pK_2} H_2A^- \xrightarrow{pK_3} HA^{2-} \xrightarrow{pK_4} A^{3-}$$
1.52 mmol

Addition of 3.50 mmol of NaOH creates 0.46 mmol of A^{3-} plus 1.06 mmol of HA^{2-}

$$pH = pK_4 + \log \frac{[A^{3-}]}{[HA^{2-}]} = 10.334 + \log \frac{0.46}{1.06} = 9.97$$

$$E = \left\{ 0.339 - \frac{0.059\ 16}{6} \log \frac{1}{[Cu^{2+}(right)]^3} \right\} -$$

$$\left\{ 0.147 - \frac{0.059\ 16}{6} \log \frac{1}{[H^+(left)]^6} \right\}$$

Inserting $E = 0.418\ V$ and $[H^+(left)] = 10^{-9.97}$ gives $[Cu^{2+}(right)] = 5._{02} \times 10^{-13}\ M$

Now find $[Y^{4-}]$ in right half-cell:

	Ca^{2+}	+	EDTA	$\rightleftarrows$	CaY^{2-}
initial mmol	2.38		2.22		—
final mmol	0.16		—		2.22

$$K_f\ (for\ CaY^{2-}) = \frac{[CaY^{2-}]}{[Ca^{2+}][Y^{4-}]}$$

$$4.9 \times 10^{10} = \frac{(2.22\ mmol/150\ mL)}{(0.16\ mmol/150\ mL)[Y^{4-}]} \quad \Rightarrow\ [Y^{4-}] = 2.8_3 \times 10^{-10}\ M$$

$$K_f\ (for\ CuY^{2-}) = \frac{[CuY^{2-}]}{[Cu^{2+}][Y^{4-}]} = \frac{(0.001\ 37)(100/150)}{(5._{02} \times 10^{-13})(2.8_3 \times 10^{-10})} = 6 \times 10^{18}$$

S17. $E = -0.103\,8 - \dfrac{0.059\,16}{2} \log \dfrac{P_{CO}}{P_{CO_2}[H^+]^2}$

$E = -0.103\,8 + 0.059\,16 \log [H^+] - \dfrac{0.059\,16}{2} \log \dfrac{P_{CO}}{P_{CO_2}}$

$\underbrace{\hphantom{E = -0.103\,8 + 0.059\,16 \log [H^+]}}$
This is $E^{\circ\prime}$ when pH = 7

$E^{\circ\prime} = -0.103\,8 + 0.059\,16 \log (10^{-7.00}) = -0.517\,9\ V$

S18. $E = E^\circ - \dfrac{0.059\,16}{2} \log \dfrac{[HCO_2H]}{P_{CO_2}\,[H^+]^2}$

Substituting $[HCO_2H] = \dfrac{[H^+]\,F_{HCO_2H}}{[H^+] + K_a}$ into the Nernst equation gives

$E = -0.114 - \dfrac{0.059\,16}{2} \log \dfrac{[H^+]\,F_{HCO_2H}}{([H^+] + K_a)\,P_{CO_2}\,[H^+]^2}$

$E = -0.114 + \dfrac{0.059\,16}{2} \log \{ ([H^+] + K_a)\,[H^+] \} - \dfrac{0.059\,16}{2} \log \dfrac{F_{HCO_2H}}{P_{CO_2}}$

$\underbrace{\hphantom{E = -0.114 + \dfrac{0.059\,16}{2} \log \{ ([H^+] + K_a)\,[H^+] \}}}$
This is $E^{\circ\prime}$ when pH = 7

Inserting $K_a = 1.8 \times 10^{-4}$ for HCO_2H and $[H^+] = 10^{-7.00}$ gives

$E^{\circ\prime} = -0.114 + \dfrac{0.059\,16}{2} \log \{ (10^{-7.00} + 1.8 \times 10^{-4})(10^{-7.00}) \} = -0.432\ V$

S19. Calling anthraquinone-2,6-disulfonate A, the reaction may be written

$A + 2H^+ + 2e^- \rightleftarrows H_2A.$

$E = 0.229 - \dfrac{0.059\,16}{2} \log \dfrac{[H_2A]}{[A][H^+]^2}$

But $[H_2A] = [H^+]^2 F_{H_2A} / ([H^+]^2 + [H^+]K_1 + K_1K_2)$ and $[A] = F_A$.

Making these substitutions and rearranging the resulting Nernst equation gives

$E = 0.229 - \dfrac{0.059\,16}{2} \log \dfrac{1}{[H^+]^2 + [H^+]K_1 + K_1K_2} - \dfrac{0.059\,16}{2} \log \dfrac{F_{H_2A}}{F_A}$

$\underbrace{\hphantom{E = 0.229 - \dfrac{0.059\,16}{2} \log \dfrac{1}{[H^+]^2 + [H^+]K_1 + K_1K_2}}}$
$E^{\circ\prime} = -0.184\ V$

CHAPTER 15: SUPPLEMENTARY SOLUTIONS
ELECTRODES AND POTENTIOMETRY

S1.   (a) -0.419 V (b) 0.720 V (c) -0.479 V (d) -0.009 V (e) 0.009 V

S2. $Br_2(aq) + 2e- \rightleftarrows 2Br^-$ $E_+^\circ = 1.098$ V

$E_+ = E_+^\circ - \dfrac{0.059\ 16}{2} \log \dfrac{[Br^-]^2}{[Br_2(aq)]} = 1.098 - \dfrac{0.059\ 16}{2} \log \dfrac{[0.234]^2}{[0.002\ 17]} = 1.057$ V

$E = E_+ - E_- = 1.057 - 0.241 = 0.816$ V

S3. $V_e = 25.0$ mL. $Ag^+ + e^- \rightleftarrows Ag(s) \Rightarrow E_+ = 0.799 - 0.059\ 16 \log \dfrac{1}{[Ag^+]}$

0.1 mL: $[SCN^-] = \underbrace{\left(\dfrac{24.9}{25.0}\right)}_{\substack{\text{Fraction} \\ \text{remaining}}} \underbrace{(0.100\ \text{M})}_{\substack{\text{Original} \\ \text{concentration}}} \underbrace{\left(\dfrac{50.0}{50.1}\right)}_{\substack{\text{Dilution} \\ \text{factor}}} = 0.099\ 4$ M

$[Ag^+] = K_{sp}/[SCN^-] = (1.1 \times 10^{-12})/0.099\ 4 = 1.1 \times 10^{-11}$ M

$E = E_+ - E_- = \left\{ 0.799 - 0.059\ 16 \log \dfrac{1}{1.1 \times 10^{-11}} \right\} - 0.241 = -0.090$ V

10.0 mL: $[SCN^-] = \left(\dfrac{15.0}{25.0}\right)(0.100\ \text{M})\left(\dfrac{50.0}{60.0}\right) = 0.050\ 0$ M

$[Ag^+] = K_{sp}/[SCN^-] = (1.1 \times 10^{-12})/0.050\ 0 = 2.2 \times 10^{-11}$ M

$E = E_+ - E_- = \left\{ 0.799 - 0.059\ 16 \log \dfrac{1}{2.2 \times 10^{-11}} \right\} - 0.241 = -0.073$ V

25.0 mL: $[Ag^+] = [SCN^-] \Rightarrow [Ag^+] = \sqrt{K_{sp}} = \sqrt{1.1 \times 10^{-12}} = 1.0_5 \times 10^{-6}$ M

$E = E_+ - E_- = \left\{ 0.799 - 0.059\ 16 \log \dfrac{1}{1.0_5 \times 10^{-6}} \right\} - 0.241 = 0.204$ V

30.0 mL: This is 5.0 mL past $V_e \Rightarrow [Ag^+] = \left(\dfrac{5.0}{80.0}\right)(0.200\ \text{M}) = 0.012\ 5$ M

$E = E_+ - E_- = \left\{ 0.799 - 0.059\ 16 \log \dfrac{1}{0.012\ 5} \right\} - 0.241 = 0.445$ V

S4.   (a) $Ag^+(right) + e^- \rightleftarrows Ag$ $E_+^\circ = 0.799$ V

$^-\ \underline{Ag^+(left) + e^- \rightleftarrows Ag}$ $\underline{E_-^\circ = 0.799\ \text{V}}$

$Ag^+(right) \rightleftarrows Ag^+(left)$ $E^\circ = 0$

$E = -0.059\ 16 \log \dfrac{[Ag^+(left)]}{[Ag^+(right)]}$

(b) NH_3: $V_e = 9.8$ mL $\Rightarrow NH_3/Ag^+ = 1.99 \Rightarrow 2NH_3 + Ag^+ \rightleftarrows (NH_3)_2Ag^+$

$H_2NCH_2CH_2NH_2$: $V_e = 5.0$ mL $\Rightarrow H_2NCH_2CH_2NH_2/Ag^+ = 1.15$

$\Rightarrow H_2NCH_2CH_2NH_2 + Ag^+ \rightleftarrows (H_2NCH_2CH_2NH_2)Ag^+$

(c) At the equivalence point, $E = 0.15$ V and $[Ag^+(left)] = 0.100$ M.

$$0.15 = -0.059\ 16 \log \frac{0.100}{[Ag^+(\text{right})]} \Rightarrow [Ag^+(\text{right})] \approx 0.03\ M$$

S5. right half-reaction : $Zn^{2+} + 2e^- \rightleftarrows Zn(s)$ $E° = -0.762\ V$

$$E = \left\{-0.762 - \frac{0.059\ 16}{2} \log \frac{1}{[Zn^{2+}]} \right\} - 0.241$$

MgY^{2-} : $K_f = 6.2 \times 10^8$; ZnY^{2-} : $K_f = 3.2 \times 10^{16}$.

When 10 mL of EDTA has been added, $[Mg^{2+}] = (4/5)(0.100)(50/60) = 0.066\ 7\ M$ and $[MgY^{2-}] = (1/5)(0.100)(50/60) = 0.016\ 7\ M$.

But $\dfrac{[MgY^{2-}]}{[Mg^{2+}][Y^{4-}]} = 6.2 \times 10^8 \Rightarrow [Y^{4-}] = 4.04 \times 10^{-10}\ M$.

The concentration of $[ZnY^{2-}]$ is $(50/60)(1.00 \times 10^{-5}) = 8.33 \times 10^{-6}\ M$.

Because $\dfrac{[ZnY^{2-}]}{[Zn^{2+}][Y^{4-}]} = 3.2 \times 10^{16}$ and we know $[Y^{4-}]$ and $[ZnY^{2-}]$, we can

compute that $[Zn^{2+}] = 6.44 \times 10^{-13}\ M$.

$$E = \left\{-0.762 - \frac{0.059\ 16}{2} \log \frac{1}{6.44 \times 10^{-13}} \right\} - 0.241 = -1.364\ V$$

S6. Relative mobilities:

$H^+ \rightarrow 36.30$	$Cl^- \rightarrow 7.91$
$4.01 \leftarrow Li^+$	$8.13 \leftarrow Br^-$

Anion diffusion from each side is nearly equal, while H^+ diffuses much faster than Li^+. Positive charge builds on the right, so the <u>left</u> side will be negative.

S7. Velocity = mobility × field = $(20.50 \times 10^{-8}\ m^2/(s{\cdot}V)) \times (1000\ V/m) = 2.05 \times 10^{-4}\ m\ s^{-1}$ for OH^- and $(5.70 \times 10^{-8})(1000) = 5.70 \times 10^{-5}\ m\ s^{-1}$ for F^-. To cover 0.010 0 m will require $(0.010\ 0\ m)/(2.05 \times 10^{-4}\ m\ s^{-1}) = 48.8\ s$ for OH^- and $(0.010\ 0)/(5.70 \times 10^{-5}) = 175\ s$ for NO_3^-.

S8. (a) $E = \text{constant} + \beta \dfrac{0.059\ 16}{3} \log \mathcal{A}_{La^{3+}}(\text{outside})$.

(b) If the activity increases by a factor of 10, the potential will increase by $0.059\ 16/3 = 19.7$ mV.

(c) $E_1 = \text{constant} + (1.00)\dfrac{0.059\ 16}{3} \log (2.36 \times 10^{-4})$

$E_2 = \text{constant} + (1.00)\dfrac{0.059\ 16}{3} \log (4.44 \times 10^{-3})$

$\Delta E = E_2 - E_1 = \dfrac{0.059\ 16}{3} \log \dfrac{4.44 \times 10^{-3}}{2.36 \times 10^{-4}} = +25.1$ mV

(d) $E = \text{constant} + \dfrac{0.059\ 16}{3} \log \left\{ [La^{3+}] + \dfrac{1}{1200}[Fe^{3+}]^{1/1} \right\}$

$$0.100 = \text{constant} + \frac{0.059\ 16}{3} \log (1.00 \times 10^{-4}) \Rightarrow \text{constant} = 0.178_9 \text{ mV}$$

$$E = 0.178_9 + \frac{0.059\ 16}{3} \log \left\{ (1.00 \times 10^{-4}) + \frac{1}{1200} (0.010) \right\} = +100.7 \text{ mV}$$

S9. $\dfrac{[ML]}{[M][L]} = 3.6 \times 10^{10} = \dfrac{0.050}{[M](0.50)} \Rightarrow [M] = 2.8 \times 10^{-12} \text{ M}$

S10. For the pure $[Ca^{2+}]$ solution we can write

$$E = \text{constant} + \frac{(0.970)(0.059\ 16)}{2} \log (1.00 \times 10^{-3})$$

Putting in $E = 0.300\ 0$ V gives constant $= 0.386\ 1$ V. For the solution
containing the interfering ions we can say that

$$E = 0.386\ 1 + \frac{(0.970)(0.059\ 16)}{2} \log \left\{ (1.00 \times 10^{-3}) + (0.040)(1.00 \times 10^{-3}) \right.$$

$$+ (0.021)(1.00 \times 10^{-3}) + (0.081)(5.00 \times 10^{-4})$$

$$\left. + (6.6 \times 10^{-5}) (0.100)^2 + (1.7 \times 10^{-4})(0.500)^2 \right\} = 0.301\ 7 \text{ V}$$

At equal concentrations, Zn^{2+} interferes the most.

S11. $0.194\ 6 = \text{constant} + \dfrac{0.057\ 1}{2} \log \left\{ [M^{2+}]_o + 0.002\ 0\ (0.200)^2 \right\}$

After making the standard addition, the concentration of M^{2+} is

$$\frac{100.0\ [M^{2+}]_o + 1.00\ (1.07 \times 10^{-2})}{101.0} = (0.990_1\ [M^{2+}]_o + 1.05_9 \times 10^{-4}) \text{ M}$$

$$0.200\ 7 = \text{constant} + \frac{0.057\ 1}{2} \log \left\{ (0.990_1\ [M^{2+}]_o + 1.05_9 \times 10^{-4}) + 0.002\ 0\ (0.200)^2 \right\}$$

$$= \text{constant} + \frac{0.057\ 1}{2} \log \left\{ 0.990_1\ [M^{2+}]_o + 1.85_9 \times 10^{-4} \right\}$$

$$\Delta E = (0.200\ 7 - 0.194\ 6) = \frac{0.057\ 1}{2} \log \frac{(0.990_1\ [M^{2+}]_o + 1.85_9 \times 10^{-4})}{[M^{2+}]_o + 8.00 \times 10^{-5}}$$

$$\Rightarrow [M^{2+}]_o = 8.5_3 \times 10^{-5} \text{ M}$$

S12. For $Mg(EDTA)^{2-}$, $K_f = 6.2 \times 10^8$. This solution contains 30% more EDTA than
Mg^{2+}. Therefore $[EDTA]/[Mg(EDTA)^{2-}] = 0.30$

$$\frac{[Mg(EDTA)^{2-}]}{[Mg^{2+}][EDTA]} = \alpha_{Y^{4-}} K_f = 0.36\ K_f$$

$$[Mg^{2+}] = \frac{[Mg(EDTA)^{2-}]}{[EDTA]\alpha_{Y^{4-}} K_f} = \frac{1}{(0.30)(0.36)K_f} = 1.49 \times 10^{-8} \text{ M}$$

For $Mn(EDTA)^{2-}$, $K_f = 7.4 \times 10^{13}$ and $[Mn^{2+}] = 1.25 \times 10^{-13}$ M

S1.   (a) $2Cu^{2+} + In^+ \rightarrow 2Cu^+ + In^{3+}$

(b) $Cu^{2+} + e^- \rightleftarrows Cu^+ \qquad E° = 0.161$ V

   $In^{3+} + 2e^- \rightleftarrows In^+ \qquad E° = -0.444$ V

(c) $E = \left\{ 0.161 - 0.059\ 16 \log \dfrac{[Cu^+]}{[Cu^{2+}]} \right\} - \left\{ 0.241 \right\}$ \hfill (A)

   $E = \left\{ -0.444 - \dfrac{0.059\ 16}{2} \log \dfrac{[In^+]}{[In^{3+}]} \right\} - \left\{ 0.241 \right\}$ \hfill (B)

(d) 1.00 mL :   Use Equation B with $[In^+]/[In^{3+}] = 39.0/1.00$, since
   $V_e = 40.0$ mL $\Rightarrow E = -0.732$ V

   20.0 mL : $[In^+]/[In^{3+}] = 20.0/20.0 \Rightarrow E = -0.685$ V

   39.0 mL : $[In^+]/[In^{3+}] = 1.0/39.0 \Rightarrow E = -0.638$ V

   40.0 mL :   This is V_e, where $[Cu^+] = 2[In^{3+}]$ and $[Cu^{2+}] = 2[In^+]$.
   Adding the two Nernst equations gives

   $E_+ = 0.161 - 0.059\ 16 \log \dfrac{[Cu^+]}{[Cu^{2+}]}$

   $2E_+ = -0.888 - 0.059\ 16 \log \dfrac{[In^+]}{[In^{3+}]}$

   _____

   $3E_+ = -0.727 - 0.059\ 16 \log \dfrac{[In^+][Cu^+]}{[In^{3+}][Cu^{2+}]}$

   $3E_+ = -0.727 - 0.059\ 16 \log \dfrac{[In^+]2[In^{3+}]}{[In^{3+}]2[In^+]}$

   $\Rightarrow E_+ = -0.242$ V and $E = -0.483$ V.

   41.0 mL :   Use Equation A with $[Cu^+]/[Cu^{2+}] = 40.0/1.0 \Rightarrow E = -0.175$ V

   80.0 mL : $[Cu^+]/[Cu^{2+}] = 40.0/40.0 \Rightarrow E = -0.080$ V

S2.   (a) $Fe^{3+} + Cu^+ \rightarrow Fe^{2+} + Cu^{2+}$

(b) $Fe^{3+} + e^- \rightleftarrows Fe^{2+} \qquad E° = 0.771$ V

   $Cu^{2+} + e^- \rightleftarrows Cu^+ \qquad E° = 0.161$ V

(c) $E = \left\{ 0.771 - 0.059\ 16 \log \dfrac{[Fe^{2+}]}{[Fe^{3+}]} \right\} - \left\{ 0.197 \right\}$ \hfill (A)

   $E = \left\{ 0.161 - 0.059\ 16 \log \dfrac{[Cu^+]}{[Cu^{2+}]} \right\} - \left\{ 0.197 \right\}$ \hfill (B)

(d) 1.0 mL :   Use Equation A with $[Fe^{2+}]/[Fe^{3+}] = 1.0/24.0$, since
   $V_e = 25.0$ mL $\Rightarrow E = 0.656$ V

   12.5 mL : $[Fe^{2+}]/[Fe^{3+}] = 12.5/12.5 \Rightarrow E = 0.574$ V

   24.5 mL : $[Fe^{2+}]/[Fe^{3+}] = 24.5/0.5 \Rightarrow E = 0.474$ V

25.0 mL :   This is V_e, where $[Cu^{2+}] = [Fe^{2+}]$ and $[Cu^+] = [Fe^3]$.

Adding the two Nernst equations gives

$$E_+ = 0.771 - 0.059\ 16 \log \frac{[Fe^{2+}]}{[Fe^{3+}]}$$

$$E_+ = 0.161 - 0.059\ 16 \log \frac{[Cu^+]}{[Cu^{2+}]}$$

$$2E_+ = 0.932 - 0.059\ 16 \log \frac{[Fe^{2+}][Cu^+]}{[Fe^{3+}][Cu^{2+}]}$$

$$2E_+ = 0.932 - 0.059\ 16 \log \frac{[Fe^{2+}][Fe^{3+}]}{[Fe^{3+}][Fe^{2+}]}$$

$$\Rightarrow E_+ = 0.466\ \text{V and } E = 0.269\ \text{V}.$$

25.5 mL :   Use Equation B with $[Cu^+]/[Cu^{2+}] = 0.5/25.0 \Rightarrow E = 0.065\ \text{V}$

30.0 mL : $[Cu^+]/[Cu^{2+}] = 5.0/25.0 \Rightarrow E = 0.005\ \text{V}$

S3.   (a)   Titration reaction: $Cr^{2+} + Fe^{3+} \rightarrow Fe^{2+} + Cr^{3+}$

Titrant: $Cr^{3+} + e^- \rightleftarrows Cr^{2+}$ $E^\circ = -0.42\ \text{V}$

Analyte: $Fe^{3+} + e^- \rightleftarrows Fe^{2+}$ $E^\circ = 0.767\ \text{V}$

$\tau = 10^{(-0.42 - E)/0.059\ 16}$ $\alpha = 10^{(0.767 - E)/0.059\ 16}$

$\phi = \dfrac{\alpha\,(1 + \tau)}{(1 + \alpha)}$ $V_e = 50.0\ \text{mL}$

	A	B	C	D	E	F	G
1	E°(T) =	E (vs S.H.E.)	Tau	Alpha	Phi	E(vs AgAgCl)	Volume (mL)
2	-0.42	0.885	8.73E-23	1.01E-02	0.01002	0.688	0.501
3	E°(A) =	0.767	8.63E-21	1.00E+00	0.50000	0.570	25.000
4	0.767	0.649	8.52E-19	9.88E+01	0.98998	0.452	49.499
5	Nernst =	0.590	8.47E-18	9.81E+02	0.99898	0.393	49.949
6	0.05916	0.174	9.11E-11	1.06E+10	1.00000	-0.023	50.000
7	Ve =	-0.302	1.01E-02	1.17E+18	1.01013	-0.499	50.506
8	50	-0.361	1.01E-01	1.17E+19	1.10062	-0.558	55.031
9		-0.420	1.00E+00	1.16E+20	2.00000	-0.617	100.000
10							
11	C2 = 10^(($A$2-B2)/$A$6)			E2 = D2*(1+C2)/(1+D2)			
12	D2 = 10^(($A$4-B2)/$A$6)			F2 = B2-0.197		G2 = $A$8*E2	

(b)   Titration reaction: $5Fe^{2+} + MnO_4^- + 8H^+ \rightarrow 5Fe^{3+} + Mn^{2+} + 4H_2O$

Titrant: $Fe^{3+} + e^- \rightleftarrows Fe^{2+}$ $E^\circ = 0.68\ \text{V}$

Analyte: $MnO_4^- + 8H^+ + 5e^- \rightarrow Mn^{2+} + 4H_2O$ $E^\circ = 1.507\ \text{V}$

$\tau = 10^{(0.68 - E)/0.059\ 16}$ $\alpha = 10^{\{[5(1.507 - E)/0.059\ 16] - 8\ pH\}}$

$\phi = \dfrac{\alpha\,(1 + \tau)}{(1 + \alpha)}$ $V_e = 50.0\ \text{mL}$

	A	B	C	D	E	F	G
1	E°(T) =	E (vs S.H.E.)	Tau	Alpha	Phi	E(vs AglAgCl)	Volume (mL)
2	0.68	1.559	1.39E-15	1.01E-02	0.01002	1.362	0.501
3	E°(A) =	1.535	3.53E-15	1.08E+00	0.51930	1.338	25.965
4	1.507	1.512	8.64E-15	9.49E+01	0.98958	1.315	49.479
5	Nernst =	1.499	1.43E-14	1.19E+03	0.99916	1.302	49.958
6	0.05916	1.390	9.97E-13	1.94E+12	1.00000	1.193	50.000
7	Ve =	0.798	1.01E-02	2.10E+62	1.01013	0.601	50.506
8	50	0.739	1.01E-01	2.04E+67	1.10062	0.542	55.031
9	pH =	0.680	1.00E+00	1.97E+72	2.00000	0.483	100.000
	-0.3						
	C2 = 10^(($A$2-B2)/$A$6)			F2 = B2-0.197		G2 = $A$8*E2	
10	D2 = 10^(5*($A$4-B2)/$A$6-8*$A$10)			E2 = D2*(1+C2)/(1+D2)			

(c)   Titration reaction: dehydro. + $2Fe^{2+}$ + $2H^+$ $\rightarrow$ ascorbic acid + $2Fe^{3+}$ + H_2O

Titrant: dehydro. + $2H^+$ + $2e^-$ $\rightarrow$ ascorbic acid + H_2O E° = 0.390 V

Analyte: Fe^{3+} + e^- $\rightleftarrows$ Fe^{2+} E° = 0.732 V

$$\tau = 10^{\{[2(0.390 - E)/0.059\,16] - 2\,pH\}} \qquad \alpha = 10^{(0.732 - E)/0.059\,16}$$

$$\phi = \frac{\alpha\,(1 + \tau)}{(1 + \alpha)} \qquad\qquad V_e = 30.0_6 \text{ mL}$$

	A	B	C	D	E	F	G
1	E°(T) =	E (vs S.H.E.)	Tau	Alpha	Phi	E(vs AglAgCl)	Volume (mL)
2	0.39	0.850	2.81E-16	1.01E-02	0.01002	0.653	0.301
3	E°(A) =	0.732	2.74E-12	1.00E+00	0.50000	0.535	15.030
4	0.732	0.614	2.67E-08	9.88E+01	0.98998	0.417	29.759
5	Nernst =	0.554	2.86E-06	1.02E+03	0.99902	0.357	30.031
6	0.05916	0.504	1.40E-04	7.14E+03	1.00000	0.307	30.060
7	Ve =	0.449	1.01E-02	6.08E+04	1.01011	0.252	30.364
8	30.06	0.420	9.68E-02	1.88E+05	1.09678	0.223	32.969
9	pH =	0.390	1.00E+00	6.04E+05	2.00000	0.149	60.120
	0						
	C2 = 10^(2*($A$2-B2)/$A$6-2*$A$10)			E2 = D2*(1+C2)/(1+D2)			
	D2 = 10^(($A$4-B2)/$A$6)			F2 = B2-0.197		G2 = $A$8*E2	

(d)   Titration reaction: UO_2^{2+} + Sn^{2+} + $4H^+$ $\rightarrow$ U^{4+} + Sn^{4+} + $2H_2O$

Titrant: UO_2^{2+} + $4H^+$ + $2e^-$ $\rightarrow$ U^{4+} + $2H_2O$ E° = 0.273 V

Analyte: Sn^{4+} + $2e^-$ $\rightleftarrows$ Sn^{2+} E° = 0.139 V

$$\tau = 10^{\{2(0.273 - E)/0.059\,16 - 4\,pH\}} \qquad \alpha = 10^{[2(0.139 - E)/0.059\,16]}$$

$$\phi = \frac{(1 + \tau)}{\tau\,(1 + \alpha)} \qquad\qquad V_e = 25.0 \text{ mL}$$

	A	B	C	D	E	F	G
1	E°(T) =	E (vs S.H.E.)	Tau	Alpha	Phi	E(vs Ag|AgCl)	Volume (mL)
2	0.273	0.080	3.35E+06	9.88E+01	0.01002	-0.117	0.251
3	E°(A) =	0.139	3.39E+04	1.00E+00	0.50001	-0.058	12.500
4	0.139	0.195	4.33E+02	1.28E-02	0.98965	-0.002	24.741
5	Nernst =	0.205	1.99E+02	5.87E-03	0.99916	0.008	24.979
6	0.05916	0.206	1.84E+02	5.43E-03	1.00000	0.009	25.000
7	Ve =	0.217	7.82E+01	2.31E-03	1.01046	0.020	25.261
8	25	0.243	1.03E+01	3.05E-04	1.09645	0.046	27.411
9	pH =	0.273	1.00E+00	2.95E-05	1.99994	0.076	49.999
10	0						
11	C2 = 10^(2*($A$2-B2)/$A$6-4*$A$10)				E2 = (1+C2)/(C2*(1+D2))		
12	D2 = 10^(2*($A$4-B2)/$A$6)			F2 = B2-0.197		G2 = $A$8*E2	

S4.

	A	B	C	D	E	F	G
1	E°(T) =	E (vs S.H.E.)	Tau	Alpha	Phi	E(vs Ag|AgCl)	Volume (mL)
2	1.36	0.653	5.06E+57	9.88E+01	0.01002	0.456	0.100
3	E°(A) =	0.735	2.44E+49	4.06E+00	0.19763	0.538	1.976
4	0.771	0.761	5.63E+46	1.48E+00	0.40391	0.564	4.039
5	Nernst =	0.781	5.27E+44	6.78E-01	0.59609	0.584	5.961
6	0.05916	0.807	1.22E+42	2.46E-01	0.80237	0.610	8.024
7	Ve =	0.827	1.14E+40	1.13E-01	0.89840	0.630	8.984
8	10	0.889	5.87E+33	1.01E-02	0.98998	0.692	9.900
9	pH =	1.177	3.63E+04	1.37E-07	1.00000	0.980	10.000
10	1	1.224	6.21E-01	2.20E-08	1.00990	1.027	10.099
11	V(Fe) =	1.234	6.01E-02	1.49E-08	1.10159	1.037	11.016
12	120	1.237	2.98E-02	1.33E-08	1.20313	1.040	12.031
13	[Cr]o =						
14	0.04		C2 = 10^(6*($A$2-B2)/$A$6-14*$A$10)				
15			D2 = 10^(($A$4-B2)/$A$6)				
16			E2 = (0.5+Sqrt(0.25+2*E2*$A$8*$A$14/				
17			(C2*E2*$A$8+C2*$A$12)))/(1+D2)				
18			F2 = B2-0.197				
19			G2 = $A$8*E2				

S5.

	A	B	C	D	E	F	G	H
1	E°(T) =	E(S.HE)	Tau	Alpha1	Alpha2	Phi	E(S.C.E)	Volume
2	1.24	0.126	2.1E+75	2.8E+00	5.9E+21	0.27	-0.115	19.995
3	E°(A1) =	0.139	2.8E+74	1.0E+00	2.1E+21	0.50	-0.102	37.500
4	0.139	0.194	5.3E+70	1.4E-02	3.0E+19	0.99	-0.047	73.977
5	E°(A2) =	0.462	4.0E+52	1.2E-11	2.6E+10	1.00	0.221	75.000
6	0.77	0.729	3.6E+34	1.1E-20	2.4E+01	1.01	0.488	75.987
7	Nernst =	0.770	6.0E+31	4.7E-22	1.0E+00	1.17	0.529	87.500
8	0.05916	1.077	1.0E+11	1.9E-32	4.2E-11	1.33	0.836	100.000
9	Ve =	1.226	8.8E+00	1.8E-37	3.8E-16	1.48	0.985	111.309
10	75							
11	pCl =							
12	0	C2 = 10^(4*($A$2-B2)/$A$8-2*$A$12-6*$A$14)						
13	pH =							
14	0	D2 = 10^(2*($A$4-B2)/$A$8)						
15	Tl(total)=	E2 = 10^(2*($A$6-B2)/$A$8)						
16	0.005	F2 = ((1+C2)/C2)*((1/(1+D2))+(($A$16/$A$18)/(1+E2)))						
17	Sn(total)=	G2 = B2-0.241						
18	0.015	H2 = $A$10*F2						

S6. $3MnO_4^- + 5Mo^{3+} + 4H^+ \rightarrow 3Mn^{2+} + 5MoO_2^{2+} + 2H_2O$

$(22.11 - 0.07) = 22.04$ mL of 0.012 34 M $KMnO_4$ = 0.272 0 mmol of MnO_4^-

which will react with $(5/3)(0.272\ 0) = 0.453\ 3$ mmol of Mo^{3+}.

$[Mo^{3+}] = 0.453\ 3$ mmol/50.00 mL = 9.066 mM.

S7. $2MnO_4^- + 5H_2C_2O_4 + 6H^+ \rightarrow 2Mn^{2+} + 10CO_2 + 8H_2O$

12.34 mL of 0.004 321 M $KMnO_4$ = 0.053 32 mmol of MnO_4^- which reacts with

$(5/2)(0.053\ 32) = 0.133\ 3$ mmol of $H_2C_2O_4$ which came from $(2/3)(0.133\ 3) =$

0.088 87 mmol of La^{3+}. $[La^{3+}] = 0.088\ 87$ mmol/25.00 mL = 3.555 mM.

S8.

$$C_3H_8O_3 \quad + \quad 3H_2O \quad \rightleftarrows \quad 3HCO_2H \quad + \quad 8e^- \quad + \quad 8H^+$$

glycerol   formic acid
(average oxidation   (oxidation number
number of C = -2/3)   of C = +2)

$$\underline{8Ce^{4+} + 8e^- \rightleftarrows 8Ce^{3+}}$$

$$C_3H_8O_3 + 8Ce^{4+} + 3H_2O \rightleftarrows 3HCO_2H = 8Ce^{3+} + 8H^+$$

One mole of glycerol requires eight moles of Ce^{4+}.

50.0 mL of 0.089 9 M Ce^{4+} = 4.495 mmol

$\underline{10.05\text{ mL of }0.043\ 7\text{ M }Fe^{2+} = 0.439\text{ mmol}}$

Ce^{4+} reacting with glycerol = 4.056 mmol

glycerol = (1/8) (4.056) = 0.507_0 mmol = 46.7 mg $\Rightarrow$ original solution = 30.5

wt % glycerol

S9.

$C_6H_5{-}NO + 4H^+ + 4e^- \rightleftharpoons C_6H_5{-}NH_2 + H_2O$

$C_6H_5{-}NO_2 + 6H^+ + 6e^- \rightleftharpoons C_6H_5{-}NH_2 + 2H_2O$

One mole of nitrosobenzene reacts with 4 moles of Cr^{2+} and one mole of nitrobenzene reacts with 6 moles of Cr^{2+}. Let x = mg of C_6H_5NO and 24.43 - x = mg of $C_6H_5NO_2$. We can write

$$\frac{4x}{107.112} + \frac{6(24.43 - x)}{123.111} = \text{mmol of } Cr^{2+} = (21.57)(0.050\,00)$$

$$x = 9.84 \text{ mg} = 40.3\%$$

S10. 0.139 2 g of $KBrO_3$ = 0.833 5 mmol of BrO_3^- which generates 3(0.833 5) = 2.501 mmol of Br_2 according to the reaction

$$BrO_3^- + 5Br^- + 6H^+ \rightarrow 3Br_2 + 3H_2O$$

One mole of excess Br_2 generates one mole of I_2 by the reaction

$$Br_2 + 2I^- \rightarrow 2Br^- + I_2$$

The I_2 requires 19.18 mL of 0.050 00 M $S_2O_3^{2-}$ = 0.959 0 mmol of $S_2O_3^{2-}$. Therefore there must have been (1/2)(0.959 0) = 0.479 5 mmol of I_2 = 0.479 5 mmol of excess Br_2. The Br_2 that reacted with acetoacetanilide must have been 2.501 - 0.479 5 = 2.021 mmol of Br_2. Since one mole of Br_2 reacts with one mole of acetoacetanilide there must have been 2.021 mmol of acetoacetanilide, or (1/2)(2.021) = 1.010 mmol of Be^{2+} = 9.107 mg of Be.

S11. Denote the average oxidation number of Bi as 3+b and the average oxidation number of Cu as 2+c.

$$Bi_2^{3+b}\ Sr_2^{2+}\ Ca_{0.8}^{2+}\ Y_{0.2}^{3+}\ Cu_2^{2+c}\ O_x$$

Positive charge = 6 + 2b +4 + 1.6 + 0.6 + 4 + 2c = 16.2 + 2b + 2c

The charge must be balanced by $O^{2-} \Rightarrow x = 8.1 + b + c$

The formula weight of the superconductor is 760.37 + 15.999 4(8.1+b+c).

One gram contains 1/[770.14 + 15.999 4(8.1+b+c)] moles.

Experiment A: Initial Cu^+ = 0.100 0 mmol; final Cu^+ = 0.052 2 mmol. Therefore 110.6 mg of superconductor consumed 0.047 8 mmol Cu^+.

2×mmol Bi^{5+} + mmol Cu^{3+} in 110.6 mg superconductor = 0.047 8.

Experiment B: Initial Fe^{2+} = 0.050 0 mmol; final Fe^{2+} = 0.021 3 mmol. Therefore 143.9 mg of superconductor consumed 0.028 7 mmol Fe^{2+}.

2 × mmol Bi^{5+} in 143.9 mg superconductor = 0.028 7.

Normalizing to 1 gram of superconductor gives

Expt A: $2(\text{mmol Bi}^{5+}) + \text{mmol Cu}^{3+}$ in 1 g superconductor = 0.432 2

Expt B: $2(\text{mmol Bi}^{5+})$ in 1 g superconductor = 0.199 4

It is easier not to get lost in the arithmetic if we suppose that the oxidized bismuth is Bi^{4+} and equate one mole of Bi^{5+} to two moles of Bi^{4+}. Therefore we can rewrite the two equations above as

$\text{mmol Bi}^{4+} + \text{mmol Cu}^{3+}$ in 1 g superconductor = 0.433 2   (1)

mmol Bi^{4+} in 1 g superconductor = 0.199 4   (2)

Subtracting Equation 2 from Equation 1 gives

mmol Cu^{3+} in 1 g superconductor = 0.233 8   (3)

Equations 2 and 3 tell us that the stoichiometric relationship in the formula of the superconductor is b/c = 0.199 4/0.233 8 = 0.852 9.

Since 1 g of superconductor contains 0.789 4 mmol Cu^{3+}, we can say

$$\frac{\text{mol Cu}^{3+}}{\text{mol solid}} = 2c$$

$$\frac{\text{mol Cu}^{3+} / \text{mol solid}}{\text{gram solid} / \text{mol solid}} = \frac{2c}{760.37 + 15.999\ 4(8.1 + b + c)}$$

$$\frac{\text{mol Cu}^{3+}}{\text{gram solid}} = \frac{2c}{770.14 + 15.999\ 4(8.1 + b + c)} = 2.338 \times 10^{-4} \qquad (4)$$

Substituting $b = 0.852\ 9c$ in the denominator of Equation 4 allows us to solve for c:

$$\frac{2c}{770.14 + 15.999\ 4(8.1 + 1.852\ 9c)} = 2.338 \times 10^{-4} \Rightarrow c = 0.105_5$$

$$\Rightarrow b = 0.852\ 9c = 0.090_0$$

The average oxidation numbers are $Bi^{+3.090_0}$ and $Cu^{+2.105_5}$ and the formula of the compound is $Bi_2Sr_2Ca_{0.8}Y_{0.2}Cu_2O_{8.295_5}$, since the oxygen stoichiometry derived at the beginning of the solution is $x = 8.1 + b + c$.

S12.   Coefficients of the balanced equations are :

(a) $1 , 1 \to 1$

(b) $2 , 1 , 2 \to 1 , 2 , 5$

(c) $1 , 24 , 24 \to 6 , 6 , 42 , 48 , 1$

(d) $1 , 1 , 1 \to 1 , 1$

(e) $3 , 1 , 3 \to 1 , 6 , 6$

(f) $1 , 1 \to 1 , 1 , 1$

(g) $1 , 1 \to 2 , 1 , 2$

(h) $1 , 5 , 6 \to 3 , 3$

(i) $1 , 2 , 1 \to 1 , 1 , 1$

One mole of Be^{3+} requires 228 moles of thiosulfate.

S13. (a)

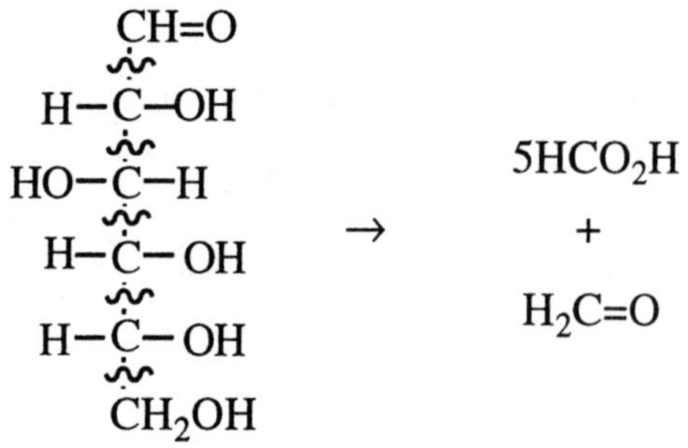

$$
\begin{array}{c}
CH{=}O \\
H{-}C{-}OH \\
HO{-}C{-}H \\
H{-}C{-}OH \\
H{-}C{-}OH \\
CH_2OH
\end{array}
\quad \rightarrow \quad
\begin{array}{c}
5HCO_2H \\
+ \\
H_2C{=}O
\end{array}
$$

To balance this reaction we can write

$$C_6H_{12}O_6 + 5H_2O \rightleftarrows \underbrace{C_6H_{12}O_{11}}_{5HCO_2H+H_2CO} + 10H^+ + 10e^-$$

glucose

$$\underline{5[IO_4^- + 2H^+ + 2e^- \rightleftarrows IO_3^- + H_2O]}$$

$$5IO_4^- + C_6H_{12}O_6 \rightarrow 5HCO_2H + H_2CO + 5IO_3^-$$

b)

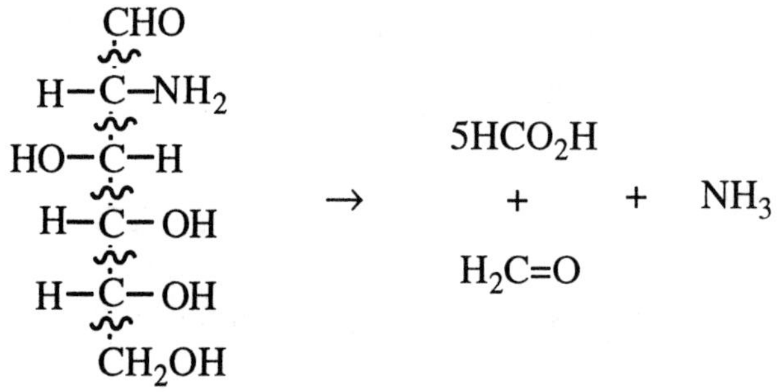

$$
\begin{array}{c}
CHO \\
H{-}C{-}NH_2 \\
HO{-}C{-}H \\
H{-}C{-}OH \\
H{-}C{-}OH \\
CH_2OH
\end{array}
\quad \rightarrow \quad
\begin{array}{c}
5HCO_2H \\
+ \\
H_2C{=}O
\end{array}
\quad + \quad NH_3
$$

$$C_6H_{13}O_5N + 6H_2O \rightleftarrows \underbrace{C_6H_{15}O_{11}N}_{5HCO_2H+H_2CO+NH_3} + 10H^+ + 10e^-$$

$$\underline{5[IO_4^- + 2H^+ + 2e^- \rightleftarrows IO_3^- + H_2O]}$$

$$C_6H_{13}O_5N + H_2O + 5IO_4^- \rightarrow 5HCO_2H + H_2CO + NH_3 + 5IO_3^-$$

(c)

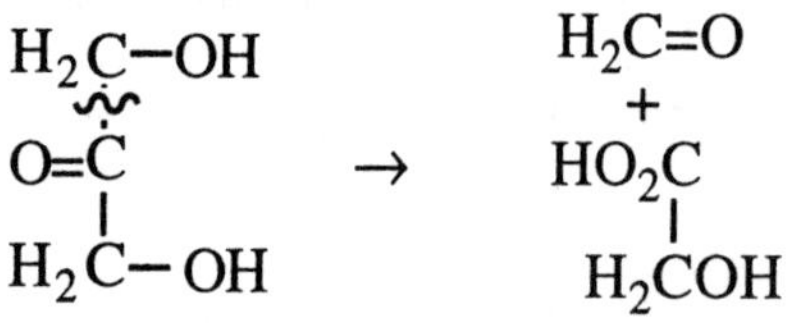

$$
\begin{array}{c}
H_2C{-}OH \\
O{=}C \\
H_2C{-}OH
\end{array}
\quad \rightarrow \quad
\begin{array}{c}
H_2C{=}O \\
+ \\
HO_2C \\
H_2COH
\end{array}
$$

$$C_3H_6O_3 + H_2O \rightleftarrows \underbrace{C_3H_6O_4}_{HCO+C_2H_4O_3} + 2H^+ + 2e^-$$

$$\underline{IO_4^- + 2H^+ + 2e^- \rightleftarrows IO_3^- + H_2O}$$

$$C_3H_6O_3 + IO_4^- \rightarrow H_2CO + HO_2CCH_2OH + IO_3^-$$

S1. $(0.100 \text{ mol})\left(96\ 485\ \dfrac{C}{\text{mol}}\right) = I \cdot (3\ 600\ \text{s}) \Rightarrow I = 2.680\ \dfrac{C}{s} = 2.680\ \text{A}$

S2. (a) $E(\text{cathode}) = -0.828 - 0.059\ 16\ \log(0.033)^{1/2}(1.0) = -0.784_2\ \text{V}$

   $E(\text{anode}) = 1.078 - 0.059\ 16\ \log\ (0.010) = 1.196_3\ \text{V}$

   (Note that $E(\text{anode})$ is calculated for a reduction half-reaction.)

   $E_{eq} = E(\text{cathode}) - E(\text{anode}) = -1.981\ \text{V}$

(b) Ohmic potential $= I \cdot R = (0.052\ \text{A})(4.3\ \Omega) = 0.224_4\ \text{V}$

   $E(\text{applied}) = E_{eq} - IR = -2.205_5$

(c) $E(\text{applied}) = E_{eq} - IR - \text{overpotentials} = -1.981 - 0.224_4 - 0.30 - 0.08 = -2.585_5\ \text{V}$

(d) concentration polarization changes the electrode potentials:

   $E(\text{cathode}) = -0.828 - 0.059\ 16\ \log\ (0.033)^{1/2}(2.0) = -0.802_0\ \text{V}$

   $E(\text{anode}) = 1.078 - 0.059\ 16\ \log\ (0.002\ 0) = 1.237_7$

   $E_{eq} = -0.802_0 - 1.237_7 = -2.040\ \text{V}$

   $E(\text{applied}) = -2.040 - 0.22 - 0.30 - 0.08 = -2.64\ \text{V}$

S3. (a) Let us assume that the left electrode is the anode:

   cathode: $\frac{1}{2}\text{Cl}_2\ (g) + e^- \rightleftharpoons \text{Cl}^-$ $E° = 1.360\ \text{V}$

   anode: $\frac{1}{2}\text{Hg}_2\text{Cl}_2\ (s) + e^- \rightleftharpoons \text{Hg}(l) + \text{Cl}^-$ $E° = 0.268\ \text{V}$

   $E(\text{cathode}) = 1.360 - 0.059\ 16\ \log\ \dfrac{[\text{Cl}^-]}{\sqrt{P_{\text{Cl}_2}}} = 1.360 - 0.059\ 16\ \log\ \dfrac{0.080}{\sqrt{0.10}} = 1.395\ \text{V}$

   $E(\text{anode}) = 0.241\ \text{V}$ (This is a saturated calomel electrode.)

   $E_{eq} = E(\text{cathode}) - E(\text{anode}) = 1.395 - 0.241 = 1.154\ \text{V}$

   Since $E(\text{cell})$ is positive, we guessed the direction of the reaction correctly.

   The net reaction is $\frac{1}{2}\text{Cl}_2(g) + \text{Hg}(l) + \text{Cl}^- \rightleftharpoons \text{Cl}^- + \frac{1}{2}\text{Hg}_2\text{Cl}_2\ (s)$

(b) $E = E_{eq} - I \cdot R = 1.154 - (0.025\ \text{A})(2.12\ \Omega) = 1.101\ \text{V}$

(c) $E(\text{for electrolysis}) = -E_{eq} - I \cdot R = -1.154 - (0.025\ \text{A})(2.12\ \Omega) = -1.207\ \text{V}$

(d) $E(\text{cathode}) = 1.360 - 0.059\ 16\ \log\ \dfrac{0.040}{\sqrt{0.20}} = 1.422\ \text{V}$

   $E_{eq} = E(\text{cathode}) - E(\text{anode}) = 1.422 - 0.241 = 1.181\ \text{V}$

   $E(\text{for electrolysis}) = -E_{eq} - I \cdot R = -1.181 - (0.025\ \text{A})(2.12\ \Omega) = -1.234\ \text{V}$

(e) $E(\text{for electrolysis}) = -1.234 - 0.15 = -1.38\ \text{V}$

S4. $Pb(tartrate) + 2H_2O \rightarrow PbO_2(s) + tartrate^{2-} + 4H^+ + 2e^-$
   Pb^{2+} Pb^{4+}

Oxidation occurs at the <u>anode</u>.

The mass of Pb (tartrate) (FW 355.3) giving 0.122 1 g of PbO_2 (FW = 239.2)

is $(355.3/239.2)(0.122\ 1\ g) = 0.181\ 4$ g.

$\%\ Pb = \dfrac{0.181\ 4}{0.531\ 4} \times 100 = 34.13\%$

S5. (a) $Fe^{2+} + e^- \rightleftarrows Fe(s)$ $E° = -0.44$ V

   $E(cathode) = -0.44 - 0.059\ 16 \log \dfrac{1}{1.0 \times 10^{-7}} = -0.854$ V

 (b) $E(cathode,\ vs\ Ag\,|\,AgCl) = -0.854 - 0.197 = -1.05$ V

 (c) Concentration polarization means that Fe^{2+} cannot diffuse to the cathode as fast
   as it is consumed. The concentration of Fe^{2+} at the electrode surface would be
   $< 0.10\ \mu M$, so the potential would be more negative.

S6. When 99% of Hg(II) is reduced, the formal concentration will be 1.0×10^{-5} M, and
the predominant form is $Hg(NH_3)_4^{2+}$.

$\beta_4 = \dfrac{[Hg(NH_3)_4^{2+}]}{[Hg^{2+}][NH_3]^4} = \dfrac{(1.0 \times 10^{-5})}{[Hg^{2+}](1.0)^4} \Rightarrow [Hg^{2+}] = 5 \times 10^{-25}$ M

$Hg^{2+} + 2e^- \rightleftarrows Hg(l)$ $E° = 0.852$

$E(cathode) = 0.852 - \dfrac{0.059\ 16}{2} \log \dfrac{1}{5 \times 10^{-25}} = 0.133$ V

S7. Relevant information :

   $Fe^{2+} + 2e^- \rightleftarrows Fe(s)$ $E° = -0.44$ V

   $Co^{2+} + 2e^- \rightleftarrows Co(s)$ $E° = -0.282$ V

   $CoY^{2-}\ \ K_f = 2.0 \times 10^{16}$

   $FeY^{2-}\ \ K_f = 2.1 \times 10^{14}$

   $\alpha_{Y4-} = 3.8 \times 10^{-9}$ at pH 4.0

When 99% of FeY^{2-} is removed, $[FeY^{2-}] = 1.0 \times 10^{-8}$ M.

$[Fe^{2+}] = \dfrac{[FeY^{2-}]}{K_f\ \alpha_{Y4-}\ [EDTA]} = \dfrac{1.0 \times 10^{-8}}{(2.1 \times 10^{14})(3.8 \times 10^{-9})(0.010)} = 1.3 \times 10^{-12}$

The cathode potential required to reduce FeY^{2-} to this level is

  $E(cathode) = -0.44 - \dfrac{0.059\ 16}{2} \log \dfrac{1}{1.3 \times 10^{-12}} = -0.79$ V

Will this cathode potential reduce Co^{2+}?

$$\alpha_{Y^{4-}} \ K_f \ (\text{for } CoY^{2-}) = \frac{[CoY^{2-}]}{[Co^{2+}][EDTA]} \Rightarrow [Co^{2+}] = 1.3 \times 10^{-8} \ M$$

$$E(\text{cathode, } Co^{2+}) = -0.282 - \frac{0.059 \ 16}{2} \ \log \frac{1}{1.3 \times 10^{-8}} = -0.515 \ V$$

Since $E(\text{cathode}) < -0.515 \ V$, CoY^{2-} will be reduced. The separation is not feasible.

S8. (a) $\text{mol } e^- = \dfrac{I \cdot t}{F} = \dfrac{(4.11 \times 10^{-3} \ C/s)(834 \ s)}{96 \ 485 \ C/mol} = 3.55 \times 10^{-5} \ \text{mol}$

  (b) One mol e^- reacts with $\frac{1}{2}$ mol Br_2, which reacts with $\frac{1}{2}$ mol cyclohexene

  $\Rightarrow 1.78 \times 10^{-5}$ mol cyclohexene.

  (c) $1.78 \times 10^{-5} \ \text{mol}/3.00 \times 10^{-3} \ L = 5.92 \times 10^{-3} \ M$

S9. Step 1: Total Ti $= \dfrac{2.03 \ \text{mg Ti}/47.88 \ \text{mg/mmol}}{42.37 \ \text{mg unknown}} = 1.00_0 \ \dfrac{\mu\text{mol Ti}}{\text{mg unknown}}$

  Step 2: reducing equivalents of Ti $= \dfrac{9.27 \ C/96 \ 485 \ C/mol}{51.36 \ \text{mg unknown}} = \dfrac{1.87 \ \mu\text{mol}}{\text{mg unknown}}$

  Reducing equivalents per mol of Ti $= \dfrac{1.87 \ \mu\text{mol/mg unknown}}{1.00 \ \mu\text{mol Ti/mg unknown}}$

  $= 1.87$ equivalents/mol Ti

This represents the degree of reduction below Ti^{4+}.

The average oxidation state is $Ti^{+2.13}$ $(= 0.87 \ TiCl_2 + 0.13 \ TiCl_3)$

S1.   Slope = 11.91   standard deviation = 0.12

intercept = -0.012   standard deviation = 0.017

σ_y = 0.033 = standard deviation of current

The concentration of Al^{3+}, when I_d = 0.904 μA, is found as follows :

$$I(mA) = m[Al^{3+}] + b$$

$$[Al^{3+}] = \frac{I-b}{m}$$

$$[Al^{3+}] = \frac{[0.904\,(\pm 0.033)] - [-0.012\,(\pm 0.017)]}{11.91\,(\pm 0.12)}$$

$$[Al^{3+}] = \frac{0.916\,(\pm 0.037\,1)]}{11.91\,(\pm 0.12)} = \frac{0.916\,(\pm 4.05\%)}{11.91\,(\pm 1.01\%)}$$

$$[Al^{3+}] = 0.076\,9\,(\pm 4.17\%) = 0.076\,9 \pm 0.003\,2\ \text{mM}$$

Equation 4-24 gives an uncertainty of $\pm 0.002\,9$ mM

S2. $$\frac{[X]_i}{[S]_f + [X]_f} = \frac{I_X}{I_{S+X}}$$

$$\frac{x(ppm)}{2.65\left(\dfrac{0.500}{3.50}\right) + x\left(\dfrac{3.00}{3.50}\right)} = \frac{152\ nA}{206\ nA} \Rightarrow x = 0.760\ ppm$$

S3. $$\frac{\text{Wave height of }CHCl_3}{\text{Wave height of DDT}} = 1.40 \text{ when } \frac{[CHCl_3]}{[DDT]} = \frac{1.00\ mM}{1.00\ mM} = 1.00$$

$$\frac{\text{Wave height of }CHCl_3}{\text{Wave height of DDT}} = 0.86 \text{ when } \frac{[CHCl_3]}{[DDT]} = x$$

$$\frac{x}{0.86} = \frac{1.00}{1.40} \Rightarrow x = 0.61_4$$

$$0.61_4 = \frac{0.500\ mM}{[DDT]} \Rightarrow [DDT] = 0.81\ mM$$

S4.   If the conditions were perfectly reproducible, the diffusion current for Tl^+ in experiment B would be (1.21/1.15)(6.38) = 6.71 μA. The observed current in experiment B is only 6.11/6.71 = 91.1% of the expected value. That is, in experiment B the response is only 91.1% as great as in experiment A. Therefore, the responses to Cd^{2+} and Zn^{2+} in experiment B are expected to be only 91.1% as great as they are in experiment A

$$[Cd^{2+}] = \frac{(4.76/6.48)}{0.911}\,(1.02) = 0.82\ mM$$

$$[Zn^{2+}] = \frac{(8.54/6.93)}{0.911}\,(1.23) = 1.66\ mM$$

S5. $\dfrac{[Cd^{2+}]_{unknown}}{[Cd^{2+}]_{unknown} + [Cd^{2+}]_{standard}} = \dfrac{(stripping\ time)_{unknown}}{(stripping\ time)_{unknown\ +\ standard}}$

$\dfrac{[Cd^{2+}]_{unknown}}{[Cd^{2+}]_{unknown} + 0.5\ \mu g/L} = \dfrac{1.0\ s}{3.7\ s} \Rightarrow [Cd^{2+}]_{unknown} = 0.19\ \mu g/L = 1.6 \times 10^{-9}\ M$

S6. The equation of the straight line is time (s) = 8.00 $[Cu^{2+}]$ ($\mu g/L$) + 8.57

The x-intercept gives the concentration in the unknown = 1.07 $\mu g/L$

S7. The graph of current vs. scan rate gives a straight line with an intercept reasonably near zero. The analyte is confined to the electrode surface. Otherwise, the graph of current vs. $\sqrt{scan\ rate}$ would give the better straight line.

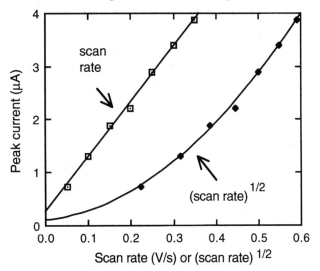

S8. (a) Slope = +0.049 V = 0.059(q-p) $\Rightarrow$ q-p $\approx$ +1. That is, the oxidized species has one more imidazole ligand than the reduced species. The chemistry is either $ML^+ + L + e^- \rightarrow ML_2$ or $M^+ + L + e^- \rightarrow ML$.

Intercept = 0.029V = $E^{free}_{1/2}$ - 0.059 log ($\beta^{ox}_p / \beta^{red}_q$). Putting in $E^{free}_{1/2}$ = -0.18 V gives log ($\beta^{ox}_p / \beta^{red}_q$) = -3.5.

(b) Since the slope is zero, the reaction is either $ML_2^+ + e^- \rightarrow ML_2$ or $ML^+ + e^- \rightarrow ML$. It cannot be $M^+ + e^- \rightarrow M$ because the product has one fewer ligand at lower free ligand concentration (-3.4 < log [L] < -2.1), and the product cannot have a negative number of ligands.

The reaction sequence in parts (a) and (b) has been interpreted as

(a) $ML^+ + L + e^- \rightarrow ML_2$

(b) $ML_2^+ + e^- \rightarrow ML_2$

S9. The end point (0.675 mL) in the figure below is where the linear current increase extrapolates back to zero. Two moles of NH_3 require 3 moles of OBr^-

$$[OBr^-] = (3/2)\frac{(30.0\ \text{mL})(4.43 \times 10^{-5}\ \text{mmol/mL})}{0.675\ \text{mL}} = 2.95\ \text{mM}.$$

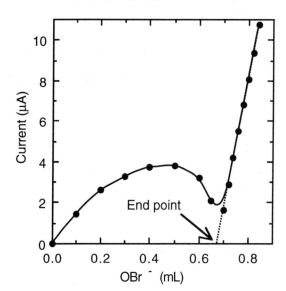

S10. 34.61 mL of methanol with 4.163 mg of H_2O/mL contains 144.08 mg H_2O = 7.997 8 mmol of H_2O. The titration of "dry" methanol tells us that 25.00 mL of methanol reacts with 3.18 mL of reagent. Therefore, 34.61 mL of methanol will react with (34.61/25.00)(3.18) = 4.40 mL of Karl Fischer reagent. The titer of the reagent is

$$\frac{7.997\ 8\ \text{mmol}\ H_2O}{(25.00 - 4.40)\ \text{mL reagent}} = 0.388\ 24\ \frac{\text{mmol}\ H_2O}{\text{mL reagent}}$$

Reagent needed to react with 1.000 g of salt in 25.00 mL of methanol = (38.12 − 3.18) = 34.94 mL. H_2O in 1.000 g of salt = (0.388 24)(34.94) = 13.565 mmol = 244.38 mg of H_2O = 24.44 wt % of the crystal.

S11.

	A	B	C	D	E	F	G	H	I	J
1	n	x	y	std dev	weight(w)	w*x*y	w*x	w*y	w*x^2	w*d^2
2		1	1.00E-04	6995	112	8E-05	6E-05	8E-09	6E-01	8E-13	6E-02
3		2	5.00E-05	3510	74	2E-04	3E-05	9E-09	6E-01	5E-13	1E-01
4		3	1.00E-05	698	11	8E-03	6E-05	8E-08	6E+00	8E-13	2E-02
5		4	5.00E-06	345	18	3E-03	5E-06	2E-08	1E+00	8E-14	3E-02
6		5	1.00E-06	64.4	3.9	7E-02	4E-06	7E-08	4E+00	7E-14	2E+00
7		6	5.00E-07	32.4	1.8	3E-01	5E-06	2E-07	1E+01	8E-14	1E+00
8		7	1.00E-07	6.88	0.64	2E+00	2E-06	2E-07	2E+01	2E-14	2E-01
9		8	5.00E-08	3.17	0.32	1E+01	2E-06	5E-07	3E+01	2E-14	2E-02
10		9	2.00E-08	1.03	0.2	3E+01	5E-07	5E-07	3E+01	1E-14	8E-04
11	n=									
12		9								
13		1.667E-4	11656		4E+01	2E-04	2E-06	1E+02	2E-12	3E+00
14		|<---------------------------------Column sums (B-J)--------------------------------->|								
15	D=	sigma(y)=			Example: E13 = Sum(E2:E10)					
16	8.623E-11	6.93E-01								
17	m=	sigma(m)=			E2 = 1/D2^2		H2 = E2*C2			
18	6.967E+07	4.58E+05			F2 = E2*B2*C2		I2 = E2*B2*B2			
19	b=	sigma(b)=			G2 = E2*B2		J2 = E2*(C2-$A$18*B2-$A$20)^2			
20	-3.578E-01	1.15E-01								
21										
22	D = E13*I13-G13*G13				sigma(y) = Sqrt(J13/(A12-2))					
23	m = (E13*F13-G13*H13)/A16				sigma(m) = B16*Sqrt(E13/A16)					
24	b = (H13*I13-G13*F13)/A16				sigma(b) = B16*Sqrt(I13/A16)					

Unweighted parameters:
$m = 7.002 (\pm 0.005) \times 10^7$
$b = -1.5 (\pm 1.9)$
$s_y = 5.0$

Weighted parameters:
$m = 6.967 (\pm 0.046) \times 10^7$
$b = -0.36 (\pm 0.12)$
$s_y = 0.693$

S12. (a) $A = 0.97 \pm 0.95$ $B = 1.385 \pm 0.027$ $C = -0.001\ 23 \pm 0.000\ 14$ $s_y = 0.919$

(b)

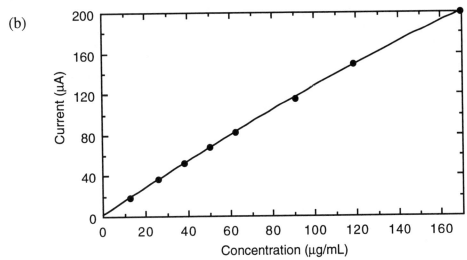

(c) $Current = A + B[\alpha\text{-tocopherol}] + C[\alpha\text{-tocopherol}]^2$

$170 = 0.967 + 1.385\ [\alpha\text{-tocopherol}] - 0.001\ 23\ [\alpha\text{-tocopherol}]^2$

$$\Rightarrow [\alpha\text{-tocopherol}] = 139.3 \ \mu g/mL$$

$$171 = 0.967 + 1.385 \ [\alpha\text{-tocopherol}] - 0.001 \ 23 \ [\alpha\text{-tocopherol}]^2$$

$$\Rightarrow [\alpha\text{-tocopherol}] = 140.2 \ \mu g/mL$$

$$169 = 0.967 + 1.385 \ [\alpha\text{-tocopherol}] - 0.001 \ 23 \ [\alpha\text{-tocopherol}]^2$$

$$\Rightarrow [\alpha\text{-tocopherol}] = 138.3 \ \mu g/mL$$

Answer: $[\alpha\text{-tocopherol}] = 139 \pm 1 \ \mu g/mL$

S1. Using Equations 19-7 and setting $b = 0.100$ cm, we find

$$[X] = \frac{\begin{vmatrix} 0.282 & 387 \\ 0.303 & 642 \end{vmatrix}}{\begin{vmatrix} 1\,640 & 387 \\ 399 & 642 \end{vmatrix}} = \frac{(0.282)(642) - (387)(0.303)}{(1\,640)(642) - (387)(399)} = 7.10 \times 10^{-5} \text{ M}$$

$$[Y] = \frac{\begin{vmatrix} 1\,640 & 0.282 \\ 399 & 0.303 \end{vmatrix}}{\begin{vmatrix} 1\,640 & 387 \\ 399 & 642 \end{vmatrix}} = 4.28 \times 10^{-4} \text{ M}$$

S2. We can use the spreadsheet from Problem 7 in Chapter 19 if we divide the absorbances by 2 to change from a 2.000-cm cell to a 1.000-cm cell. Putting absorbances of 0.333, 0.249 and 0.180 into the spreadsheet gives $[X] = 2.086 \times 10^{-5}$ M, $[Y] = 2.387 \times 10^{-5}$, and $[Z] = 6.317 \times 10^{-6}$ M.

S3. A graph of $\Delta A/[X]$ versus ΔA is a scattered straight line with a slope of -1.401 5 and an intercept of 5.943 2.

$$K = -\text{slope} = 1.40. \qquad \Delta\varepsilon = \frac{\text{intercept}}{KP_0} = \frac{5.943\,2}{1.40 \times 0.001\,00} = 4\,240 \text{ M}^{-1} \text{ cm}^{-1}$$

[X] (M)	ΔA	$\Delta A/[X]$
0	0	---
0.005 09	0.030	5.894
0.008 52	0.050	5.869
0.017 3	0.100	5.780
0.029 5	0.170	5.763
0.038 7	0.220	5.685
0.050 9	0.280	5.501
0.065 0	0.350	5.385
0.077 9	0.420	5.392
0.093 2	0.490	5.258
0.106 2	0.550	5.179

S4. For Zn^{2+}, the maximum absorbance occurs at a mole fraction of metal of 0.33, indicating formation of a 2:1 ligand:metal complex. For Ga^{3+}, the maximum absorbance at a mole fraction of 0.25, indicates formation of a 3:1 ligand:metal complex.

S5. 340 nm: $E = h\nu = h\dfrac{c}{\lambda} = (6.626\ 1 \times 10^{-34}\ \text{J·s})\ \dfrac{2.997\ 9 \times 10^8\ \text{m/s}}{340 \times 10^{-9}\ \text{m}} = 5.84 \times 10^{-19}\ \text{J}$

To convert to J/mol, multiply by Avogadro's number:

5.84×10^{-19} J/molecule $\times\ 6.022 \times 10^{23}$ molecules/mol $= 352$ kJ/mol.

613 nm: $E = (6.626\ 1 \times 10^{-34}\ \text{J·s})\dfrac{2.997\ 9 \times 10^8\ \text{m/s}}{613 \times 10^{-9}\ \text{m}} = 3.24 \times 10^{-19}$ J

$= 195\ \dfrac{\text{kJ}}{\text{mol}}$. The difference between the irradiation energy and the fluorescence

energy is $352 - 195 = 157$ kJ/mol.

S6.

	A	B	C
1	E(excitation) =	Concentration	Relative Intensity
2	2120	0.0000001	1.95E-04
3	E(emission) =	0.000001	1.95E-03
4	810	0.00001	1.89E-02
5	b1 =	0.0001	1.40E-01
6	0.3	0.001	7.81E-02
7	b2 =	0.01	3.89E-11
8	0.4		
9	b3 =		
10	0.5		

S7. (a) The equation for intensity can be solved for K, giving

$$K = \frac{[S]\ (1 - I_s/I_{s+x})}{I_s/I_{s+x} - [S]/([S] + [X])}$$

Putting in $I_s/I_{s+x} = 58.7/74.5$, $[S] = 250\ \mu M$ and $[X] = 200\ \mu M$
gives $K = 228\ \mu M$.

(b) Solving the intensity equation for [X] gives

$$[X] = \frac{K[S]}{(I_s/I_{s+x})(K + [S]) - [S]} - [S]$$

Using the values $I_s/I_{s+x} = 63.5/74.6$, $[S] = 250 + 200$
$= 450\ \mu M$ and $K = 228\ \mu M$ gives $[X] = 357\ \mu M$.

SPECTROPHOTOMETERS

S1. $T + a + R = 0.70 + a + 0.08 = 1 \Rightarrow a = 0.22$

S2. $P_2/P_1 = e^{-\alpha b} = e^{-(1.40 \text{ cm}^{-1})(0.200 \text{ cm})} = 0.756$

S3.   (a) $R = \left(\dfrac{1-n}{1+n}\right)^2 = \left(\dfrac{1-1.62}{1+1.62}\right)^2 = 0.056\,0$

   $T = \dfrac{1-R}{1+R} = 0.894$ (independent of thickness if absorption = 0)

   (b) $T = \dfrac{(1-R)^2 e^{-\alpha b}}{1-R^2 e^{-2\alpha b}} = \dfrac{(1-0.056\,0)^2 e^{-(0.85 \text{ cm}^{-1})(0.500 \text{ cm})}}{1-0.056\,0^2 e^{-2(0.85 \text{ cm}^{-1})(0.500 \text{ cm})}} = 0.583$

S4.   (a) The critical angle, θ_c, is such that $(n_1/n_2)\sin\theta_c = 1$. For $n_1 = 2.7$ and $n_2 = 2.0$, $\theta_c = 48°$. That is θ must be $\geq 48°$ for total internal reflection.

   (b) $0.012\,\dfrac{dB}{m} = \dfrac{-10\log\left(\dfrac{\text{power out}}{\text{power in}}\right)}{\text{length}}$

   For length = 0.50 m, we find $\dfrac{\text{power out}}{\text{power in}} = 0.998\,6$

S5.   At 13 μm the slope $dn/d\lambda$ is greater for NaCl than for KBr. Therefore a NaCl prism will have greater dispersion than a KBr prism.

S6. $n_1 \sin\theta_1 = n_2 \sin\theta_2$, where $n_1 = 1.46$ and $n_2 = 1.63$

   (a) If $\theta_1 = 30°$, $\theta_2 = 26.6°$

   (b) If $\theta_1 = 0°$, $\theta_2 = 0°$ (no refraction)

S7.   (a) For incident light: $n_{air} \sin\theta = n_{prism} \sin\alpha \Rightarrow \sin\theta = \dfrac{n_{prism} \sin\alpha}{n_{air}}$

   For exiting light: $n_{prism} \sin\alpha = n_{air} \sin\theta \Rightarrow \sin\theta = \dfrac{n_{prism} \sin\alpha}{n_{air}}$

   Therefore entrance and exit angles are the same.

   (b) $\alpha = 30°$ if light is parallel to the prism base.

   Therefore $\sin\theta = \dfrac{(1.500) \sin 30°}{1.000} \Rightarrow \theta = 48.59°$

S8.   Since exitance is proportional to T^4, $\dfrac{\text{exitance at 900 K}}{\text{exitance at 300 K}} = \left(\dfrac{900}{300}\right)^4 = 81$

   Exitance at 900 K $= \left(5.669\,8 \times 10^{-8}\,\dfrac{W}{m^2 K^4}\right)(900 \text{ K})^4 = 3.72 \times 10^4\,\dfrac{W}{m^2}$

S9.   (a)   The mass of a 1-cm length of the cylinder is: (volume) (density) =

$$\pi r^2(\text{length})(7.89 \text{ g/cm}^3) = \pi(0.32 \text{ cm})^2(1 \text{ cm})(7.86 \text{ g/cm}^3) = 2.5_3 \text{ g}$$

The surface area of the 1-cm length of cylinder is π(diameter)(length)

$$= \pi(0.64 \text{ cm})(1 \text{ cm}) = 2.0_1 \text{ cm}^2$$

$$\text{Exitance} = \left(5.669\ 8 \times 10^{-8} \frac{W}{m^2 K^4}\right)(1\ 373 \text{ K})^4 = 2.01 \times 10^5 \frac{W}{m^2}$$

$$= 20._1 \frac{W}{cm^2} = 20._1 \frac{J}{s \cdot cm^2}$$

$$\text{Energy loss rate (J/s)} = \left[\text{exitance}\left(\frac{J}{s\ cm^2}\right)\right][\text{surface area (cm}^2)]$$

$$= \left(20._1 \frac{J}{s \cdot cm^2}\right)(2.0_1 \text{ cm}^2) = 40._4 \text{ J/s}$$

Heat capacity of 1-cm length is [mass (g)]$\left[\text{heat capacity}\left(\frac{J}{g\ K}\right)\right]$ =

$$(2.5_3 \text{ g})\left(0.606 \frac{J}{g \cdot K}\right) = 1.5_3 \text{ J/K}$$

$$\text{Cooling rate} = \frac{\text{energy loss rate}}{\text{heat capacity}} = \frac{40._4 \text{ J/s}}{1.5_3 \text{ J/K}} = 26 \text{ K/s}$$

(b)   The ratio of surface area/volume is 10 times greater so the cooling rate is also 10 times greater = 260°C/s.

S10. (a) $n\lambda = d(\sin\theta - \sin\phi)$

$$1 \cdot 400 \times 10^{-9} \text{ m} = d(\sin 20° - \sin 10°) \Rightarrow d = 2.38 \times 10^{-6} \text{ m}$$

$$\text{Lines/cm} = 1/(2.38 \times 10^{-4} \text{ cm}) = 4.21 \times 10^3 \text{ lines/cm}$$

(b) $\lambda = 1/(1\ 000 \text{ cm}^{-1}) = 10^{-3} \text{ cm} \Rightarrow d = 5.94 \times 10^{-3} \text{ cm} \Rightarrow 168 \text{ lines/cm}$

S11. (a) $\text{Resolution} = \dfrac{\lambda}{\Delta\lambda} = \dfrac{443.531}{0.072} = 6.2 \times 10^3$

For $\lambda = 443.495$ nm, $\tilde{v} = 1/\lambda = 2.254\ 82 \times 10^6 \text{ m}^{-1} = 2.254\ 82 \times 10^4 \text{ cm}^{-1}$.

For $\lambda = 443.567$ nm, $\tilde{v} = 2.254\ 45 \times 10^4 \text{ cm}^{-1}$ Difference = 3.7 cm^{-1}.

For $\lambda = 443.495$ nm, $v = c/\lambda = 6.759\ 77 \times 10^{14}$ Hz.

For $\lambda = 443.567$ nm, $v = 6.758\ 67 \times 10^{14}$ Hz   Difference = 1.10×10^{11} Hz.

(b) $\Delta\lambda = \dfrac{\lambda}{10^4} = \dfrac{443.495}{10^4} = 0.04$ nm

(c)   Resolution = nN = (2)(6.00 cm × 2 120 cm^{-1}) = 2.54×10^4

(d)   200 lines/mm = 5 μm/line = d

$$\frac{\Delta\phi}{\Delta\lambda} = \frac{2}{d \cos \phi} = \frac{1}{(5 \ \mu m) \cos 10°} = 0.406 \ \frac{\text{radians}}{\mu m} = 23.3°/\mu m$$

For $\Delta\lambda = 0.072$ nm, $\Delta\phi = (23.3°/\mu m)(7.2 \times 10^{-5} \ \mu m) = 1.7 \times 10^{-3}$ degrees

For 20th order diffraction, the dispersion will be 10 times greater, or 0.017°.

S12. True transmittance $= 10^{-1.26} = 0.055$. With 0.4% stray light, the apparent transmittance is $0.055 + 0.004 = 0.059$. The apparent absorbance is $- \log 0.036 \ 6 = 1.23$.

S13. (a) $\Delta = \pm 4$ cm

(b) Resolution $\approx 1/\Delta = 0.25$ cm^{-1}

(c) $\delta = 1/(2\Delta v) = 1/(2 \cdot 4 \ 000 \ \text{cm}^{-1}) = 1.25 \ \mu m$

S14. To increase the ratio from 3/1 to 9/1 (a factor of $9/3 = 3$) requires $3^2 = 9$ scans.

S15.

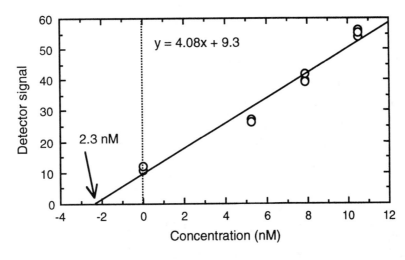

S1. The Doppler linewidth is given by

$$\Delta\nu = \nu\,(7 \times 10^{-7})\,\sqrt{T/M} = (c/\lambda)\,(7 \times 10^{-7})\,\sqrt{T/M}$$

$= 6.2$ GHz for Fe at 3 000 K (with $\lambda = 248 \times 10^{-9}$ m and $M = 56$)

$= 8.8$ GHz for Fe at 6 000 K (with $\lambda = 248 \times 10^{-9}$ m and $M = 56$)

$= 3.2$ GHz for Hg at 3 000 K (with $\lambda = 254 \times 10^{-9}$ m and $M = 201$)

$= 4.5$ GHz for Hg at 6 000 K (with $\lambda = 254 \times 10^{-9}$ m and $M = 201$)

S2. (a) $\Delta E = h\nu = \dfrac{hc}{\lambda} = \dfrac{(6.626\ 1 \times 10^{-34}\ \text{J·s})(2.997\ 9 \times 10^{8}\ \text{m/s})}{327 \times 10^{-9}\ \text{m}}$

$= 6.07 \times 10^{-19}$ J/molecule $= 366$ kJ/mol

(b) $\dfrac{N^*}{N_o} = \dfrac{g^*}{g_o}\,e^{-\Delta E/kT} = 3e^{-(6.07\times10^{-19}\ \text{J})/(1.381\times10^{-23}\ \text{J/K})(2\ 400\text{K})} = 3.29 \times 10^{-8}$

(c) At 2 415 K, $N^*/N_o = 3.69 \times 10^{-8} \Rightarrow$ 12% increase from 2 400 to 2 415 K

(d) At 6 000 K, $N^*/N_o = 0.002\ 0$

S3. Sensitivity = concentration of Mo giving $A = 0.004\ 36$ $(= 99\%\ T)$

$$\dfrac{[\text{Mo}]\ \text{giving}\ A = 0.004\ 36}{[\text{Mo}]\ \text{giving}\ A = 0.025} = \dfrac{0.004\ 36}{0.025}$$

$$\dfrac{x}{23.6\ \mu\text{g/mL}} = \dfrac{0.004\ 36}{0.025} \Rightarrow x = 4.1\ \mu\text{g/mL}$$

S4. (a)

μg/mL	$A_{422.7}/A_{324.7}$	
1.00	0.6056	
2.00	0.6062	average = 0.6019
3.00	0.5913	
4.00	0.6045	

(b) When μg Ca/mL = μg Cu/mL, $A_{422.7}/A_{324.7} = 0.6019$

When μg Ca/mL = x and μg Cu/mL = 2.47, $A_{422.7}/A_{324.7} =$
$0.218/0.269 = 0.8104$.

We can set up a ratio to solve for Ca :

$$\dfrac{\text{Concentration ratio in unknown}}{\text{Concentration ratio in standard mixture}} = \dfrac{\text{Absorbance ratio in unknown}}{\text{Absorbance ratio in standard mixture}}$$

$$\dfrac{[\text{Ca}]/2.47}{1} = \dfrac{0.8104.}{0.6019} \Rightarrow [\text{Ca}] = 3.33\ \mu\text{g/mL}$$

S5. (a) The graph below shows the quadratic fit to the calibration data.

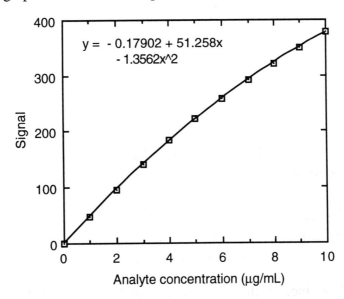

$$y = -0.17902 + 51.258x - 1.3562x^2$$

(b) $344.0 = -0.179\ 02 + 51.258x - 1.356\ 2x^2 \Rightarrow x = 8.73\ \mu g/mL$

(by solving quadratic equation and using the root that lies on this graph)

$341.8 = -0.179\ 02 + 51.258x - 1.356\ 2x^2 \Rightarrow x = 8.81\ \mu g/mL$

$346.2 = -0.179\ 02 + 51.258x - 1.356\ 2x^2 \Rightarrow x = 8.81\ \mu g/mL$

$341.8 = -0.179\ 02 + 51.258x - 1.356\ 2x^2 \Rightarrow x = 8.65\ \mu g/mL$

Concentration of unknown $= 8.73 \pm 0.08\ \mu g/mL$

S1. (a) Fraction remaining $= \dfrac{V_1}{V_1 + KV_2}$

$0.005 = \dfrac{100}{100 + (610)V_2} \Rightarrow V_2 = 32.6 \text{ mL}$

(b) $0.005 = \left(\dfrac{100}{100 + (610)V_2}\right)^4 \Rightarrow V_2 = 0.453 \text{ mL}. \quad 4 \times V_2 = 1.81 \text{ mL}$

S2. (a) $D \equiv \dfrac{[HA]_{ether} + [A^-]_{ether}}{[HA]_{aq} + [A^-]_{aq}} \approx \dfrac{[HA]_{ether}}{[HA]_{aq} + [A^-]_{aq}}$

$K \equiv \dfrac{[HA]_{ether}}{[HA]_{aq}}$

(b) $D = \dfrac{K \cdot [H^+]}{[H^+] + K_a} = \dfrac{92 \cdot 10^{-4.00}}{10^{-4.00} + 4.2 \times 10^{-4}} = 17.7$

(c) At pH 3.50, more of the acid will be in the form HA, which is the extractable species. Therefore the distribution coefficient will be greater than at pH 4.00.

(d) $1 = \dfrac{92 \cdot [H^+]}{[H^+] + 4.2 \times 10^{-4}} \Rightarrow [H^+] = 4.62 \times 10^{-6} \text{ M} \Rightarrow \text{pH} = 5.34$

S3. (a) $t'_r = 13.81 - 1.06 = 12.75 \text{ min}. \quad k' = 12.75/1.06 = 12.03$

(b) $w_{1/2} = 0.172 \text{ min} \Rightarrow N = \dfrac{5.55\, t_r^2}{w_{1/2}^2} = \dfrac{5.55\,(13.81)^2}{0.172^2} = 3.58 \times 10^4 \text{ plates}$

(c) $w = 0.302 \text{ min} \Rightarrow N = \dfrac{16\, t_r^2}{w^2} = \dfrac{16\,(13.81)^2}{0.302^2} = 3.35 \times 10^4 \text{ plates}$

(d) $H = (30.0 \times 10^3 \text{ mm})/(3.35 \times 10^4 \text{ plates}) = 0.896 \text{ mm}$

(e) Volume of 30.0 m of column $= \pi r^2 \times \text{length} = \pi(0.026\ 2 \text{ cm})^2 \times 3\ 000 \text{ cm}$
$= 6.47 \text{ mL}.$
Volume flow rate $= (6.47 \text{ mL}) / (1.06 \text{ min}) = 6.10 \text{ mL/min}$

S4. (a) $N = \dfrac{16\, t_r^2}{w^2} = \dfrac{16\,(12.83 \text{ min})^2}{(0.307 \text{ min})^2} = 2.80 \times 10^4 \text{ plates}$

(b) $H = (158 \text{ mm})/(2.80 \times 10^4 \text{ plates}) = 5.64 \text{ μm}$

S5. (a) Volume of mobile phase per cm of column $= (0.426)\pi r^2 \times (1 \text{ cm}) =$
$(0.426)\pi(0.960 \text{ cm})^2 \times (1 \text{ cm}) = 1.23 \text{ mL}$
Linear flow rate $= (2.22 \text{ mL/min}) / (1.23 \text{ mL/cm}) = 1.80 \text{ cm/min}$

(b) $t_m = (50.6 \text{ cm}) / 1.80 \text{ cm/min}) = 28.1_1 \text{ min}$

$$k' = \frac{t_r - t_m}{t_m} \Rightarrow 8.04 = \frac{t_r - 28.11}{28.11} \Rightarrow t_r = 254 \text{ min}$$

S6. $\alpha = (k_2') / (k_1') = 1.02; \quad k_2' = 5.10; \quad k_{av}' = 5.05$

$$\text{Resolution} = \frac{\sqrt{N}}{4} \left(\frac{\alpha-1}{\alpha}\right) \left(\frac{k_2'}{1 + k_{av}'}\right)$$

$$1.00 = \frac{\sqrt{N}}{4} \left(\frac{1.02 - 1}{1.02}\right) \left(\frac{5.10}{1 + 5.05}\right) \Rightarrow N = 5.86 \times 10^4 \text{ plates}$$

$(5.86 \times 10^4 \text{ plates})(155 \ \mu m/ \text{ plate}) = 9.08 \text{ m}$

To double the resolution requires four times as many plates = 36.3 m.

S7. (a) Heptane: $t_r' = 14.56 - 1.06 = 13.50$ min. $k' = 13.50/1.06 = 12.74$

$C_6H_4F_2$: $t_r' = 14.77 - 1.06 = 13.71$ min. $k' = 13.71/1.06 = 12.93$

(b) $\alpha = 12.93/12.74 = 1.015$

(c) $w_{1/2}$ (heptane) $= 0.126$ min; $w_{1/2}$ $(C_6H_4F_2) = 0.119$ min

$$N = \frac{5.55 \ t_r^2}{w_{1/2}^2} = \frac{5.55 \ (14.56)^2}{0.126^2} = 7.41 \times 10^4 \text{ plates for heptane}$$

$$N = \frac{5.55 \ (14.77)^2}{0.119^2} = 8.55 \times 10^4 \text{ plates for } C_6H_4F_2$$

(d) w (heptane) $= 0.216$ min; w $(C_6H_4F_2) = 0.196$ min

$$N = \frac{16 \ t_r^2}{w^2} = \frac{16 \ (14.56)^2}{0.216^2} = 7.27 \times 10^4 \text{ plates for heptane}$$

$$N = \frac{16 \ (14.77)^2}{0.196^2} = 9.09 \times 10^4 \text{ plates for } C_6H_4F_2$$

(e) $\text{Resolution} = \frac{\Delta t_r}{w_{av}} = \frac{14.77 - 14.56}{0.206} = 1.019$

(f) $N = \sqrt{(7.27 \times 10^4)(9.09 \times 10^4)} = 8.13 \times 10^4 \text{ plates}$

$$\text{Resolution} = \frac{\sqrt{N}}{4} \left(\frac{\alpha-1}{\alpha}\right) \left(\frac{k_2'}{1 + k_{av}'}\right)$$

$$= \frac{\sqrt{8.13 \times 10^4}}{4} \left(\frac{1.015-1}{1.015}\right) \left(\frac{12.93}{1 + 12.84}\right) = 0.984$$

(g) Fraction of time in stationary phase $= \dfrac{t_s}{t_s + t_m} = \dfrac{k't_m}{k't_m + t_m} =$

$$\frac{k'}{k' + 1} = \frac{12.74}{12.74 + 1} = 0.927$$

S8. $\dfrac{\text{Large load}}{\text{Small load}} = \left(\dfrac{\text{large column radius}}{\text{small column radius}}\right)^2$

$\dfrac{72.4 \text{ mg}}{10.0 \text{ mg}} = \left(\dfrac{3.00 \text{ cm}}{\text{small column diameter}}\right)^2 \Rightarrow \text{diameter} = 1.11 \text{ cm}$

Use the same concentration $\Rightarrow$ volume $= \dfrac{10.0}{72.4}(0.500 \text{ mL}) = 0.069\ 1 \text{ mL}$

S9. Flux $= -D\dfrac{dc}{dx} = -\left(2.0 \times 10^{-9} \dfrac{\text{m}^2}{\text{s}}\right)\left(-3.4 \dfrac{\text{mol}}{\text{L·cm}}\right)$

$= \left(2.0 \times 10^{-9} \dfrac{\text{m}^2}{\text{s}}\right)\left(\dfrac{100 \text{ cm}}{\text{m}}\right)^2\left(3.4 \dfrac{\text{mol}}{\text{L·cm}}\right)\left(\dfrac{0.001 \text{ L}}{\text{cm}^3}\right) = 6.80 \times 10^{-8} \dfrac{\text{mol/s}}{\text{cm}^2}$

$= 4.09 \times 10^{16} \dfrac{\text{ions/s}}{\text{cm}^2}$

In 5.5 s, 2.3×10^{17} ions will cross each square centimeter.

CHAPTER 23: SUPPLEMENTARY SOLUTIONS
GAS AND LIQUID CHROMATOGRAPHY

S1. $I = 100\left[6 + (7-6)\dfrac{\log(6.45) - \log(6.12)}{\log(7.20) - \log(6.12)}\right] = 632$

S2. (a): hexane < benzene < heptane < nitropropane < pyridine < octane < nonane

(b): hexane < heptane < benzene < octane < pyridine < nitropropane < nonane

(c): hexane < heptane < octane < nonane < benzene < pyridine < nitropropane

S3. $$\frac{[C]/[D] \text{ in unknown}}{[C]/[D] \text{ in standard}} = \frac{\text{area ratio in unknown}}{\text{area ratio in standard}}$$

$$\frac{[C]/(1\,230\ \mu g/25.00\ mL)}{(236/337)} = \frac{3.33/2.22}{4.42/5.52} \Rightarrow [C] = 64.5\ \mu g/mL$$

$$[C] \text{ in original unknown} = \frac{25}{10}(64.5\ \mu g/mL) = 161\ \mu g/mL$$

S4. The more polar the solvent, the greater the eluent strength on a polar column, and the shorter the retention time. Since 60 vol % acetonitrile is less polar than 40 vol % acetonitrile, the retention time will be greater for the 60 vol % eluent.

S5. $$k' = \frac{t_r - t_m}{t_m} \Rightarrow 3.50 = \frac{t_r - 1.33}{1.33} \Rightarrow t_r = 5.99 \text{ min}$$

Number of plates $= N = (0.150 \text{ m})/(10.6 \times 10^{-6} \text{ m/plate}) = 1.42 \times 10^4$

$$N = 5.55\left(\frac{t_r}{w_{1/2}}\right)^2 \Rightarrow 1.42 \times 10^4 = 5.55\left(\frac{5.99}{w_{1/2}}\right)^2 \Rightarrow w_{1/2} = 0.118 \text{ min}$$

Raise the temperature to get the same separation efficiency at a faster linear flow rate, as in Figure 22-16.

S1. $N = \dfrac{5.55\, t_r^2}{w_{1/2}^2} = \dfrac{5.55\,(17.80\ \text{min})^2}{(0.17_4\ \text{min})^2} = 5.8 \times 10^4$ plates

Plate height $= (0.60\ \text{m}) / (5.8 \times 10^4\ \text{plates}) = 10.3\ \mu\text{m}$

S2. $V_t = \pi r^2 l = \pi(1.27\ \text{cm})^2(28.4\ \text{cm}) = 143._9\ \text{mL}$

$K_{av} = \dfrac{V_r - V_0}{V_t - V_0} = \dfrac{98.6 - 43.4}{143.8 - 43.4} = 0.550$

S3.   The calibration curve follows the equation log (MW) = 12.590 − 0.607 6 t. For t = 13.00 min, MW = 4.9×10^4.

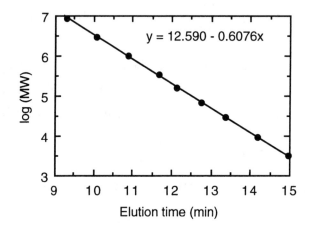

S4.   Observed current − extrapolated current $= 0.05$(extrapolated current)

$(3V + 0.01V^2 + 0.000\,2V^3) - (3V) = (0.05)3V \;\;\Rightarrow\;\; V = 12.08\ \text{kV}$

S5.   The 13-*cis* isomer is a slightly more compact molecule, and ought to have a smaller effective hydrodynamic radius.

S6.   (a) $\mu_{app} = \dfrac{u_{net}}{E} = \dfrac{L_d / t}{V / L_t} = \dfrac{0.82\ \text{m} / 408\ \text{s}}{25.4 \times 10^3\ \text{V} / 0.96\ \text{m}} = 7.6 \times 10^{-8}\ \dfrac{\text{m}^2}{\text{s·V}}$

(b) Moles injected $= \mu_{app}\left(E\,\dfrac{\kappa_b}{\kappa_s}\right) t\pi r^2 C$

$= \left(7.6\times 10^{-8}\ \dfrac{\text{m}^2}{\text{s·V}}\right)\left(\dfrac{5.0\times10^3\ \text{V}}{0.96\ \text{m}}\right)(2)(3.0\ \text{s})(\pi)(25\times10^{-6}\ \text{m})^2\left(2.4\times10^{-3}\ \dfrac{\text{mol}}{\text{m}^3}\right)$

$= 11\ \text{fmol}$

S7.   For the acid HA, the charge is α_{A^-}, where α is the fraction of dissociation:

$$\alpha_{A^-} = \frac{K_A}{[H^+] + K_A}$$

Let's designate acetic acid as HA and formic acid as HF with acid dissociation constant K_F. The difference in charge between the two acids is

$$\Delta = \alpha_{A^-} - \alpha_{F^-} = \frac{K_A}{[H^+] + K_A} - \frac{K_F}{[H^+] + K_F}$$

We are seeking the maximum D, for which $d\Delta/d[H^+] = 0$:

$$\frac{d\Delta}{d[H^+]} = 0 = \frac{-K_A}{([H^+] + K_A)^2} + \frac{K_F}{([H^+] + K_F)^2}$$

$$K_A([H^+] + K_F)^2 = K_F([H^+] + K_A)^2$$

$$K_A[H^+]^2 + 2K_AK_F[H^+] + K_AK_F^2 = K_F[H^+]^2 + 2K_AK_F[H^+] + K_FK_A^2$$

$$(K_A - K_F)[H^+]^2 = K_FK_A^2 - K_AK_F^2$$

$$(K_A - K_F)[H^+]^2 = K_FK_A(K_A - K_F) \Rightarrow [H^+] = \sqrt{K_FK_A}$$

Inserting the values $K_A = 1.75 \times 10^{-5}$ and $K_F = 1.80 \times 10^{-4}$ gives pH = 4.25

S8.   (a) At pH 12, the mixture contains RCO_2^-, RNH_2, Na^+ and OH^-. All of these species pass directly through the cation-exchange column (which is in the Na^+ form). There is no retention of any component.

   (b) At pH 2, the mixture to be separated contains RCO_2H, RNH_3^+, H^+ and Cl^-. The RNH_3^+ cation will be retained by the column and the other species will be eluted

S1. One mole of product (206.243 g) comes from one mole of piperazine (86.137 g).

Grams of piperazine in sample =

(0.793 3 g of piperazine / g of sample) × (0.123 4 g of sample) = 0.097 89 g.

Mass of product $= \left(\dfrac{206.243}{86.137}\right)(0.097\ 89\ \text{g}) = 0.234\ 4\ \text{g}.$

S2. 2.188 g bis(dimethylglyoximate)nickel (II) = 7.573×10^{-3} mol Ni = 0.444 5 g Ni = 39.44% Ni.

S3. Mol K^+ in unknown = mol $K^+B(C_6H_5)_4^- = \dfrac{1.003\ \text{g}}{358.33\ \text{g/mol}} = 2.799$ mmol

= 0.109 4 g K = 8.665 wt % K.

S4. (a) mmol C = mmol $CO_2 = \dfrac{16.432\ \text{mg}}{44.010\ \text{mg/mmol}} = 0.373\ 37$ mmol = 4.484 5 mg C

$\dfrac{4.484\ 5\ \text{mg C}}{8.732\ \text{mg unknown}} \times 100 = 51.36$ wt % C

mmol H = 2 × mmol $H_2O = \dfrac{2 \times 2.840\ \text{mg}}{18.015\ \text{mg/mmol}} = 0.315\ 29$ mmol

= 0.317 80 mg H = 3.639 wt % H

(b) Atomic ratio (C:H) = 0.373 37 / 0.315 29 = 1.184

A small integer ratio is 6:5.

S5. In step (1) we make 0.726 8 g of lead piperazine dithiocarbamate, which equals (0.726 8 / 527.77) = $1.377\ 1 \times 10^{-3}$ mol of Pb. The $PbCl_2$ in 100.00 mL will be twice as much, since only 50.00 mL was treated in step (1). Moles of $PbCl_2$ in 100 mL = 2.754×10^{-3} mol.

In step (2), the grams of $Pb(IO_3)_2$ formed will be

$$\frac{1}{4}(\underbrace{2.754 \times 10^{-3}\ \text{mol}}_{\text{moles } PbCl_2})(557.0\ \text{g/mol}) = 0.383\ 5\ \text{g}$$

The $Cu(IO_3)_2$ from 25.00 mL is 0.838 8 - 0.383 5 = 0.455 3 g.

The moles of Cu in 0.455 3 g of $Cu(IO_3)_2$ are (0.455 3/413.35) = 1.101×10^{-3} mol. The moles of Cu in 100.0 mL are 4 (1.101×10^{-3}) = 4.404×10^{-3} mol = 0.279 9 g = 13.99%.

S6. (a) $2NaHCO_3 \xrightarrow{100°C} Na_2CO_3 + H_2O + CO_2$

$2KHCO_3 \xrightarrow{150°C} K_2CO_3 + H_2O + CO_2$

(b) 2 mol $NaHCO_3$ (168.014 g) would produce 1 mol (105.989 g) of Na_2CO_3.

Therefore 50.0 g of $NaHCO_3$ produces $\dfrac{105.989}{168.014} \times 50.0 = 31.5$ g Na_2CO_3,

losing 18.5 g in the process.

2 mol $KHCO_3$ (200.230 g) would produce 1 mol (138.206 g) of K_2CO_3.

Therefore 50.0 g of $KHCO_3$ produces $\dfrac{138.206}{200.230} \times 50.0 = 34.5$ g K_2CO_3,

losing 15.5 g in the process.

(c) Starting with 100 mg, the mixture will lose 18.5 mg (18.5 wt %) of its mass near 100°C and another 15.5 mg (15.5 wt %) at 150°C. There will be a plateau between 100 and 150 °C at 81.5 wt % (= 100 - 18.5) and a second plateau above 150°C at 66.0 wt % (= 100 - 15.5 - 18.5).

S7. The atomic ratio H:C is

$$\frac{\left(\dfrac{6.84 \pm 0.10 \text{ g}}{1.008 \text{ g/mol}}\right)}{\left(\dfrac{71.22 \pm 1.1 \text{ g}}{12.011 \text{ g/mol}}\right)} = \frac{6.786 \pm 0.099}{5.930 \pm 0.091\ 6} = \frac{6.786 \pm 1.46\%}{5.930 \pm 1.54\%} = 1.144 \pm 0.024$$

If we define the stoichiometry coefficient for C to be 8, then the stoichiometry coefficient for H is $8(1.144 \pm 0.024) = 9.15 \pm 0.19$.

The atomic ratio N:C is

$$\frac{\left(\dfrac{10.33 \pm 0.13 \text{ g}}{14.007 \text{ g/mol}}\right)}{\left(\dfrac{71.22 \pm 1.1 \text{ g}}{12.011 \text{ g/mol}}\right)} = \frac{0.737\ 5 \pm 0.009\ 3}{5.930 \pm 0.091\ 6} = \frac{0.737\ 5 \pm 1.26\%}{5.930 \pm 1.54\%}$$

$$= 0.124\ 4 \pm 0.002\ 5$$

If we define the stoichiometry coefficient for C to be 8, then the stoichiometry coefficient for N is $8(0.124\ 4 \pm 0.002\ 5) = 0.995 \pm 0.020$.

The empirical formula is reasonably expressed as $C_8H_{9.15\pm0.19}N_{0.995\pm0.020}$

S8. We will use mmol and mg, instead of mol and g, for our calculations. From the formula weight of $YBa_2Cu_3O_{7-x}$, we can say

$$\text{mmol of } YBa_2Cu_3O_{7-x} \text{ in experiment} = \frac{28.19 \text{ mg}}{(666.19-16.00x) \text{ mg/mmol}}$$

All mass lost represents loss of oxygen from $YBa_2Cu_3O_{7-x}$.

$$\text{mmol of O lost} = \frac{(28.19 - 25.85) \text{ mg}}{16.00 \text{ mg/mmol}} = 0.146_{25} \text{ mmol}$$

From the reaction stoichiometry, we can say

$$\frac{\text{mmol of O lost}}{\text{mmol YBa}_2\text{Cu}_3\text{O}_{7-x}} = \frac{3.5\text{-}x}{1}$$

$$\frac{0.1462_5 \text{ mmol}}{\left(\dfrac{28.19 \text{ mg}}{(666.19\text{-}16.00x) \text{ mg/mmol}}\right)} = 3.5\text{-}x$$

This terrible expression can be solved for the dimensionless stoichiometry coefficient, x, by cross multiplication:

$$0.1462_5 = \left(\frac{(28.19)}{(666.19\text{-}16.00x)}\right)(3.5\text{-}x)$$

$$(0.1462_5)(666.19\text{-}16.00x) = (28.19)(3.5\text{-}x)$$

$$97.4_{30} - 2.34_{00}x = 98.66_{50} - 28.19x$$

$$28.19x - 2.34_{00}x = 98.66_{50} - 97.4_{30}$$

$$25.85_{00}x = 1.2_{350}$$

$$x = 0.048$$

Note that the number 3.5 is *exact*, since it comes from the stoichiometry.

S9.   mol Br = mol Ag^+ at 1st end point = $(0.405 \times 10^{-3}$ kg$)(0.093\ 84$ mol/kg$)$

   $= 3.80 \times 10^{-5}$ mol

mol Cl = mol Ag^+ at 2nd end point = $(0.787 \times 10^{-3}$ kg$)(0.093\ 84$ mol/kg$)$

   $= 7.39 \times 10^{-5}$ mol

wt % Br $= \dfrac{(3.80 \times 10^{-5} \text{ mol})(79.904 \text{ g/mol})}{8.463 \times 10^{-3} \text{ g sample}} \times 100 = 35.9\%$

wt % Cl $= \dfrac{(7.39 \times 10^{-5} \text{ mol})(35.453 \text{ g/mol})}{8.463 \times 10^{-3} \text{ g sample}} \times 100 = 30.9\%$

Br/Cl atomic ratio $= \dfrac{3.80 \times 10^{-5} \text{ mol}}{7.39 \times 10^{-5} \text{ mol}} = 0.514$

S1. Use Equation 26-7 with $s_s = 0.06$ and $e = 0.03$. The initial value of t for 95% confidence in Table 4-2 is 1.960. $n = t^2 s_s^2 / e^2 = 15.4$ For $n = 15$, there are 14 degrees of freedom, so $t \approx 2.15$, which gives $n = 18.5$. For 17 degrees of freedom, $t \approx 2.11$, which gives $n = 17.8$. It appears that the calculation will oscillate near 18 samples. For 90% confidence, the initial t is 1.645 in Table 4-2 and the same series of calculations gives $n = \underline{13\ samples}$.

S2. (a)

Mass (m, g)	Standard deviation (R, %)	$K_s = mR^2$ (g)
0.54	6.33	21.6
1.78	3.21	18.3
4.22	2.00	16.9
8.63	1.55	20.7
	average	19.4

(b) $mR^2 = K_s$. For $R = 1$ and $K_s = 19.4$ g, we find $m = 19.4$ g.

(c) Use Equation 26-7 with $s_s = 0.01$ and $e = 0.007\ 5$. The initial value of t for 90% confidence in Table 4-2 is 1.645. $n = t^2 s_s^2 / e^2 = 4.8$ For $n = 5$, there are 4 degrees of freedom, so $t = 2.132$, which gives $n = 8.1$. For 7 degrees of freedom, $t = 1.895$, which gives $n = 6.4$. Continuing, we find $t = 2.015 \Rightarrow n = 7.2$. This gives $t = 1.943 \Rightarrow n = 6.7$. Use 7 samples.

S3. (a) Volume $= (4/3)\pi r^3$, where $r = 0.100$ mm $= 0.010\ 0$ cm.

Volume $= 4.189 \times 10^{-6}$ mL.

Mass of diamond particle $= (4.189 \times 10^{-6}$ mL$)(3.51$ g/mL$) = 1.470 \times 10^{-5}$ g.

Mass of SiC particle $= (4.189 \times 10^{-6}$ mL$)(3.23$ g/mL$) = 1.353 \times 10^{-5}$ g.

Number of particles of diamond in 1.00 g of mixture

$= (0.870$ g$)/(1.470 \times 10^{-5}$ g/particle$) = 5.92 \times 10^4$.

Number of particles of SiC in 1.00 g of mixture

$= (0.130$ g$)/(1.353 \times 10^{-5}$ g/particle$) = 9.61 \times 10^3$.

(b) The fraction of each type of particle is

$p_{diamond} = (5.92 \times 10^4)/(5.92 \times 10^4 + 9.61 \times 10^3) = 0.860$

$q_{SiC} = (9.61 \times 10^3)/(5.92 \times 10^4 + 9.61 \times 10^3) = 0.140$.

(i) 1.00 g:

Number of particles in 1.00 g is $n = 5.92 \times 10^4 + 9.61 \times 10^3 = 6.88 \times 10^4$.

Expected number of diamond particles in 1.00 g is $np = 5.92 \times 10^4$.

Expected number of SiC particles in 1.00 g is $nq = 9.61 \times 10^3$.

Sampling standard deviation $= \sqrt{npq} = \sqrt{(6.88 \times 10^4)(0.860)(0.140)} = 91.0$.

Relative sampling standard deviation for diamond $= \dfrac{91.0}{5.92 \times 10^4} = 0.154\%$

Relative sampling standard deviation for SiC $= \dfrac{91.0}{9.61 \times 10^3} = 0.947\ \%.$

(ii) <u>0.100 g</u>:

Number of particles in 0.100 g is $n = 5.92 \times 10^3 + 9.61 \times 10^2 = 6.88 \times 10^3.$

Expected number of diamond particles in 0.100 g is $np = 5.92 \times 10^3.$

Expected number of SiC particles in 0.100 g is $nq = 9.61 \times 10^2.$

Sampling standard deviation $= \sqrt{npq} = \sqrt{(6.88 \times 10^3)(0.860)(0.140)} = 28.8.$

Relative sampling standard deviation for diamond $= \dfrac{28.8}{5.92 \times 10^3} = 0.486\%$

Relative sampling standard deviation for SiC $= \dfrac{28.8}{9.61 \times 10^2} = 3.00\ \%.$